Applications of Cryogenic Technology

Volume 5

CRYO–72 PROCEEDINGS...

A compilation of Papers Invited by
The Cryogenic Society of America
to the
Fifth Annual Conference of the Society
held concurrently with
The American Vacuum Society Conference
and Equipment Display
in
Chicago, Illinois
October 3, 4, and 5, 1972

CRYOGENIC SOCIETY OF AMERICA
INC.

Applications of Cryogenic Technology

Volume 5

Edited by **Robert H. Carr, Ph.D.**
Professor of Physics
California State University
Los Angeles

Scholium International, Inc. • *Whitestone, New York 11357*

ISBN 0-87936-001-1

LCN 68-57815

Editor's Preface

Readers of this series will correctly expect this volume to be something between a conference proceedings and a reference text. In publishing selected papers from its annual conferences, the Cryogenic Society of America attempts to produce books which have the relevance of proceedings, but some of the lasting power of reference works. The latter value is derived from the invited-paper format of the conferences. The typical length of an invited paper allows its topic to be presented in the depth and breadth appropriate for a book chapter. The invitation system permits the program committee to select topics which converge on an area of interest.

To the authors of the chapters I owe both thanks and apologies. Without them this book would not exist. I regret that our publishing schedule, complicated by our contracting with a new publisher for the series, did not allow time to send proofs before printing. Although I have attempted to confirm with the authors any substantive editorial changes or questions on interpretation, I must apologize for denying them opportunity to review their work in nearly final form.

In his program chairman's preface Ray Szara has acknowledged those who participated in the conference. He has, of course, modestly omitted his own work as program chairman. On behalf of the Cryogenic Society I would like to thank him for successfully coordinating the entire program and for personally obtaining several of the speakers.

The diversity of the applications of cryogenic and helium technology represented in this book would be hard to exaggerate. Temperatures discussed range from millidegrees K to tens of millions of degrees; pressures, from 10^{-7} Torr to 10^{3} psi; areas, from square angstroms to thousands of square feet; magnetic fields from zero to hundreds of kilogauss; and powers from microwatts to gigawatts. The viewpoints range from tabulation of useful data and description of operational equipment, both laboratory and industrial, to speculation about the future.

I've enjoyed reading these papers; I hope you will, too.

Robert H. Carr
Editor, Volume 5

Chairman's Preface

CRYO-72 was the fifth annual conference of The Cryogenic Society of America. It was held October 3, 4, and 5, 1972, in Chicago, Illinois concurrently with American Vacuum Society Conference and Equipment Display.

The conference was divided into four sessions. The first dealt with helium and consisted of invited papers arranged by the Helium Division of the Cryogenic Society of America. Thanks go to J. B. Reinoehl, Helium Division Program Chairman, for obtaining high caliber speakers and most interesting subject matter for this session. The other sessions included Refrigeration and Applications, Superconductivity, and lastly Cryopumping. This final session was complimentary to the program of the American Vacuum Society.

The CSA program began with a keynote address by Dr. John H. Pomeroy of the National Aeronautics and Space Administration reviewing present problems of space and giving insight into the future. In the Russell B. Scott Memorial Address, Dr. Samuel C. Collins ably described "A Half Century Quest for Improvement in the Art of Cooling." I am grateful to both of these gentlemen for honoring us by their presence, as well as delighting us with their presentations.

Sincere thanks go to all the authors for making a generous contribution to Cryogenic literature; I am also grateful to the Session Chairmen who exerted a cohesive influence on the program. Credit for the excellent job of local arrangements belongs to the officers of the Midwest Chapter of The Cryogenic Society and to their wives.

Finally, I would like to thank Dr. Leo Garwin (1972 President of CSA) and Robert W. Vance for their advice and assistance in the preparation of the program, and Dr. Robert H. Carr for editing ***Applications of Cryogenic Technology, Volume 5***.

Romuald J. Szara
Program Chairman
CRYO-72

Contributors

A. C. Anderson, Ph.D.
Paper No. 8
Professor, Department of Physics
University of Illinois
Urbana, Illinois 61801

N. W. Baer
Paper No. 18
Member Technical Staff
Department of Electrical Engr.
University of Colorado
Boulder Colorado 80302

Roger W. Boom, Ph.D.*
Paper No. 13
Professor
Department of Nuclear Engineering
University of Wisconsin
Madison, Wisconsin

R. A. Byrns
Paper No. 14
Mechanical Engineer
Lawrence Berkeley Laboratory
University of California
Berkeley, California 94720

Robert H. Carr, Ph.D.
Editor, Volume 5
Professor of Physics
California State University
Los Angeles, California

David H. Crimmins*
Paper No. 1
Plant Applications Manager
Gulf General Atomic Co.
San Diego, California 92112

F. J. DiSalvo*
Paper No. 12
Member Technical Staff
Bell Laboratories
Murray Hill, New Jersey 07974

*Delivered paper at conference.

H. Fernández-Morán, M.D. Ph.D. *
Paper No. 9
A. N. Pritzker Professor of Biophysics
Department of Biophysics &
Pritzker School of Medicine
University of Chicago
Chicago, Illinois 60637

Patricia J. Giarratano*
Paper No. 4
Mechanical Engineer, Cryogenics Division
Institute for Basic Standards, NBS
Boulder, Colorado 80302

E. F. Graham, Ph.D. *
Paper No. 10
Professor
University of Minnesota
St. Paul, Minnesota

W. D. Gregory, Ph.D.*
Paper No. 11
Department of Physics
Georgetown University
Washington, D. C. 20007

R. E. Hayes, Ph.D.
Paper No. 18
Associate Professor
Department of Electrical Engr.
University of Colorado
Boulder, Colorado 80302

L. O. Hoppie, Ph.D.*
Paper No. 18
Senior Research Engineer
GM Research Laboratories
Warren, Michigan 48090
(Work done at University of Colorado, Department of Electrical Engineering)

Jack E. Jensen*
Paper No. 5
Senior Mechanical Engineer
Brookhaven National Laboratory
Upton, New York

J. K. Jones*
Paper No. 7
Assistant Project Manager
Petrocarbon Developments Ltd.
Manchester, England

D. A. Krieger Paper No. 11	Georgetown Instruments, Inc. Washington, D. C. 20007
Gerald L. Kulcinski, Ph.D. Paper No. 13	Associate Professor Department of Nuclear Engineering University of Wisconsin Madison, Wisconsin
R. D. Ladd Paper No. 11	Georgetown Instruments, Inc. Washington, D. C. 20007
Lawrence M. Litz, Ph.D. Paper No. 6	Project Manager Sterling Forest Technical Center Union Carbide Corp. Tuxedo, New York 10987
W. N. Mathews, Jr., Ph.D. Paper No. 11	Department of Physics Georgetown University Washington, D. C. 20007
Charles W. Maynard, Ph.D. Paper No. 13	Professor Department of Nuclear Engineering University of Wisconsin Madison, Wisconsin
E. C. W. Perryman* Paper No. 3	Director of Applied Research & Development Chalk River Nuclear Laboratories Atomic Energy of Canada Ltd. Chalk River, Ontario Canada KOJ1JO
James W. Shearer, Ph.D.* Paper No. 2	Senior Scientist Lawrence Livermore Laboratory University of California/ Livermore, California
A. M. Smith, Ph.D.* Papers No. 16, 17	Supervisor, Research Section von Karman Gas Dynamics Facility ARO, Inc. Arnold Air Force Station, Tenn.

George E. Smith* Paper No. 6	Research Engineer Sterling Forest Technical Center Union Carbide Corp. Tuxedo, New York 10987
J. M. Stacey Paper No. 7	Petrocarbon Developments Ltd. Manchester, England
Bruce Strauss, Ph.D.* Paper No. 15	Staff Member National Accelerator Lab Batavia, Illinois
Coworkers on Paper 15:	
V. Bartenev A. Kuznetsov, Prof. B. Morozov V. Nikitin Y. Pilipenko V. Popov L. Zolin	Institute of High Energy Physics Serpukhov, USSR
J. Klen E. Malamud D. Sutter	National Accelerator Lab Batavia, Illinois
Romuald J. Szarza Program Chairman CRYO-72	The James Franck Institute University of Chicago 5640 Ellis Avenue Chicago, Illinois 60637
J. T. Tanabe* Paper No. 14	Mechanical Engineer Lawrence Berkeley Laboratory University of California Berkeley, California 94720
William F. Vogelsang, Ph.D. Paper No. 13	Associate Professor Department of Nuclear Engineering University of Wisconsin Madison, Wisconsin

Contents

Special Speakers

LUNCHEON SPEAKER

Dr. Samuel C. Collins

Dr. Collins is known throughout the world for the Collins Cryostat, the first machine to make liquid helium in quantities sufficient to be used as a refrigerant, and which opened up many fields of cryogenic research. Although his interests extend beyond cryogenics into physical chemistry, gas purification, medical technology, and cryobiology, Dr. Collins **limited** the topic of his Russell B. Scott Memorial Address to **"A Half Century Quest for Improvements in the Art of Cooling."** The talk began with early experiments with nonmechanical methods of freezing water, then continued with descriptions of the various turbines, rotary engines, diaphragm engines, reciprocating engines, and accompanying heat exchangers which led to a successful continuous process for helium liquefaction. In conclusion, some modern developments, such as replacing the Joule-Thomson expansion valve with a two-phase engine, were discussed.

KEYNOTE ADDRESS

Dr. John H. Pomeroy

Dr. Pomeroy began his professional career as a chemist, but he is better known as a scientific-administrator for the USAEC and NASA, and as a senior science editor for Encyclopaedia Britannica, Inc. It is in his current capacity as assistant director of the NASA Lunar Sample Program that Dr. Pomeroy delivered the CRYO-72 keynote address: **"Low Temperatures, Low Pressures, Space Travel and NASA."** The space program played a key role a decade ago in the development of cryogenics, and NASA's various branches, laboratories, spacecraft and supported research are probably still the largest investigators, producers, consumers, and explorers of low temperatures and low pressures in the solar system and beyond. Dr. Pomeroy discussed NASA's currently used vacuum and cryogenic equipment. He then offered some speculations about future developments in these areas and their relevance to future astronautical and cosmological exploration.

1

A HIGH TEMPERATURE HELIUM-GAS-COOLED REACTOR

David H. Crimmins

The High Temperature Helium-Gas-Cooled Reactor (HTGR) plant provides a number of significant advantages to electric utilities over other types of thermal power plants. Less heat must be rejected to the environment because of the efficient high temperature steam produced, made possible because of the high temperature operation of the helium-cooled graphite core structure. Maintenance is eased by the noncorrosive properties of helium, and radioactive releases to the environment are virtually eliminated because all radioactive products can be removed from the helium by means of cryogenic systems.

The general design of a 1160 MWe HTGR reactor system, completely enclosed within the prestressed concrete reactor vessel is described.

1-1. Introduction

Four years ago, in 1968, Dr. Goodjohn of Gulf General Atomic presented a paper on helium-cooled reactors at the Helium Centennial Symposium. In that presentation he described the five helium-cooled reactors operating or under construction throughout the world. He also stated, and I quote from the proceedings of that meeting, "The history of helium-cooled reactors to date is interesting, though limited, and the future holds considerable promise. As the technology progresses, it is reasonable to expect that the next two decades will see an almost universal application of helium-cooled reactor technology." This was a rather ambitious statement for him to make at that time, but it may turn out to be prophetic.

Two important trends have evolved within the past four years: (1) an increased environmental awareness; and (2), not unrelated to the first, six large commercial HTGRs have been sold in the past fourteen months.

Helium as a reactor coolant has several advantages over water. The two most notable are the almost total elimination of gaseous and liquid radioactive discharges from the plant, and the considerable reduction, and potential elimination, of thermal discharge to the surrounding bodies of water. These advantages accrue from the following operational characteristics:

1. The use of inert helium gas coolant results in negligible coolant activation, corrosion, and carryover of activation products. Fission products which might enter the coolant under the design fuel leakage assumptions are readily removed by a bypass helium purification system. This continually removes the fission products, including tritium, in a manner which results in the almost complete elimination of gaseous waste effluent from the HTGR plant.

   Liquid waste in the HTGR plant during normal operation occurs only as a result of decontamination operations. It is of such low activity level and small volume that periodic processing by demineralization or distillation can result in similarly negligible liquid release.

2. The high temperature coolant leads to an optimum cycle efficiency of close to 40%, which is comparable to the efficiencies available from the most modern fossil-fired steam-electric stations. The high efficiency minimizes the thermal discharge and reduces the cost of cooling ponds or wet cooling towers (if once-through water cooling is not available or is environmentally unacceptable).

   In the future, we expect to offer a helium-cooled plant that will eliminate thermal discharge to water bodies. This will be the direct cycle HTGR, using a helium Brayton cycle to drive a gas turbine directly. The heat rejection characteristics of this concept make air cooling, rather than water cooling, economically attractive. Elimination of the entire steam turbine plant should also result in decreased power costs.

1-2. HTGR Description

General. HTGR NSSS (Nuclear Steam Supply System) is the key element of a load-following nuclear generating plant, capable of producing up to 1160 MWe in a conventional reheat steam cycle. The NSSS produces main superheated steam at 955°F and 2500 psig and reheat steam at 1001°F and 575 psig.

The major components and systems of the HTGR NSSS include: (1) the prestressed concrete reactor vessel (PCRV) enclosing the entire primary coolant system and interconnecting ducts and plenums; (2) once-through steam generators, each containing economizer-evaporator-superheater and reheater sections; (3) steam-turbine-driven, axial-flow helium circulators; (4) hexagonally shaped graphite fuel elements incorporating coated fuel particles; and (5) top-mounted, cable-type control rod drive mechanisms. The arrangement of these components is shown in Figure 1-1.

The NSSS contains six independent primary coolant loops, each having a helium circulator and steam generator. Helium, at a pressure of about 685 psig, is circulated by means of the helium circulators downward through the reactor core and through the once-through steam generators which are located in the PCRV in separate cavities around the main cavity, before returning to the helium circulators. The main superheated steam produced in the steam generators at 955°F and 2500 psig passes to the high-pressure element of the turbine. The cold reheat steam from the high-pressure turbine exhaust is used to drive the helium circulators. It then passes to the reheat section of the steam generator and on to the intermediate- and low-pressure sections of the main turbine.

Prestressed Concrete Reactor Vessel. The overall PCRV is a right circular cylinder. The steam generators and helium circulators are located in sidewall cavities spaced symmetrically around the central cavity. Additional cavities are provided for the core auxiliary cooling circulators and heat exchangers.

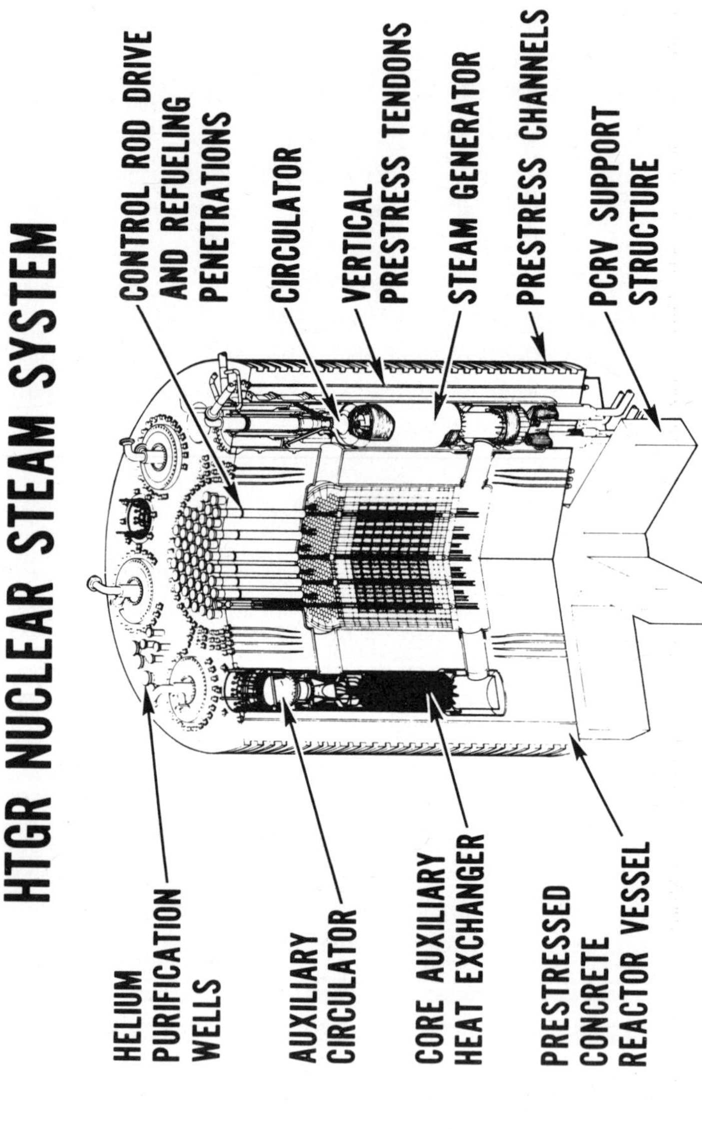
HTGR NUCLEAR STEAM SYSTEM
CONTROL ROD DRIVE AND REFUELING PENETRATIONS
CIRCULATOR
VERTICAL PRESTRESS TENDONS
STEAM GENERATOR
PRESTRESS CHANNELS
PCRV SUPPORT STRUCTURE
HELIUM PURIFICATION WELLS
AUXILIARY CIRCULATOR
CORE AUXILIARY HEAT EXCHANGER
PRESTRESSED CONCRETE REACTOR VESSEL

Fig. 1-1. HTGR Nuclear Steam System

Prestressing of the PCRV is accomplished by means of high-strength vertical wire tendons and external circumferential wire wrapping. The concrete is protected from the helium gas temperatures by means of a thermal barrier mounted on the inner wall of the PCRV liner and by cooling tubes attached to the outer surface of the liner. The metal liner acts as a barrier to the helium. The helium purification system, which continuously processes a bypass stream of the helium coolant, is contained in wells built into the PCRV head.

The PCRV is housed in a secondary containment structure designed to contain the entire primary coolant system helium under conditions postulated for the design basis accident.

Heat Transfer Systems. The steam generators are of a once-through type consisting of a helically coiled economizer-evaporator-superheater tube bundle and a reheater tube bundle.

The cold reheat steam is used as the driving fluid for the single-stage, steam-turbine-driven, axial-flow helium circulators. Output of each of the circulators can be individually regulated by bypassing cold reheat steam around their drive turbines. A helium shutoff valve is provided at the discharge of each of the circulators.

Auxiliary cooling loops, each consisting of a circulator, a helium shutoff valve, and a helium-water heat exchanger, are provided within the PCRV to cool the core after reactor shutdown if it is necessary or desirable to take the main loops out of service.

Reactor Core. The reactor core is made up of hexagonally shaped, graphite fuel elements approximately 14 inches across the flats by 31 inches high. The fuel in the form of coated particles of uranium dicarbide as the fissile material and thorium carbide or oxide as the fertile material, is contained in bonded rods located in vertical blind holes in the fuel elements. The fuel elements are stacked in columns eight blocks high; thus, the active core is about 21 feet high and 27 feet in diameter (23 feet in diameter for the MWe plant). The core is surrounded by replaceable graphite reflector elements and permanent graphite reflector blocks.

Control of the reactor is accomplished by means of control rod pairs, 73 in the 1160 MWe plant and 49 in the 770 MWe size.

Each pair of control rods is suspended by cables operated by an electric-motor-driven rod drive mechanism. The control rods, which contain neutron-absorbing material, move in vertical passages in the central column of fuel elements in each refueling region. An emergency shutdown system capable of injecting small neutron-absorbing balls into core cavities is also included in each control rod assembly.

The fuel loading is based upon a four-year cycle; that is, approximately one-fourth of the core is replaced each year. The initial fuel loading is of highly enriched uranium (93% U-235) and of thorium. The U-233 bred from the thorium can be recycled to the core within a year.

The fuel in each region is replaced through the associated refueling penetration in the PCRV top head. Spent fuel removed from the reactor is stored in dry storage wells; the fission products are allowed to decay for about four months, at which time the spent fuel is ready for shipment to the reprocessing and recycling fuel fabrication plant.

1-3. Radioactive Effluent

Helium Purification System. The helium coolant purification system for an HTGR plant consists of two identical trains of equipment, see Figure 1-2. One train is on line at all times during power operation and processes approximately 10% of the primary coolant inventory per hour.

The various sections of the system perform their functions of successively removing by filtering, condensation, or adsorption, all of the particulate, chemical, and fission product contaminants that may be present in the primary coolant helium. The high temperature filter adsorber effectively removes all of the condensible nuclides, such as metallic fission products by direct adsorption on charcoal. The volatile fission products, including iodine, are removed by chemisorption on potassium which is impregnated in the charcoal. The gaseous fission products, Krypton, xenon, and some of the tritium, are removed by the nitrogen-cooled low-temperature delay bed. The delay bed provides at least a one-year

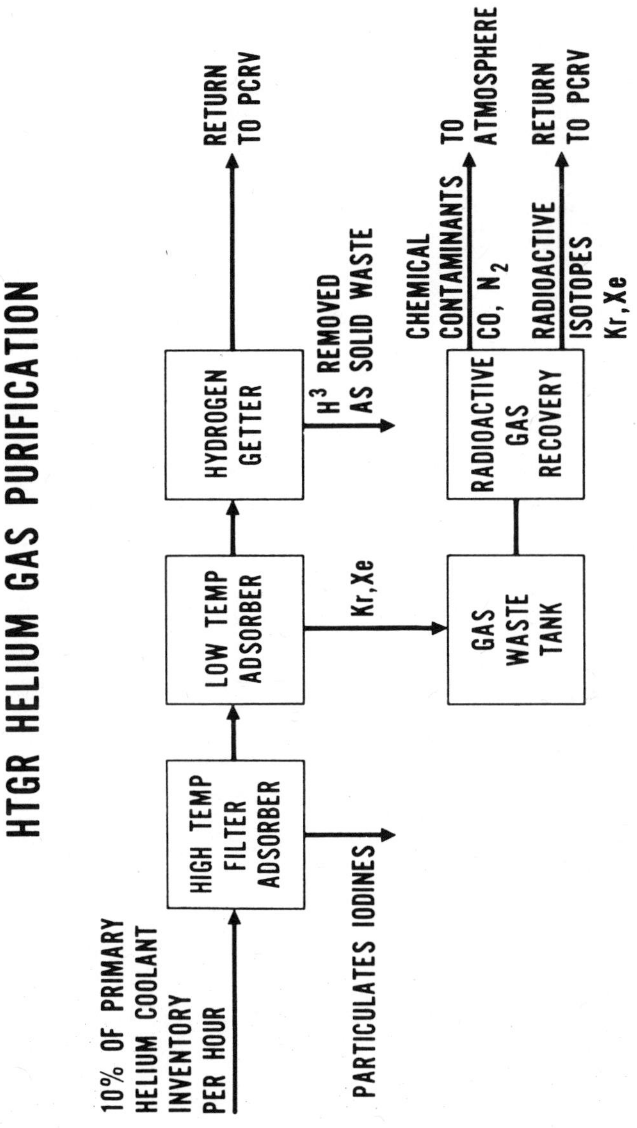

Fig. 1-2. HTGR Helium Gas Purification

delay time for Krypton. Breakthrough of Kr-85 is one of the signals used to indicate the need for regeneration of the delay bed. Tritium, which remains in the gaseous state in the nonoxidizing HTGR atmosphere, is removed by one of the two titanium sponges in the hydrogen getter unit. Following processing in the helium purification train, the purified helium is returned to a distribution header and then back into the primary coolant system.

Regeneration and Radioactive Gas Effluent. Regeneration of the low-temperature delay bed is performed by taking the purification train off the line, allowing two months' decay and then depressurizing and heating the adsorber. All effluent gas flows to the gas waste tank.

Prior to the recent controversy concerning the release of radioactivity to the environment, the design basis for the HTGR involved the release of the contents of the gas waste tank to the atmosphere at periodic intervals. The Kr-85 levels were less than 1% of the allowable limits. In an effort to reduce this release even further, Gulf General Atomic has now added additional equipment which further processes the contents of the gas waste tank and returns essentially 100% of the Kr-85 back to the primary coolant system to be adsorbed on the on-line low-temperature adsorber.

The low-temperature adsorbers are adequate for the storage of the entire Kr-85 inventory expected to be deposited in the adsorbers during the lifetime of the plant. However, as an alternate, it is also possible to periodically pump the Kr-85 into standard gas cylinders for off-site disposal. The addition of this equipment to process the gas in the gas waste tank results in only insignificant amounts of gaseous radioactivity being released from the HTGR plant. The radioactive gaseous release is now expected to be only that which arises from minor leakage in the plant systems. It should be negligible.

Regeneration of the hydrogen getter unit is accomplished by replacing the titanium sponge in the getter bed and shipping the sponge as a solid waste to off-site disposal. The life of the titanium sponge is determined by the accumulation of hydrogen from assumed water leakage into the nuclear steam system. The design replacement interval is approximately monthly. The removal of tritium in this manner circumvents the problems associated with the accumulation of tritium oxide in the plant water systems during operation and the need for ultimate disposal of large amounts of water containing tritium.

Liquid Waste Effluents. The HTGR does not generate radioactive liquid waste during normal operation. Liquid waste would be produced by infrequent decontamination operations or as a result of equipment failure such as a steam generator tube leak.

Decontamination operations involve the use of liquids for the removal of dust or other lightly adherent material from the surfaces of the fuel handling equipment, control rod drives or other equipment that might be removed from the primary circuit. The liquid from such operations is collected in a radioactive liquid waste tank. The volume is expected to be less than 2400 gallons per year with a total activity level of less than 100 curies.

Following a steam generator tube leak the volume and activity level will be temporarily higher depending on the carryover of fission products by the water that is removed in the coolers and dryers of the helium purification system.

The activity levels in the liquid waste tank are low enough that dilution to levels well below the tolerable limits is relatively easy. Likewise, a small capacity demineralizer or distillation unit can be used to reduce liquid waste effluent to as small a volume as desired.

As for all nuclear plants, solid wastes require proper packaging followed by shipment to off-site storage.

1-4. Thermal Effluent

The HTGR requires about 30% less cooling water than light-water reactor types because of the higher efficiency of the HTGR type power plant. The smaller amount of water required for a given temperature rise, or conversely, the smaller temperature rise for a given cooling water flow rate does not, of course, completely alleviate the siting problem facing many utilities. Very small temperature rises of the order of a few degrees, which require either large flow rates or equipment or provide outfall mixing, are becoming a requirement. Wet cooling towers or cooling ponds alleviate the temperature rise problem but result in increased costs for the heat rejection system. Again, the HTGR with the more efficient steam cycle requires a smaller heat rejection system for a given electrical output. In many areas of the country the use of wet cooling towers may also face environmental restrictions. The most serious of these restrictions in the future would appear to be the consumption of fresh water, of the order of 25 million gpd for a nominal 1100 MWe station, to replace evaporation and spray losses.

1-5. Future Developments

Direct Cycle HTGR. Use of a closed cycle is mandatory to secure the compactness of machinery necessary for practical application of the gas turbine to large-scale power production. Such a cycle needs a heat rejection system with a specification differing in a very important way from a steam cycle. Whereas the heat from a steam cycle is rejected at virtually one temperature, heat rejection from a gas-turbine cycle is spread over a range of temperatures of the same order of magnitude as the compressor temperature rise, around 350°F or around 250°F if an intercooler is used. The temperature entropy diagrams (Figure 1-3) illustrate the difference noted for the two cycles. As a result, cooling by ambient air is a practical means for rejecting heat, whereas air cooling of a steam plant would require about ten times the air flow. This eliminates any requirements for a water supply.

The advantages of the closed-cycle gas-turbine power plant were considerable even in the absence of the environmental factors discussed in this paper. In comparison with a steam plant, these advantages included:

1. Two orders of magnitude increase in working fluid exhaust density, greatly reducing the size of equipment, ducting and piping.
2. Reduction in expansion ratio from around 2500 to 1 for steam to about 2.5 to 1 for helium, allowing higher

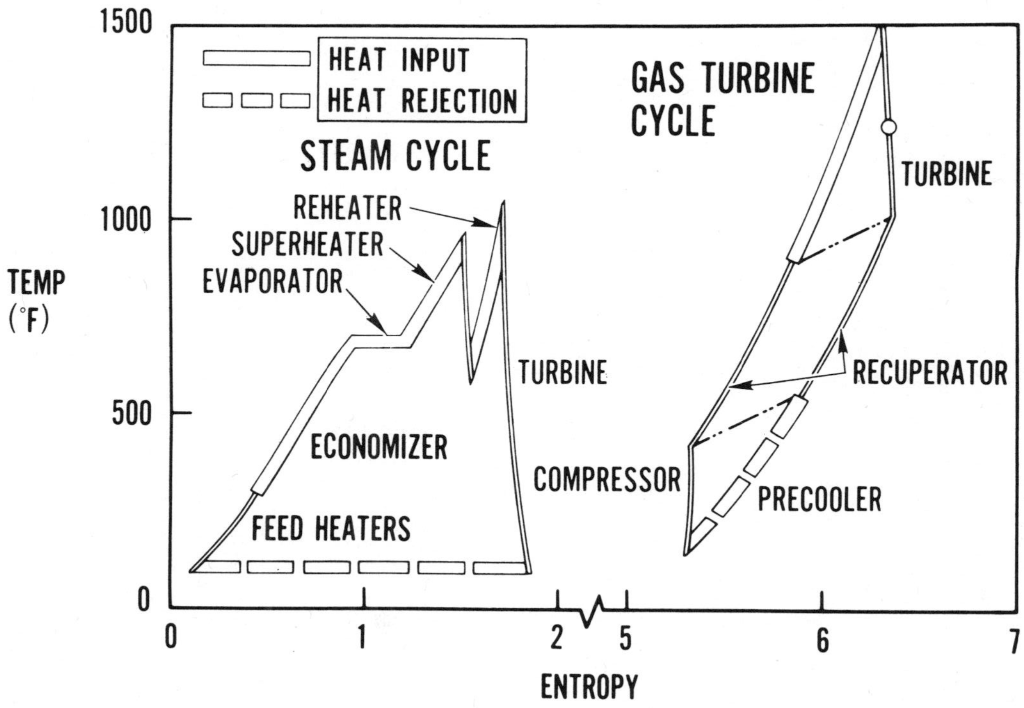

Fig. 1-3. Temperature-Entropy Diagrams

expansion efficiency and much more effective structural materials deployment.

3. Very substantial reduction in complexity, by eliminating equipment and instrumentation concerned with water treatment, boiler feed pumping, heating and de-aeration.
4. Elimination of wetness and corrosion problems.

One might wonder why the gas turbine is not already with us. The reasons are rather simple: the advent of nuclear power was needed to avoid the corrosion and temperature problems of the input heat exchanger necessary to employ fossil fuel, and the light water reactors operated at too low a temperature to provide any solution by means of a water-to-gas heat exchanger.

The proportions shown in Figure 1-4 are based on stresses consistent with 30-year life, and in this respect are a very conservative departure from what has come to be the normally accepted figures for aircraft gas turbines. This 30-year life, coupled with the restriction of the very much lower temperature levels currently available from the HTGR-type reactors, removes the design of this kind of system from the developmental needs of the aircraft gas turbine, with which it should not be confused.

It is, of course, difficult at this time to estimate the costs with great accuracy, but the basic advantages already mentioned lead to the conservative anticipation of at least a 10% to 15% reduction in the total plant cost as compared with a steam plant. In addition, the necessary dry cooling equipment is expected to cost about one-third of that needed for steam.

Usage of Helium by HTGRs. The total helium inventory in one 1160 MWe HTGR is approximately 1.7×10^6 SCF, and make-up requirements are estimated at 200,000 SCF per year. Our projected HTGR economy through the year 2000 indicates that the annual helium consumption by HTGRs in this country would be of the order of 100,000,000 SCF. This represents about 2% of the annual production forecast for the year 2000, or about 3% of the helium presently in storage under the Helium Conservation Program.

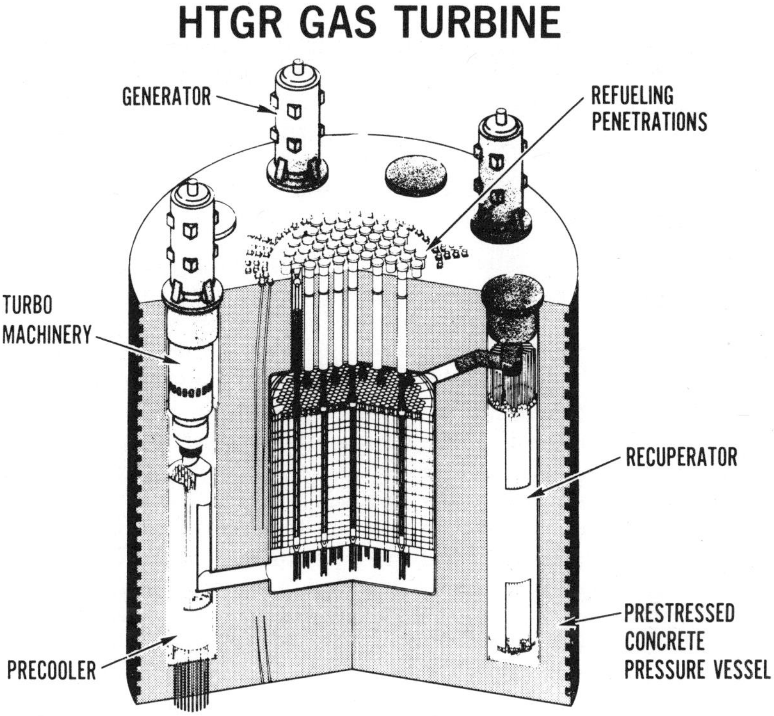

Fig. 1-4. HTGR Gas Turbine

2

LASER POSSIBILITIES IN PULSED CONTROLLED THERMONUCLEAR FUSION

James W. Shearer

Two possibilities for using lasers in controlled fusion reactors are discussed. In the first, the laser would be used to heat a deuterium plasma in a uniform magnetic field similar to "conventional" fusion machine concepts. Large cryogenically cooled coils would be needed to provide fields of a hundred kilogauss or greater.

In the second proposal, small thermonuclear "microexplosions" would be created in a thick-walled vacuum chamber by means of a spherical array of laser beams focused on small spherical pellets of solid heavy hydrogen. These pellets must be prepared cryogenically and propelled into the vacuum chamber. This controlled fusion concept does not require a magnetic field for containment because the pellet plasma is so dense that sufficient nuclear reactions occur before it comes apart. The proposal involves questions of explosion magnitude and repetition rate, efficiencies of laser operation and light absorption, and energy recovery.

2-1. The Controlled Fusion Problem

The demand for energy and power continues to climb while resources of coal and oil are beginning to dwindle.[1] This energy crisis has spurred the search for other sources of power; one such possibility is the energy release obtained from nuclear fusion reactions in the heavy isotopes of hydrogen—deuterium and tritium.

The hydrogen bomb is the only way in which appreciable amounts of nuclear fusion energy have been released by man. For obvious reasons, this is an unsatisfactory source of continuous useful power. For over twenty years, research has been underway throughout the world, searching for an alternative method of releasing the energy of nuclear fusion in a manner that can be controlled on an industrial scale. Thus far the achievement of this goal has eluded all efforts.[2]

The expansion of laser technology in the last ten years now allows us to consider some additional possibilities for the controlled release of fusion energy. However, there are some formidable technical challenges in these new possibilities, and much remains to be learned. This paper is intended as a brief, introductory review of these concepts.

2-2. High Power Pulsed Laser Systems

Two different lasers dominate the high power pulse field today: the neodymium-glass laser, and the carbon dioxide laser. Some "typical" characteristics of the light output from the largest of these lasers are indicated in Table 2-1.

Table 2-1. "Typical" High Power Pulsed Laser Characteristics

Laser	*Nd-Glass*	*CO_2*
Wavelength (cm)	1.06×10^{-4}	10.6×10^{-4}
Cutoff Density (see Sec. 2-5)	10^{21}	10^{19}
Largest Output Energy (joules)	100-1000	100-1000
Pulse Length (nanoseconds)	.01 - 10	1 - 100
Efficiency	~0.1%	~10%
Cost	~10^3 \$/joule	~10-100 \$/joule

The near infrared neodymium glass laser[3] consists of glass that is doped with 1% to 5% neodymium oxide. The Nd^{+3} lasing transition is excited by optical pumping from xenon-filled flashlamps. Its low efficiency arises from the fact that most of the flashlamp light is at the wrong wavelength for excitation of the excited level.

The far-infrared carbon dioxide molecular laser has been built in many forms; the most promising for high pulsed power applications is the recently developed "TEA" laser[4] (Transverse Electric Atmospheric). In this laser the excited molecular vibrational level is created by an electric discharge at atmospheric pressure whose electric field is transverse to the axis of the laser cavity. The optimum discharge is in the abnormal glow region, and considerable technical ingenuity is required to prevent arc breakdown and maintain optical homogeneity. Nevertheless, the TEA laser is undergoing rapid development at this time because its high efficiency and low cost offer the promise of many future applications. One workable design is pictured in Figure 2-1.

2-3. Controlled Thermonuclear Reaction Conditions

The lowest temperature self-sustained thermonuclear burn is the deuterium-tritium (DT) reaction ($\sim$100 million degrees in a 50:50 mixture of DT)

$$D + T \rightarrow He^4 + n^1 \tag{1}$$

where the alpha particle (He^4) carries off an energy of 3.5 MeV per particle, and the neutron (n^1) has 14.1 MeV per particle. Additional energy ($\simeq$ 5 MeV) is released when the neutron is captured by surrounding material. The total energy release by all the reactions must exceed the energy necessary to raise the fuel to ignition temperature if excess energy is to become available as useful power. This condition can be expressed mathematically by the "Lawson criterion" for the DT reaction;

$$n\tau > 5 \times 10^{13} \frac{a}{\epsilon} \text{ sec cm}^{-3} \tag{2}$$

where n is the ion density, τ is the confinement time of the plasma, a is the ratio of nuclear reaction energy output to laser energy input, and ϵ is the efficiency of absorption of laser light.[5]

An important practical limitation of "conventional" thermonuclear machines is the achievable containment pressure, P_c, whether that pressure be gas pressure or magnetic pressure within a coil. Choosing $P_c \leq 2000$ atmospheres, we then find an upper

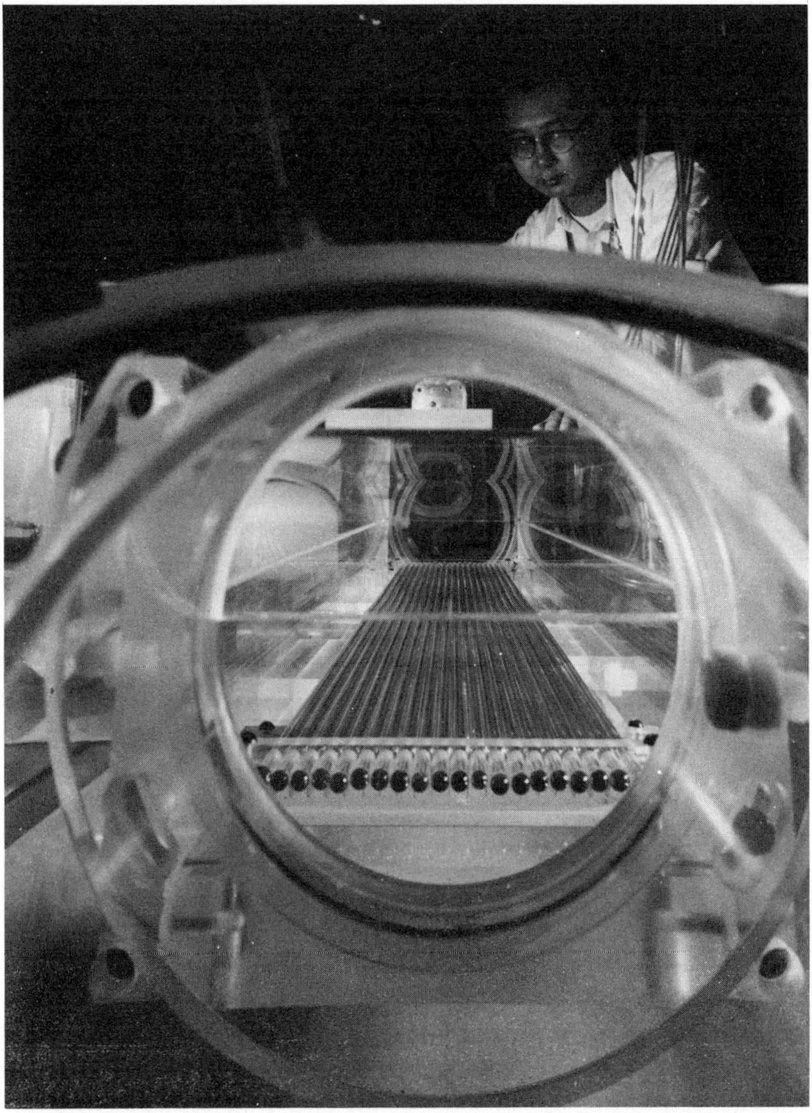

Fig. 2-1. A view taken looking down the axis of the laser beam into a TEA CO_2 laser. The polished anode is above the beam axis; a "double discharge" type of cathode is below the beam axis.

limit for the ion density n which can be contained at the ignition temperature T mentioned above:

$$n = P/2kT < 10^{17}\ \text{cm}^{-3} \tag{3}$$

where we have assumed that the electron pressure equals the ion pressure. When we now compare equation (2) and (3), we obtain a condition on the confinement time τ:

$$\tau > 5 \times 10^{-4}\ (a/\epsilon)\ \text{sec} \tag{4}$$

An additional requirement of such fusion machines is thermal insulation of the plasma from the walls. Large magnetic fields are used to provide this insulation and to contain the plasma; the magnetic field which is required to contain 2,000 atmospheres pressure is $\cong$220 kilogauss.

It would take us too far afield to catalogue here the variety of proposed fusion machines and the problems of plasma instability that have plagued fusion research for so many years.* Instead, we shall describe two laser/fusion proposals: first, a "conventional" fusion machine concept which might be improved by laser heating; second, a radically different fusion proposal involving laser-induced "microexplosions" of solid DT pellets.

2-4. Laser Heating of Long Confined Plasma Column

An important parameter for laser/plasma heating is the absorption coefficient of the laser light. The best known process for this is collisional absorption, whose mean free path ℓ_c is approximately:

$$\ell_c \cong \frac{10^{22}\, T^{3/2}}{Z^3\, n^2\, \lambda^2}\ \text{cm} \tag{5}$$

where T is the temperature in degrees Kelvin, Z is the atomic number of the ion, and λ is the vacuum wavelength of the light in cm. Using the previous discussion we set $T \cong 10^8\,°K$, $n \leq 10^{17}$, $Z = 1$, and $\lambda = 10.6\ \mu m$ (CO_2 laser light), and we find:

*For a discussion of one "conventional" fusion reactor design see paper 13 in this book.

$$\ell_c \cong 10 \text{ km} \tag{6}$$

An even longer mean free path is found for neodymium light at the same density.

This large result might be reduced in two ways. First, at high laser light intensities other absorption processes involving plasma instabilities can be excited. These promise to lower the mean free path by several orders of magnitude.[6] Second, one can hope that efficient longer wavelength lasers may be developed in the future. This would also lower ℓ_c.

In Figures 2-2a and 2-2b we compare the "conventional" theta pinch fusion machine concept with a possible laser-heated plasma column machine, first proposed three years ago.[7] The expensive high energy low inductance capacitor bank that both heats the theta pinch plasma and provides the containment magnetic field is replaced by two items: an axial laser beam that heats the plasma, and a large magnetic field coil that surrounds both the plasma and the one meter liquid lithium blanket. In both machines the lithium is needed to capture the heat from the neutrons and to regenerate tritium (T) by means of the reaction:

$$Li^6 + n^1 \rightarrow He^4 + T \tag{7}$$

The plasma column must be long enough so that end losses do not quench it. These losses can be roughly estimated from the deuterium plasma's sound velocity, which is approximately 10^8 cm/sec at T = 5×10^7 °K. Then the length L is, approximately:

$$L > 2v_a \tau \cong .7\ (a/\epsilon) \text{ Km} \tag{8}$$

This distance is comparable to the absorption mean free path (Eq. 6); it might be lowered with more complicated magnetic field shapes, to inhibit the axial flow.

The light beam must stay on axis, and not strike the walls over this long distance. Calculations have shown, however, that if the heated plasma column has a hollow density distribution its refractive index profile forms a light pipe which channels the laser light down the axis.[8]

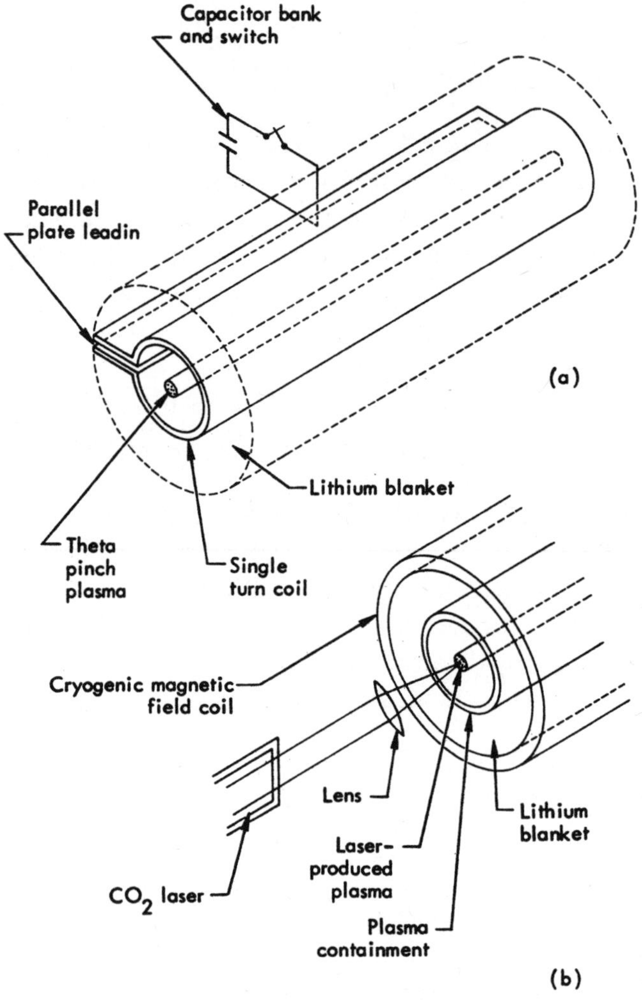

Fig. 2-2. Two fusion machine concepts at a plasma density of $\sim 10^{17}$ ions/cc. (a) The fast discharge "theta pinch" concept; the Los Alamos "SCYLLA" machine is an example. (b) A laser heated plasma inside a magnetic field, as described in the text.

If the 10 km plasma column has a cross-sectional area of 0.1 cm^2, then 50 megajoules of heat is required to reach 100 million degrees. This implies a very large laser, compared to current practice. The minimum required laser power, obtained by dividing this energy by the decay time, is of the order of 10 GW—a figure which is now being approached in practice at much shorter pulse lengths (100 nanoseconds) by TEA lasers.[9]

It can be seen that many formidable technical challenges are posed by this fusion machine concept, not the least of which is the cryogenic coil! It must be 2 meters in diameter, a few kilometers long, and produce a uniform 220 kilogauss field. Alternatively, if the coil were wound inside the lithium blanket it could be smaller, but it would have to withstand intense heating from the neutron flux.

2-5. Laser-Triggered Fusion Microexplosion Concept

Figure 2-3 illustrates a radically different proposal for useful fusion energy release.[10,11] A spherical array of laser beams is focused in vacuum on a solid spherical pellet of DT. The laser light heats and compresses the pellet until it ignites and burns in a "microexplosion" sufficiently small to be contained by the walls of the vacuum vessel. The lithium blanket serves the same purpose as in the previous proposal, but in this scheme no magnetic field is needed.

This proposal is actually more than ten years old, and it has an obscure, complicated, and controversial history. It has recently attracted much publicity because of a patent dispute and the revelation of previously classified computer calculations.[12]

In order to better understand the theoretical basis for this scheme, consider a sphere of DT whose ion density n is given by:

$$n \equiv \eta\, n_o \cong 5 \times 10^{22}\, \eta \qquad (9)$$

where n_o is the solid density and η is the "compression factor." If this sphere is at temperature T, the containment time τ can be written as:

$$\tau = \frac{1}{K} \frac{R}{V_s(T)} \qquad (10)$$

where R is the radius and $V_s(T)$ is the velocity of sound. K is a normalization factor, which is about 4-5. Equation (10) is a

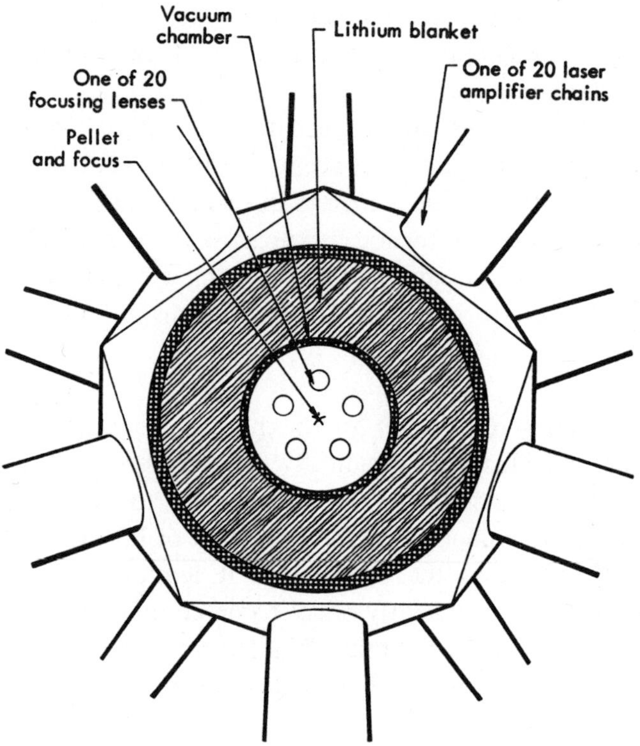

Fig. 2-3. Cutaway view of twenty-beam version of laser-triggered fusion microexplosion concept. Not shown is the DT pellet machine, the vacuum pumps or the laser optics.

mathematical expression of the principle of inertial containment—that is, the reacting material will hold together for a time τ before the rarefaction wave "eats in" from the surface, and it flies apart. Inertial containment is being substituted for magnetic field containment in this proposal.

The next important question is: how large a "microexplosion" is contemplated? This can be calculated from the ignition energy E_i for the sphere:

$$E_i = \left(\frac{4}{3}\pi R^3\right)\left(n_o \eta\right)\left[\frac{3}{2} k (T_e + T_i)\right] \tag{11}$$

where k is Boltzmann's constant, and T_e and T_i are the electron and ion temperatures, respectively.

If we substitute numbers into this equation (choosing $T_e = T_i \cong 10^8\,°K$), and apply the Lawson criterion for containment (Equation 2), we can show that[5]

$$E_i \cong K^3\, a^3/\epsilon^3\eta^2 \text{ megajoules} \tag{12}$$

Since the laser energy E_L and the nuclear yield energy E_N are given by:

$$E_L = E_i/\epsilon \qquad\qquad E_N = a\, E_L \tag{13}$$

we can write that the "microexplosion" energy is:

$$E_N \cong 100\, a^4/\epsilon^4\, \eta^2 \text{ megajoules} \tag{14}$$

where we have put $K^3=100$. For comparison, 100 megajoules corresponds to the energy yield of ~ 50 pounds of high explosive, which may be difficult to control. Furthermore, there are no megajoule lasers in existence or immediately contemplated today. Thus, we would like to make the energy E_N as low as possible.

Now $\epsilon < 1$ by definition, and $a > 1$ for practical application; therefore, the only way to reduce E_N is by increasing the compression factor η. This is the principle that is now emphasized by several groups. Values of η up to as high as 10^4 have been calculated.[12]

How could laser light produce such a high compression? The calculations predict that it would be created by the high pressure (millions of atmospheres or more) produced at the surface of the pellet by laser heating. The outer surface ablates into the vacuum, crushing the inner material by "rocket reaction." However, a high degree of spherical symmetry would be required to obtain the highest compressions. Whether such high values as quoted above could ever be achieved in practice is not known.

The microexplosion energy (Equation 14) is an even more sensitive function of the light absorption efficiency ϵ at the surface of the solid DT pellet. The ablating material forms a density gradient which the light can penetrate only as far as the "cutoff density" N_c, where the light frequency equals the plasma frequency. (This is the same cutoff density that appears in the theory of radio waves in the ionosphere, translated to higher frequencies and densities.) The effective absorption is then the sum of the absorption over the light path in and out of the plasma. Table 2-1 shows that the cutoff density for available lasers is less than solid density, so all the light absorption will be in the outer ablation plume around the solid core. Thermal conduction carries the energy toward the center. All of these phenomena involve difficult plasma physics problems that are still under study. However, it appears at present that high absorption efficiencies would be much easier to obtain if a shorter wavelength, ultraviolet pulsed laser were available whose light could penetrate closer to the core.[13]

Table 2-2 gives some examples of possible values of the parameters discussed above. Note that low values of microexplosion energy correspond to very short containment and heating times for the experiment (and for the laser pulse).

This proposal obviously poses a large number of formidable technical problems. One requirement of cryogenic interest is the DT pellet machine. It must be able to reproducibly manufacture spherical DT solid pellets of the order of $\sim$100 μm radius (barely visible specks). Then, like a baseball pitcher with exceptional control, the machine must propel each pellet into the center of a $\sim$1 meter vacuum chamber so that it intersects with the laser focus to an accuracy of the order of $\sim$10 μm. A typical pellet injection rate might be 10 pellets per sec. In order to meet competitive energy prices, the individual cost per pellet must be a fraction of a cent.

Another technical challenge common to both laser-fusion proposals is the enormous vacuum pumping capabilities that are required to evacuate the spent fuel after each pulse.[14] In a power-producing fusion reactor this will probably impose a requirement that the pumping time be only a fraction of a second.

Another problem, probably involving cryogenic techniques, is the recovery of unburned deuterium and tritium from these vacuum pumps in order to recycle this fuel back into the reactor.

2-6. Conclusions

Whatever becomes of these imaginative proposals for laser-triggered fusion, it is obvious that they are not a short-term panacea for the energy crisis. Optimists may differ, but in this author's opinion the only realistic assessment is that these are long-shot

Table 2-2. Various Hypothetical Microexplosions

Purpose	Energy Multiplication a	Light Absorption Efficiency ϵ	Compression Factor η	Pellet Radii Uncompressed R_o(μm)	Pellet Radii Compressed R (μm)	Containment Time τ(psec)	Laser Pulse Energy E_L(MJ)	Nuclear Output Energy E_N(MJ)
Breakeven Experiment	1	1	1	5,000	5,000	1100	100	100
"	1	.5	1	10,000	10,000	2200	1600	1600
"	1	.5	10^2	464	100	22	.16	.16
"	1	.5	10^3	100	10	2.2	.0016	.0016
"	1	.25	10^3	200	20	4.4	.0256	.0256
"	1	.25	10^4	43.2	2	.44	.000256	.000256
Fusion Power	10	.5	10^2	4,640	1,000	220	160	1600
"	10	.5	10^3	1,000	100	22	1.6	16
"	10	.25	10^3	2,000	200	44	25.6	256
"	10	.25	10^4	432	20	4.4	.256	2.56
"	20	.25	10^4	864	40	8.8	2.05	41
"	40	.25	10^4	1,728	80	17.6	16.4	656

shot possibilities for useful energy sources in the next century. On the other hand, they do represent the kind of imaginative thinking that may stimulate other ideas in the more immediate future.

References

1. John Holdren and Philip Herrera, *Energy,* Sierra Club, San Francisco and New York, 1971.
2. David J. Rose, "Controlled Nuclear Fusion: Status and Outlook," *Science* **172**, 797-808, 21 May 1971.
3. Michel A. Duguay, John W. Hansen, and Stanley L. Shapiro, "Study of the Nd:Glass Laser Radiation," *IEEE Journal of Quantum Electronics,* QE-6, 725-743, Nov. 1970.
4. Colin S. Willett and Thomas J. Gleason, "Gas Lasers at Room Pressure," *Laser Focus* **7**, 30-34, June 1971.
5. Ray E. Kidder, "Some Aspects of Controlled Fusion By Use of Lasers," Lawrence Livermore Laboratory Rept. UCRL-73500 (1971), to appear in the Proceedings of the Esfahan Symposium on Fundamental and Applied Laser Physics, 1971.
6. P. K. Kaw and J. M. Dawson, "Laser-Induced Anomalous Heating of a Plasma," *The Physics of Fluids* **12**, 2586-2591, December 1969.
7. John M. Dawson, Abraham Hertzberg, George C. Vlases, Harlow G. Ahlstrom, Loren C. Steinhauer, Ray E. Kidder, and W. L. Kruer, "Controlled Fusion Using Long Wavelength Laser Heating with Magnetic Confinement," Princeton Plasma Physics Laboratory Report MATT 782, 1971, revised in 1972.
8. Loren C. Steinhauer and Harlow G. Ahlstrom, "The Propagation of Radiation in a Cylindrical Plasma Column," *The Physics of Fluids* **13**, 1109-1114, June 1971.
9. M. C. Richardson, A. J. Alcock, K. Leopold, and P. Burtyn, "A High-Power High-Energy TEA CO_2 Laser," Paper Q-10 in VII International Quantum Electronics Conference, Montreal, Canada (May 8-11, 1972).
10. Moshe J. Lubin and Arthur P. Fraas, "Fusion by Laser," *Scientific American* **224**, No. 6, 21-33, June 1971.
11. Lowell Wood and John Nuckolls, "Fusion Power," *Environment* **14**, No. 4, 29-33, May 1972.
12. Gloria B. Lubkin, "AEC Opens Up on Laser Fusion Implosion Concept," *Physics Today* **25**, No. 8, 17-20, August 1972.
13. J. W. Shearer and J. J. Duderstadt, "Wavelength Dependence of Laser Light Absorption by a Solid Deuterium Target," Paper G-10 in VII International Quantum Electronics Conference, Montreal, Canada (May 8-11, 1972).
14. Harold L. Davis, "The Importance of Empty Space," *Physics Today* **25**, No. 8, 88, August 1972.

Acknowledgment

Work performed under the auspices of the U.S. Atomic Energy Commission.

3

NUCLEAR POWER IN CANADA

E. C. W. Perryman

A brief historical review of the Canadian Power Reactor Program is given, covering heavy-water moderated reactors cooled with heavy water, light water and organic liquid. The experience obtained from two prototypes, NPD (Nuclear Power Demonstration) and Douglas Point, is discussed in relation to the first year's successful operation of the Pickering Nuclear Power Station. Future improvements and trends in the CANDU family of reactors are described.

3-1. Introduction

In 1954 when Canada decided to develop nuclear power reactors she had behind her considerable experience in handling heavy water through the operation of NRX, which is now 25 years old. With this experience, the knowledge that heavy water is the most efficient moderator, and that Canada had plenty of uranium reserves but no U-235 enrichment experience, it was perhaps natural to choose a system based on heavy-water moderation and natural fuel.

The first design for the 25 MWe Nuclear Power Demonstration (NPD) reactor used a steel pressure vessel. During the design it was recognized that as reactor output increased with years, this vessel would become very large and thick and would be stretching the capabilities of Canadian industry. At that time the uncertainties that neutron irradiation introduced were unknown. In 1957 probably the most important decision in Canada's nuclear power development program was made. The design of NPD was changed from dependence on the steel pressure vessel to the use of Zircaloy pressure tubes, with the heavy-water moderator separated from the heavy-water coolant. In retrospect this was a decision of fantastic proportions. Not only was there very little experience with Zircaloy as a structural material, but the decision put Canada alone in its nuclear power development. The remainder of this paper shows what an excellent decision this was, not only because of the resulting successes in power reactor development, but also because the use of the Zircaloy pressure tubes has two very important advantages. First, it provides flexibility in choice of coolant and design. Second, it makes it possible to test a complete vessel before commitment and then to follow its behaviour and performance during the reactor's life and, if necessary, to replace it with a new vessel without destroying the plant.

3-2. The CANDU Concept

The CANDU concept can be considered as a family of reactors, each reactor being distinguished by the type of coolant. Table 3-1 and Figure 3-1 show the different characteristics of the three reactor types that are operating or under development. The dominant factor in thinking has always been neutron economy. This, together with low fabrication costs made possible by fuel simplicity (Figure 3-2), gives low fuel costs (about half that of the light-water reactors), a large amount of power per unit mass of uranium ore, and the highest plutonium production rate per unit of energy produced of any thermal reactor. This is why CANDU is sometimes called an advanced converter. Table 3-2 compares the uranium utilization of the CANDU heavy-water reactors and the USA light-water reactors. The materials of construction for the three reactor types are the same, and only in the case of the organic-cooled reactor is the fuel different. Thus the research and development program is in the main relevant to all three types. This has allowed development of all three with a minimum of effort and expenditure.

Table 3-1. CANDU–A Family of Reactor Concepts

DISTINGUISHING CHARACTERISTICS

1) Heavy-water moderation

   Fuel economy–freedom to use natural uranium
   Spreads out fuel–allows pressure tubes

2) Pressure tubes of neutron economic material

   East of proof testing
   Ease of scale-up
   Allows evolutionary improvements

3) On-power fueling

MEMBERS OF FAMILY CHARACTERIZED BY COOLANT

PHW – Pressurized heavy-water coolant
BLW – Boiling light-water coolant
OCR – Organic-cooled reactor

Table 3-2. U and Th Utilization

Fuel Cycle	*Reactor Type*	*Net Running Consumption (g/ekw-yr)*		
		U	*Pu*	*Th.*
PuSALE	LWR*	171	-0.29	–
	HWR* (nat.)	110	-0.41	–
	HWR (enr.)	120	-0.24	–
PuRECYCLE	LWR	114	–	–
	HWR	60	–	–
	FB*	1	-0.23	–
Th/U-233	LWR	65	–	2
	HWR	8	–	2

* LWR = light water coolant
HWR = heavy water coolant
FB = fast breeder

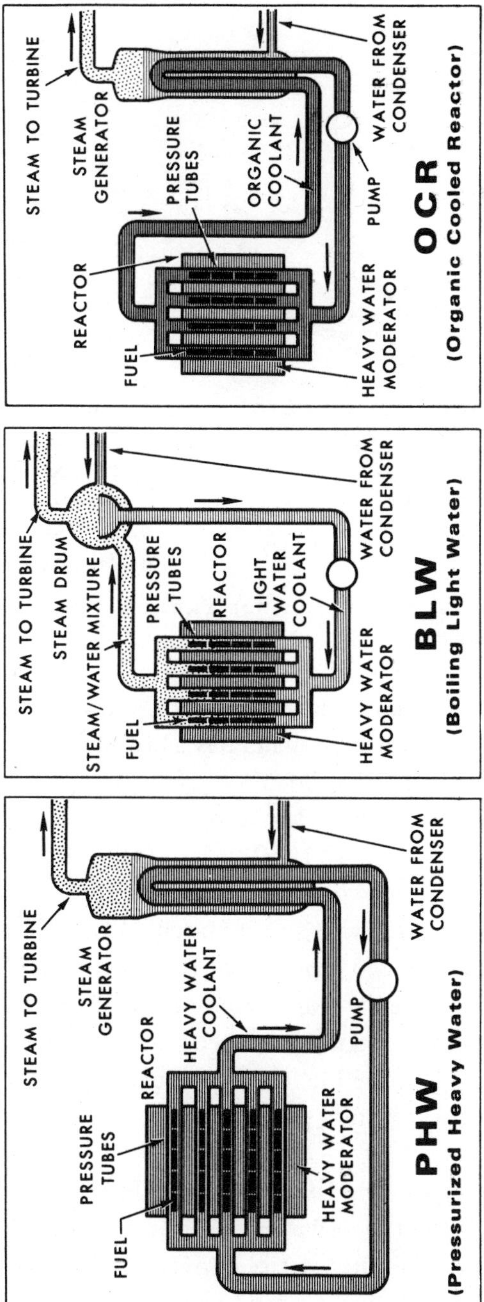

Fig. 3-1. Schematics of CANDU-PHW, -BLW, and -OCR

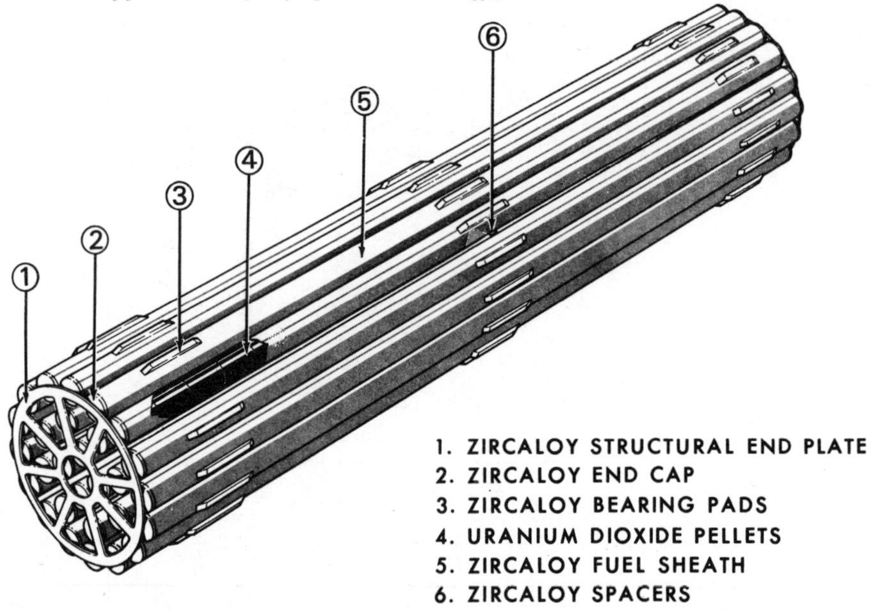

Fig. 3-2. Fuel bundle for Pickering reactor assembled from six basic components.

3-3. Canada's Nuclear Power Commitments and Requirements

At present Canada has 2115 MWe of nuclear power operating and a further 3500 MWe under construction. Over the next 20 years installed nuclear capacity is expected to grow rapidly, reaching 8000 MWe by 1980 and 35,000 MWe by 1990. In 1990, annual nuclear fuel expenditure will be about $181 million, which should be compared with an estimated $840 million per year if this power were produced from fossil fuel. Table 3-3 lists the CANDU power reactors that are operating or under construction. The major commitments to date have been made by Ontario Hydro, and today their operating nuclear power plants are producing about 14% of their total system generation.

3-4. Operating Experience–CANDU-PHW

NPD (25 MWe) came into operation in 1962 and demonstrated the viability of the CANDU-PHW concept. It is still operating and, in addition to producing power for Ontario Hydro, serves as

Table 3-3. CANDU Power Reactors—Operating Or Under Construction

Type	*MWe*	*Name*	*Start-Up*
BHW	22	NPD Rolphton	1962
PHW	208	Douglas Point	1967
PHW	125	KANUPP	1971
PHW	508	Pickering-1	1971
BLW	250	Gentilly	1971
PHW	508	Pickering-2	1971
PHW	203	RAPP-1	1972
PHW	508	Pickering-3	1972
PHW	508	Pickering-4	1973
PHW	203	RAPP-2	1974
PHW	750	Bruce-1	1976
PHW	750	Bruce-2	1977
PHW	750	Bruce-3	1978
PHW	750	Bruce-4	1979

a training ground for future electrical utility operating staff and also as a vehicle for carrying out part of the experimental program. For example, a few years ago it was converted from single-phase coolant to a two-phase boiling coolant (indirect cycle). The purpose was to determine whether boiling would introduce problems in the areas of control, coolant chemistry, radiation field build-up, or fuel. The reactor is now operating again in the single-phase coolant mode, and experiments on U-Si-A1 and UO_2-PuO_2 fuels are in progress.

Before NPD was operational, Douglas Point (200 MWe) was committed. Basically Douglas Point was the same design as NPD except that the power density was considerably higher and the on-power fueling machines were of a different design. Douglas Point became operational in 1967 and its performance has been similar to most other prototype nuclear plants in the world. For the last two years its capacity factor has averaged about 50%, but during the winter months it has been a reliable power source, last winter achieving an 80% capacity factor. It is worth noting that at least half of the unreliability has been associated with standard power station components such as valves, seals, turbines and generators. Initial difficulties experienced with the fueling machines have now been overcome, and on-power fueling is now a routine operation.

The important points learned from Douglas Point are:

1. A natural uranium reactor with very low fueling cost is practical.
2. On-power fueling is not only practical but offers many operational advantages.
3. It is necessary to design reactor buildings for complete recovery of the heavy water that escapes from the primary heat transport system.
4. Plant lay-out should be uncongested to provide easier maintenance, even though this may mean a larger heavy water inventory.
5. Heavy water and light water circuits should be separated to avoid downgrading of escaping heavy water.
6. The number of valves and mechanical joints should be minimized.

When the design of the 2,000 MWe Pickering station near Toronto began in 1965, three years' experience had accumulated from NPD (but none from Douglas Point). However, this limited experience did enable some major changes. The light-water systems were better segregated from the heavy-water systems; the number of valves and mechanical joints were reduced by more than a factor of two; air driers were included. All of these were expected to give significantly lower heavy water upkeep costs. As experience from Douglas Point accumulated, it was possible to inject some of it into Pickering during the later stages of construction and commissioning.

In February 1971 Canada's first commercial size 540 MWe unit at Pickering went critical and within three months had reached full power. Table 3-4 shows the significant dates for the three Pickering reactors that have been brought into service. Figures 3-3 and 3-4 show the operating experience on units 1 and 2,

Table 3-4. Significant Dates for the Three Pickering Reactors

Reactor	*Critical*	*First Steam*	*First Electricity*	*Full Power*	*In Service*	*On-Power Fuelling*
Pickering-1						
- Date	Feb 25/71	Mar 16/71	Apr 4/71	May 30/71	July 29/71	Nov 3/71
- Days after Critical		20	39	95	155	252
Pickering-2						
- Date	Sep 15/71	Sep 29/71	Oct 6/71	Nov 7/71	Dec 30/71	Apr 19/72
- Days after Critical		14	21	53	106	217
Pickering-3						
- Date	Apr 24/72	Apr 28/72	May 3/72	May 12/72	June 1/72	
- Days after Critical		4	9	18	38	

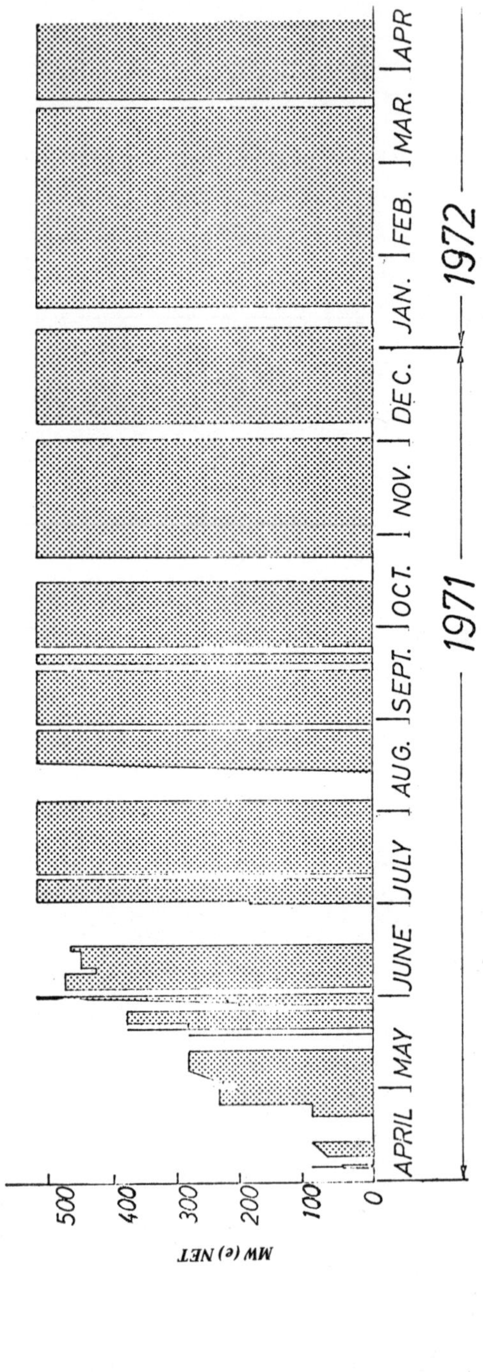

Fig. 3-3. Pickering G. S. Unit 1 Performance Summaries (Ref. 2)

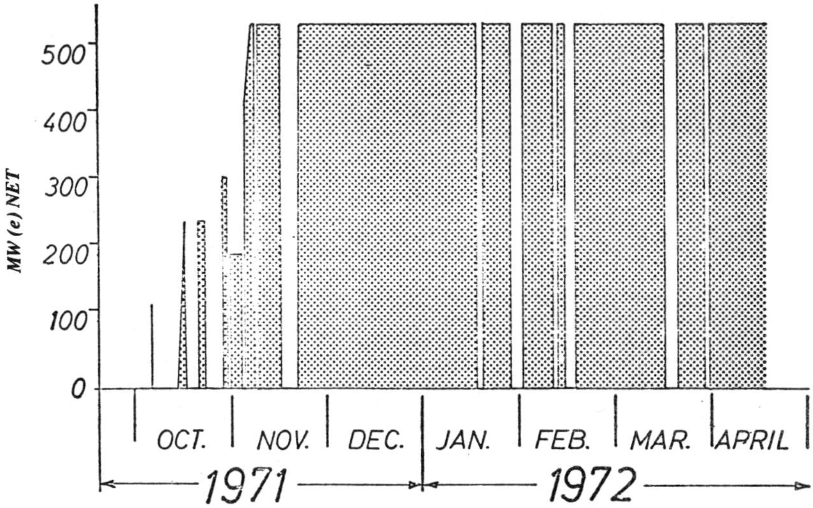

Fig. 3-4. Pickering G. S. Unit 2 Performance Summaries (Ref. 2)

respectively. To bring such complex plants to maturity in periods of 100 days or less is a very significant achievement. Since the units 1 and 2 were declared "in-service" by Ontario Hydro, they achieved an average capacity factor of 85.2% and 87.0% respectively up to the end of June 1972.* The experience gained in Douglas Point and NPD, together with the design changes and equipment modifications made in Pickering, have brought about the required low heavy water upkeep cost. Experience at Pickering to date has shown that the design target of 1.3 kg/h of heavy water can be accomplished. The results to date are given in Figure 3-5. The total upkeep decreased quickly as commissioning was completed and both units have operated well below the design figure, the heavy water upkeep cost being about 0.1 mills/kWh. The Pickering station is shown in Figure 3-6. The Bruce site containing Douglas

* At the time this paper was given (October 3, 1972), all three units had been shut down since the end of June because of the Ontario Hydro strike.

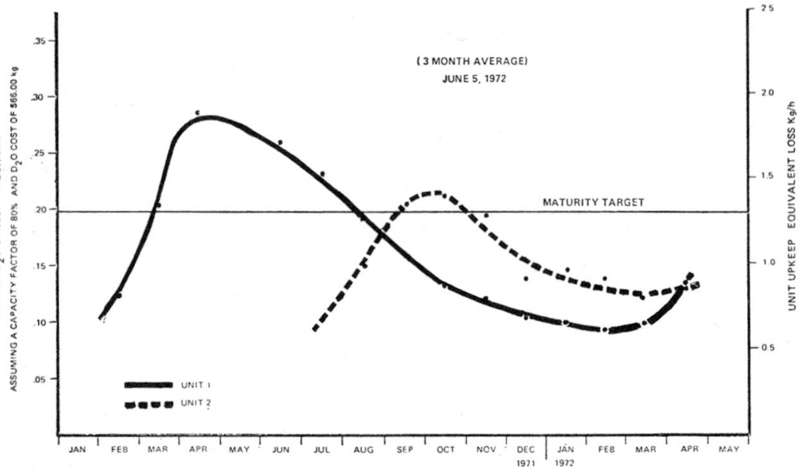

Fig. 3-5. Trends in Heavy-Water Upkeep Cost for Pickering Units 1 and 2

Fig. 3-6. Pickering Generating Station

Point, an 800 ton/year heavy-water plant, and the 3,000 MWe generating station under construction is shown in Figure 3-7.*

Fig. 3-7. Bruce Generating Station

Although the Pickering fuel machines are of the same general design as those at Douglas Point, they had to be redesigned in detail to make them compatible with the larger Pickering pressure tube. These on-power fueling machines have been brought into operation without any major difficulties and on-power fueling is now a routine operation. Although on-power fueling was chosen to obtain increased average fuel burn-up, it has demonstrated other advantages. For example, on-power fueling machines have provided

*Since this paper was written the fourth 500 MWe unit at Pickering has been brought into service and reached full power 12 days after first criticality. The operating performance of all units has been exceptionally good, giving an average station capacity factor better than 80% since the units were declared "in-service."

increased operational flexibility when problems have arisen. Fuel defects have been observed and the fueling machines have allowed the operators to remove such failed fuel after it has been detected. Moreover, these machines have given the operators flexibility in fuel management to help avoid fuel failures.

In Canada, fuel failure has not been a significant economic problem because the number of failed bundles has been less than 1% of those fed into the reactor. The problem first occurred at Douglas Point in March 1970, and since that time Chalk River has been attempting to find a solution. It quickly became apparent that the failures were associated with increasing the power output from the fuel after the Zircaloy cladding had been damaged significantly by neutron radiation. As discussed earlier, one possible answer was to change the fueling schedule and this was done at Douglas Point by moving eight bundles at a time rather than four. Although this reduces the average burn-up slightly, it does avoid large power changes on the fuel and had reduced the failure rate in Douglas Point to insignificant proportions. Since the beginning of 1971 only one bundle has failed in the reactor. In parallel with this, the development program has been working toward a permanent solution to the problem. Because of excellent loop irradiation facilities in the NRX and NRU test reactors, it has been possible to simulate the power excursions experienced by power reactor fuel and to test many fuel design modifications. More than one solution to the problem appears to be available. Modified fuel, which goes under the name CANLUB, has already been loaded into one of the Pickering reactors and is now displacing the earlier fuel in our fuel production lines.

3-5. Operating Experience–CANDU-BLW

The 250 MWe Gentilly prototype direct-cycle boiling light water reactor was built because it showed promise of lower capital and total unit energy costs. The lower capital arises from a smaller heavy water inventory, no steam generators and slightly higher efficiency. In addition, the use of light-water coolant eradicates any heavy-water leakage problem from the high-temperature, high-pressure circuits. To use natural uranium fuel necessitates reducing the coolant density. This reactor is owned by AECL and operated by Hydro-Québec at Trois Rivières between Montreal and Quebec City.

At the outset it was recognized that this reactor would have a positive void coefficient of reactivity, i.e., as the density of the coolant decreased, more reactivity and hence more power would be produced. Such changes in density can arise from perturbations in pressure, temperature, flow rate and neutron flux. Since the control of reactor power might therefore be more difficult, a prototype reactor was necessary.

Construction of the reactor started in 1966 and criticality was achieved November 1970. This prototype plant has been no different from earlier prototypes in that hardware problems associated with pumps, valves, heat exchangers and turbines have been experienced. The first electricity was produced in April 1971 and full power was achieved May 18, 1972. This long commissioning period was intentional since it was very important to fully understand the operational characteristics of the plant in view of the positive void coefficient. During the commissioning period many experiments were performed to determine what happens when a perturbation in reactivity is made. Experience has shown that the positive feedback associated with the positive void coefficient is not an instantaneous effect but has a time constant of many seconds. This enables the relatively slow-acting control rods to control such perturbations.

Like the CANDU-PHW reactors, Gentilly also has on-power fueling. The main difference is that a complete string of fuel is removed and the fuel shuffling done outside the reactor. The fueling machine has been brought into operation without any significant problems.

3-6. Future CANDU Reactor Concepts

CANDU-OCR (Organic-Cooled Reactor). Canada, like many other countries, has been interested in a reactor moderated with heavy water and using an organic such as Terphenyl to take the heat away from the fuel. The advantage of such a coolant is that it can be used at high temperature and low pressure. This should result in improved station efficiency, low corrosion product inventories leading to low radiation fields, and avoidance of heavy water leakage. A power reactor based on this coolant would be similar to the other two reactor types discussed, the major difference being in the operating temperature of the fuel cladding (500°C instead of 300°C) and pressure tubes (400°C instead of 280°C). Because of the higher coolant neutron absorption, a higher density fuel such as uranium carbide or uranium metal is necessary for a natural fuel cycle.

For the past seven years a 60 MW test reactor (WR-1) with a heavy-water moderator and organic coolant has been in operation at the Whiteshell Nuclear Research Establishment in Manitoba. When this reactor was designed, little information existed on the behaviour of zirconium alloys in high temperature organic coolant and there were major uncertainties on whether deposits of carbonaceous material on fuel element surfaces would be a real problem. Experiments in small organic loops at Chalk River showed that deposition could be avoided provided the chlorine content of the organic was maintained at a low level. Fuel experiments in the same loops also showed that zirconium alloys could be considered as fuel cladding as long as a controlled water content was maintained in the organic coolant to keep the zirconium oxide film in good repair. Insufficient evidence existed to show that zirconium alloy pressure tubes could be considered, so the reactor was built with stainless-steel pressure tubes. Four zirconium alloy tubes were installed for experimental purposes. The reactor has operated steadily for seven years with enriched UO_2 fuel clad with a zirconium-2½% niobium alloy, this being used instead of Zircaloy-4 because of its lower hydrogen pick-up. Additional experimental zirconium alloy pressure tubes have been irradiated in WR-1 and in the two organic loops at Chalk River over the last few years. These tubes have been monitored for creep, and many fracture experiments have been performed on them after irradiation. This work has shown that at least a 10-year life for a zirconium alloy pressure tube in an organic-cooled power reactor is realistic and all the stainless steel tubes in WR-1 have now been replaced by zirconium alloy tubes. Experiments in our organic loops with full-size fuel bundles have shown that zirconium alloy clad uranium carbide will operate satisfactorily at the fuel ratings required and for burn-ups that the natural fuel cycle will demand.

A very important fact that has been learned from the operation of WR-1 is that there is little transport of radioactive material in the primary heat transport system. Indeed, the radiation fields in the primary heat exchanger room are low enough to allow a man to work all day without exceeding his allowable exposure. Thus contact maintenance is possible. This is orders of magnitude different from the water-cooled power reactor and is a significant economic plus for the organic-cooled reactor. This is probably the major reason for the 84% availability achieved since start-up of WR-1.

Although the moderator, moderator tank heat-transport circuits, pressure tubes and calandria tubes of WR-1 are fully representative of a power reactor, a prototype station will be needed before the reactor concept gains commercial acceptance. At present a design study is in progress on a 500 MWe OCR fueled with natural uranium carbide.

Enriched CANDU-BLW. All CANDU reactors discussed so far are designed to burn natural uranium and the fuel design has been such that sufficiently low fabrication costs have been achieved to give the lowest fueling cost of any nuclear power reactor in the world. The fueling cost achieved, 0.7 mills/kWh, is about half that of the USA light-water reactors even assuming no value for the plutonium contained within the discharged fuel.

With the presently predicted nuclear power requirements for Canada, by 1980 about 7,600 kg of plutonium will exist in discharged fuel and in the same year the plutonium production rate will be about 1,600 kg/year. The economic value of this plutonium is most uncertain; it will lie between the lowest value defined by the cost of separating the plutonium from the fuel and the maximum value which a given reactor system can afford to pay for mixed uranium oxide and plutonium oxide fuel elements.

The possible options for the disposal of this plutonium are as follows:

1. Sell the discharged fuel.
2. Reprocess the discharged fuel and sell the separated plutonium.
3. Reprocess the discharged fuel and use the plutonium in operating reactors or in reactors designed to burn it most efficiently.

The fast reactor people state that they can afford to burn plutonium at 20 $/g. Our own staff at Whiteshell Establishment say that using the amine process for reprocessing, they can recover the plutonium at 6 $/g. Thus if the market allows, we could obtain a net of 14 $/g, which would mean that the net fuel cost for a CANDU-PHW would be very close to zero. This is unlikely to happen because those countries developing the fast reactor will ensure that they install in their power system the mix of thermal and fast reactors which will ensure that they are independent of outside sources for their plutonium requirements. There could, however, be a few years when they may have to buy from an outside source.

Over the last two years we have been investigating the third option, namely, how to best use the plutonium in CANDU reactors. We have addressed ourselves to the following questions:

1. How can we best use plutonium to reduce the capital cost of the CANDU reactor and at the same time reduce the total unit energy cost?
2. In what form should the plutonium be used?

Both these questions are extremely complicated and in the space available only a short summary of the results of our study is possible.

First we compared CANDU-PHW and CANDU-BLW designs optimized for burning UO_2-PuO_2 fuel. This showed that the CANDU-BLW reactor gave the best economics, i.e., we could afford to pay more for plutonium when burnt in a BLW than when burnt in a PHW. Our studies were then extended to optimizing the plutonium-burning BLW and we found that a reactor having a capital cost some 20% lower than a natural PHW and a TUEC some 11% lower might be possible. Such a reactor would have the same physical size as the Gentilly 250 MWe unit, but it would produce three times as much power. Unlike a natural uranium reactor, this enriched design would be undermoderated; that is, the moderator-to-coolant ratio would be considerably lower than a natural reactor. This would be achieved by spacing the pressure tubes closer together and increasing the calandria tube diameter. The reactor would have a higher power density which our heat transfer and fuel development work indicates is possible. The capital cost savings arise from:

1. Fewer fuel channels;
2. Smaller calandria;
3. No heat exchanger;
4. Reduced pumping power;
5. Lower heavy water inventory;
6. Only one fueling machine; and
7. Shorter construction time.

Earlier it was noted that a natural uranium CANDU-BLW such as Gentilly has a positive void coefficient. This coefficient is a strong function of the moderator-to-coolant ratio and the moderator-to-fuel ratio. As these ratios decrease, the void coefficient becomes less positive. The optimized plutonium reactor design would have an almost zero coefficient and therefore would be easier to control than the natural version.

This enriched reactor can be designed to be self-sufficient in plutonium. The plutonium for the first fuel charge would come from inventory, but during operation as much plutonium would be produced as would be burned. A 750 MWe reactor of this design would need 500 kg of plutonium. One can easily show that 5000 MWe of natural CANDU-BLW reactors would provide enough plutonium to support the installation of about 19,000 MWe of the enriched BLW type. Furthermore, the natural uranium requirements for 24,000 MWe of natural CANDU-BLW plus plutonium enriched BLWs would be about 60% of that for 24,000 MWe of all natural CANDU-BLW reactors.

Although, in general, plutonium and U-235 are interchangeable in an enriched reactor of the type under discussion, there are detailed differences which can have important effects on the controllability of the reactor. For example, the variation of reactivity with coolant density is a function of both the Pu-239 and Pu-240 content, which in turn is a function of the burn-up that the UO_2 fuel has received. For a given Pu-240 content, such as exists in discharged fuel from CANDU reactors, the void coefficient is less positive for plutonium fuel than for fuel enriched with U-235. However, a reactor with a near zero power coefficient can be designed with U-235 enriched fuel. Then later on in life, plutonium fuel can be used. At present we have a study in progress on a 500 MWe U-235 enriched CANDU-BLW which will have a low enough enrichment, about 1.4 w/o U-235, that reprocessing will be unnecessary. Our expectation is that the capital cost will be significantly lower than the natural CANDU-PHW; the fuel cost will be slightly higher, and the total unit energy cost will be lower. The pressure tube reactor gives the designer a great deal of flexibility. One can trade capital cost and fuel cost by changing the design, while still preserving the fundamental rule of good neutron economy to the extent that a throw-away fuel cycle can be used. Thus, the reactor can respond to the economic climate that prevails at any time. Moreover the uncertainties associated with the cost of reprocessing the plutonium, and the cost of U-235, can be avoided.

Advanced Breeders. A very reasonable question to ask is "Why has Canada's nuclear power program ignored the fast breeder reactor?" The answer is complex, but the essential reason is that we have a thermal reactor system with a high conversion ratio, good fuel utilization and a low fueling cost. The expectations in the USA are that the capital cost of the fast breeder will be about 10% higher than thermal reactors and the fuel cost will decrease with time from about 1 mill/kWh to 0.5 mill/kWh providing very high fuel burn-ups (100,000 MWd/tonne) can be achieved. The CANDU reactor fuel costs are already in this range and with potential for further reduction. Thus it seems clear that there is no economic reason for Canada to embark on fast breeder reactors. If there is any argument, it must be based on conservation of uranium resources.

To consider this we must consider the fundamentals of the fission process. To produce a chain reaction at least one neutron is required from each neutron absorbed in order to induce the succeeding fission. Of the remaining neutrons, as many as possible should be absorbed in fertile material such as U-238 or Th-232 to produce further fissile atoms of Pu-239 or U-233. As stated earlier, heavy water moderated reactors have a high conversion ratio: each fissile atom destroyed produces between 0.8 and 0.9 new fissile atoms, compared with 0.5 and 0.6 new fissile atoms for light-water reactors. When more than one new fissile atom is produced the system is said to "breed," in which case all the world's resources of fissile and fertile material can, in principle, be burned to produce energy. This means that η, the ratio of neutrons emitted to neutrons absorbed, must be greater than two and for practical purposes must exceed about 2.2 to take into account neutron losses by (1) capture in structural materials and fission products; (2) leakage; and (3) loss of fissile material during fabrication and reprocessing. Figure 3-8 shows how η varies with neutron energy. It can be seen that marginal breeding is possible with U-233 at low neutron energies typical of thermal reactors. However, thorium, from which U-233 is produced, has no natural fissile isotope and therefore the thorium/U-233 reactor cycle must be initiated by added fissile material. The fuel resource argument is summarized in Table 3-2, taken from an international study. This shows that there is no question that the fast breeder is a desirable objective for the conservation of uranium resources. However, at the anticipated growth rates for nuclear power over the next half-century or more, one thermal reactor will be required for every two or three fast breeders in order to supply the initial plutonium inventory. This implies a net running consumption of uranium for the combined breeder converter system of from 25 to 50 g/ekW-yr, depending on the breeding ratio achieved. Table 3-2 also shows that for thermal reactors operating on the thorium/U-233 cycle, heavy-water reactors have a real advantage over light-water reactors.

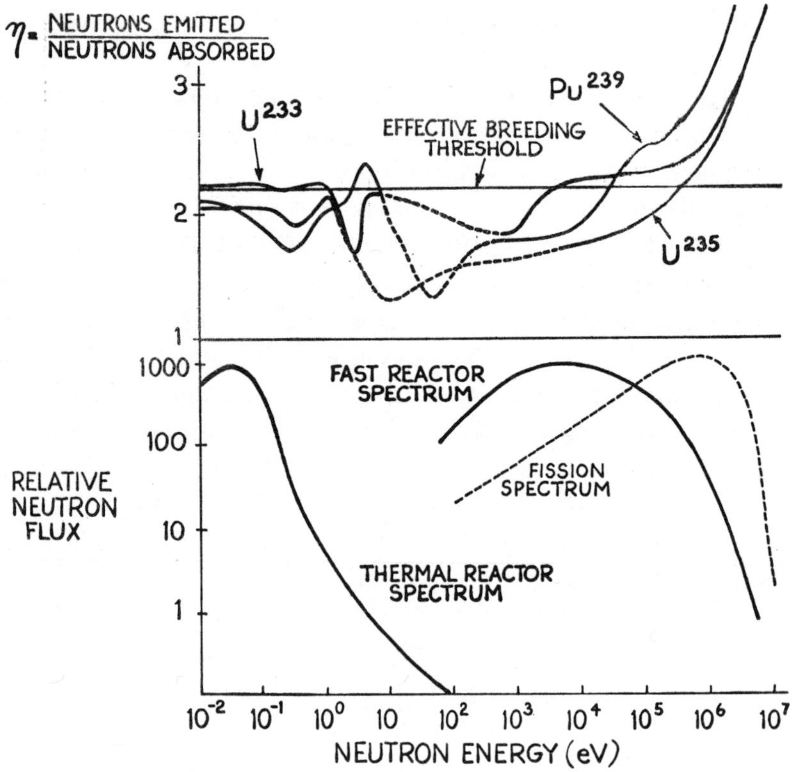

Fig. 3-8. How η Varies with Neutron Energy

As far as uranium requirements are concerned, they are not very different from the fast breeders. Like the fast breeder, an expanding power system using thorium will require extra uranium to provide the required inventory. From an overall system viewpoint, both the fast breeder and the Th/U-233 thermal reactor are near-breeder systems, both requiring extra fissile material obtained from the conversion of fertile material by neutron capture. Coupling this argument with Canada's uranium and thorium reserves, we cannot see any good reason for Canada to develop the fast reactor. The evolution of the heavy-water reactor to operate on the Th/U-233 cycle provides a logical future development of heavy-water reactors which would complement the fast reactors in the total utilization of the world's nuclear fuel resources.

Heavy Water Supply. At the present time Canada has one 400 ton/yr heavy-water plant operating at Port Hawkesbury, Nova Scotia, an 800 ton/yr plant under construction at the same site as the Douglas Point reactor, and the 3000 MWe Bruce Generating Station. The Glace Bay 400 ton/yr plant in Nova Scotia is being rehabilitated. The Port Hawkesbury, Bruce and Glace Bay plants are expected to reach maturity in 1975, 1977 and 1979 respectively. Figure 3-9 shows the commited heavy water production and our expected domestic demand. The Bruce plant is nearing completion and during 1973 we expect 530 metric tons from the Port Hawkesbury and Bruce plants.

All the plants committed so far are based on the process of isotopic exchange between the water feed and hydrogen sulphide

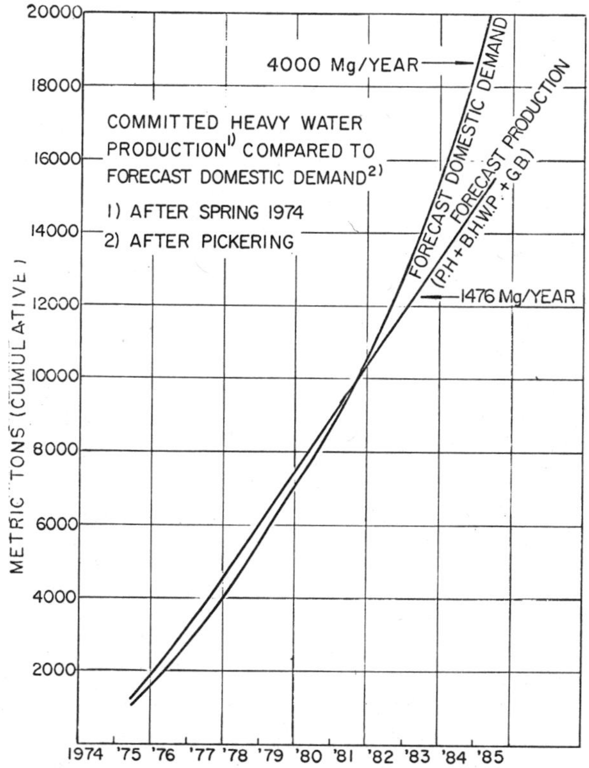

Fig. 3-9. Committed Heavy Water Production Compared to Forecast Domestic Demand. (Ref. 17)

gas. Development work is proceeding on two promising alternative processes, but it will be close to ten years before this development leads to an operating commercial unit. Heavy water production plants like U-235 enrichment plants are capital intensive. For example, about 60% of the product cost is associated with the capital investment. Both new processes that are under development are expected to give significantly lower capital costs.

3-7. Conclusions

In this short paper it has been possible to give only a very brief description of the various CANDU reactor concepts and the Canadian nuclear power program. A bibliography is included which will allow the reader to obtain more details.

References

1. J. H. Armstrong, "Operating Experience at Douglas Point GS," Proceedings of the 1972 Annual Conference of the Canadian Nuclear Association, 72-CNA-301.
2. E. P. Horton, "Early Operating Performance at Pickering Generating Station," 72-CNA-302.
3. L. W. Woodhead, "Review of Ontario Hydro's Nuclear Program and Experience," 72-CNA-303.
4. P. A. Léger and L. F. Monier, "État de la Centrale Nucléaire de Gentilly," 72-CNA-304.
5. L. W. Woodhead, D. C. Milley, K. E. Elston, E. P. Horton, A. Dahlinger and R. C. Johnston, "Commissioning and Operating Experience with Canadian Nuclear-Electric Stations," 4th U. N. International Conference on the Peaceful Uses of Atomic Energy, Paper No. A/Conf. 49/A/148.
6. D. L. S. Bate, P. F. Mayes and W. S. Philip, "Costing of Canadian Nuclear Power Plants," 4th U. N. International Conference, Paper No. A/Conf. 49/A/149.
7. G. Hake, P. J. Barry and F. C. Boyd, "Canada Judges Power Reactor Safety on Component Quality and Reliable System Performance," 4th U. N. International Conference, Paper No. A/Conf. 49/A/150.
8. R. G. Hart, L. R. Haywood, and G. A. Pon, "The CANDU Nuclear Power System: Competitive for the Foreseeable Future," 4th U.N. International Conference, Paper No. A/Conf. 49/A/151.
9. A. J. Mooradian, E. C. W. Perryman, T. J. Kennett, "Irradiation Facilities: Their Importance in Developing Canadian Nuclear Competence," 4th U. N. International Conference, Paper No. A/Conf. 49/A/152.

10. L. R. Haywood, J. A. L. Robertson, J. Pawliw, J. Howieson, L. L. Bodie, "Fuel for Canadian Power Reactors," 4th U. N. International Conference, Paper No. A/Conf. 49/A/156.
11. W. B. Lewis, M. F. Duret, D. S. Craig, J. J. Veeder, A. S. Bain, "Large-Scale Nuclear Energy from the Thorium Cycle," 4th U. N. International Conference, Paper No. A/Conf. 49/A/157.
12. J. R. MacEwan, A. S. Bain, M. J. F. Notley and R. W. Jones, "Irradiation Experience with Fuel for Power Reactors," 4th U. N. International Conference, Paper No. A/Conf. 49/A/158.
13. W. Evans, P. A. Ross-Ross, J. E. LeSurf and H. E. Thexton, "Metallurgical Properties of Zirconium-Alloy Pressure Tubes and Their Steel End-Fittings for CANDU Reactors," 4th U. N. International Conference, Paper No. A/Conf. 49/A/159.
14. A. M. Marko, P. J. Barry, R. Wilson, K. Wong, P. O. Perron, and J. L. Weeks, "Nuclear Power and the Environment," 4th U. N. International Conference, Paper No. A/Conf. 49/A/160.
15. H. K. Rae et al., "Heavy Water," AECL-3866.
16. A. R. Bancroft, "The Canadian Approach to Cheaper Heavy Water," AECL-3044, 1967.
17. L. R. Haywood, "Heavy Water," 72-CNA-305.

4

SUPERCRITICAL HELIUM HEAT TRANSFER

Patricia J. Giarratano

This paper reports part of the National Bureau of Standards Cryogenics Division's program to provide helium heat transfer information to designers of helium cooling systems. An experiment on supercritical helium heat transfer is described, and its results are compared with various standard and modified correlation expressions.

Extensive appendices give tables of helium viscosity, thermal conductivity, and Prandtl numbers from 3 to 300 K.

4-1. Introduction

The past few years have witnessed the development of reliable high field superconductors whose use as current conductors at low temperatures has led to new technological systems of increased performance and efficiency. Examples of working systems include superconducting magnets for beam control in high-energy physics and for plasma containment in controlled thermonuclear reaction studies. Serious consideration is being given to superconducting motors, generators, transmission lines, magnetic levitation systems for high-speed ground transport, and magnetic ore filtration techniques.

Helium technology must keep pace with this materials development. For relatively small laboratory systems it is common simply to immerse the conductor in a container supplied with liquid helium vented at atmospheric pressure, and thus maintained at 4 K. However, as the system size, cost, and complexity increase, it is important to take a closer look at the means for maintaining the low-temperature environment. In brief summary, the selection of an optimum system requires comparison of the merits of (1) the common pool boiling heat-transfer mode; (2) cooling by forced flow of supercritical or boiling helium through flow channels; (3) the use of superfluid helium, below 2.17 K; and (4) operation of the device in conjunction with its own closed-cycle refrigeration system, especially as the size increases.

Since 1968 the Cryogenics Division of the National Bureau of Standards has been engaged in a study to provide the reliable data on helium properties and helium heat transfer which are necessary for serious engineering studies of superconducting systems. At the start of the program such information was seriously lacking.

In this paper a brief review of forced convection supercritical helium heat transfer is presented. More complete information on the total helium program is given in references 1 through 13.

4-2. Thermodynamic and Transport Properties

The primary goal of the helium heat-transfer program was to measure forced convection heat-transfer coefficients for supercritical helium flowing in a straight tube and to determine a correlation which best represented the data of that study.

Knowledge of thermodynamic and transport properties of supercritical helium is a necessary prerequisite to heat-transfer correlation studies. An equation of state that is a substantial improvement over existing equations has been developed by McCarty, and tabulated thermodynamic properties calculated from this equation are given in reference 13 (which is available from the Cryogenics Division of the National Bureau of Standards). The viscosity of helium in the region of interest has been measured by Steward,[12] and an empirical correlation for helium thermal conductivity based on a comprehensive review and compilation of experimental data has been developed by Roder.[16] These are regarded as the best available values for the relevant thermodynamic and transport properties of helium.

Tables of viscosity, thermal conductivity and Prandtl numbers for helium are given in Tables A1 through A3 in the Appendix.

4-3. Discussion of Correlations

In convective heat transfer the quantity of heat transferred from a solid wall to a liquid or gaseous medium in direct contact with the wall is described by the following equation:

$$Q/A = h\,(T_w - T_b) \tag{1}$$

where

Q/A = heat flux, W/cm^2
h = heat transfer coefficient, W/cm^2 - K
T_w = Wall temperature, K
T_b = bulk temperature of the fluid, K.

Predictive correlations for the fully developed heat transfer coefficient h, during turbulent forced convection of a single-phase fluid flowing inside a tube, have been well-established; e.g., the correlation of Dittus-Boelter:[14]

$$\frac{hD}{\kappa} = 0.023\ Re^{0.8}\ Pr^{0.4} \tag{2}$$

Where

D = inside tube diameter, cm
κ = thermal conductivity of the fluid, W/cm-K

$$Re = \frac{DG}{\mu}$$

$$Pr = C_p \frac{\mu}{\kappa}$$

G = Mass velocity, $g/s\text{-}cm^2$
μ = viscosity, g/cm-s
C_p = specific heat, J/g-K

All properties are evaluated at the bulk temperature and pressure of the fluid.

The Dittus-Boelter equation is generally applicable for moderate ΔT situations where there is negligible property variation across the boundary layer. However, for heat transfer with large ΔT's or for heat transfer in the near critical region, where fluid properties may change significantly with small changes in temperature, the Dittus-Boelter equation is inadequate.

Many correlations have been developed for predicting heat-transfer coefficients for fluids with widely ranging properties. These generally are modifications of Equation (2). Typical modifications may involve any combination of the following.

1. Selection of some temperature other than bulk temperature for properties evaluation.
2. Inclusion of a T_w/T_b and/or L/D parameter or some function of these parameters.
3. Inclusion of a wall-to-bulk property ratio or some function of that parameter.

Examples of the variable property correlations are given by Taylor.[15]

4-4. Experimental Measurement of Supercritical Helium Heat-Transfer Coefficients

A more detailed description of the following is contained in reference 4 and the interested reader is referred to that publication.

Description of Apparatus. The basic experimental system, as shown in Figure 4-1,* consists of a closed flow loop in which supercritical helium is circulated by a miniature centrifugal pump. A schematic of the pump and its performance characteristics are shown in Figures 4-2* and 4-3 respectively. A complete discussion of the pump is given by Sixsmith and Giarratano.[6] A high-pressure helium cylinder supplies the test fluid which enters the flow loop through a port in the pump cap. Liquid helium, which is circulated through two heat exchangers, provides the cooling for the test fluid as it flows around the loop.

At two points along the 0.2128 cm i.d. x 10 cm long stainless steel test section (L/D of 20 and 40) the local heat-transfer

---

*Use of trade names in this paper is for the sake of clarity and does not in any way imply a recommendation or endorsement by the National Bureau of Standards.

1. Germanium Resistance Thermometer (Wall temperature)
2. Germanium Resistance Thermometer (Downstream bulk temperature)
3. Germanium Resistance Thermometer (Upstream bulk temperature)
4. Electrical Terminal
5. Mixing Chamber
6. Stainless Steel – Pyrex Seal
7. Flow Orifice
8. Pressure Tap
9. Pump
10. Heat Exchanger
11. Inner Copper Radiation Shield
12. Outer Copper Radiation Shield
13. Stainless Steel Dewar
14. Vacuum Can
15. Liquid Nitrogen
16. Length Of Test Section (Resistance heated)

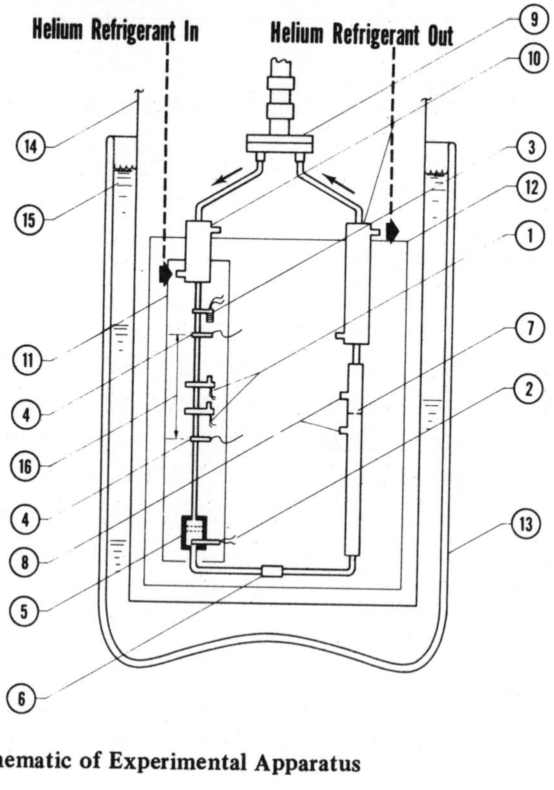

Fig. 4-1. Schematic of Experimental Apparatus

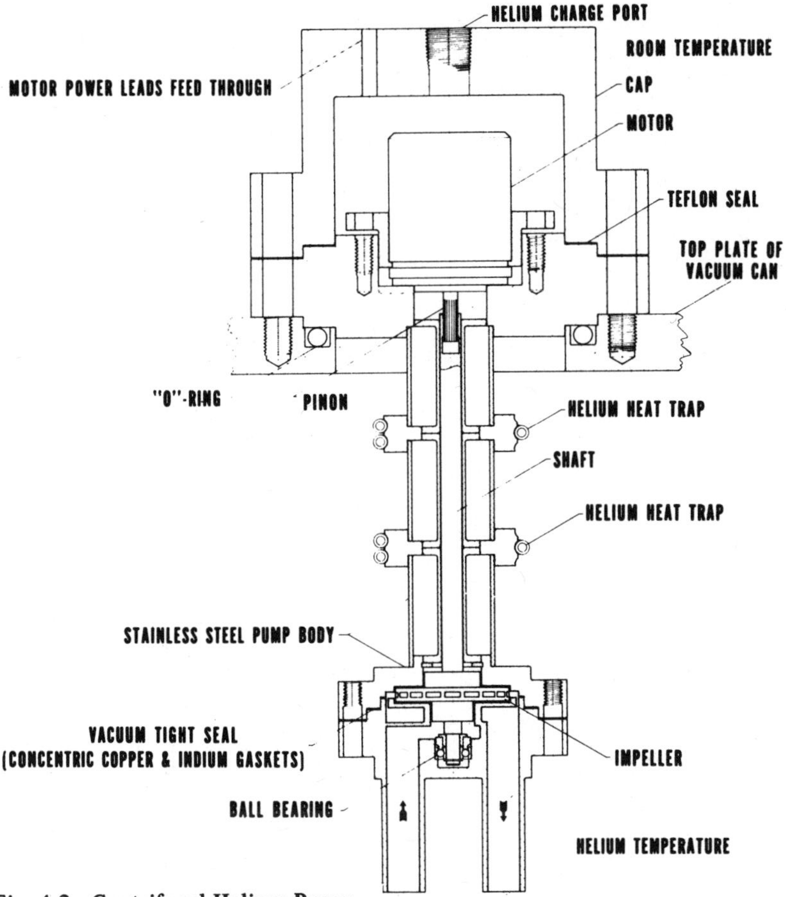

Fig. 4-2. Centrifugal Helium Pump

coefficients, h, in Equation (1), are experimentally determined by measurement of the electrical power input to the tube, Q, the outer wall temperature T_{w_o}, and the bulk temperature T_b. The inside wall temperatures, T_w, are obtained by calculation of the temperature drop through the tube wall, and the average fluid (bulk) temperatures at these points are determined from a thermal energy balance and measurement of the inlet and outlet fluid temperatures of the test section. The mass flow rate, used in correlating the data, is measured calorimetrically and by measurement of the pressure drop across a previously calibrated flow orifice.

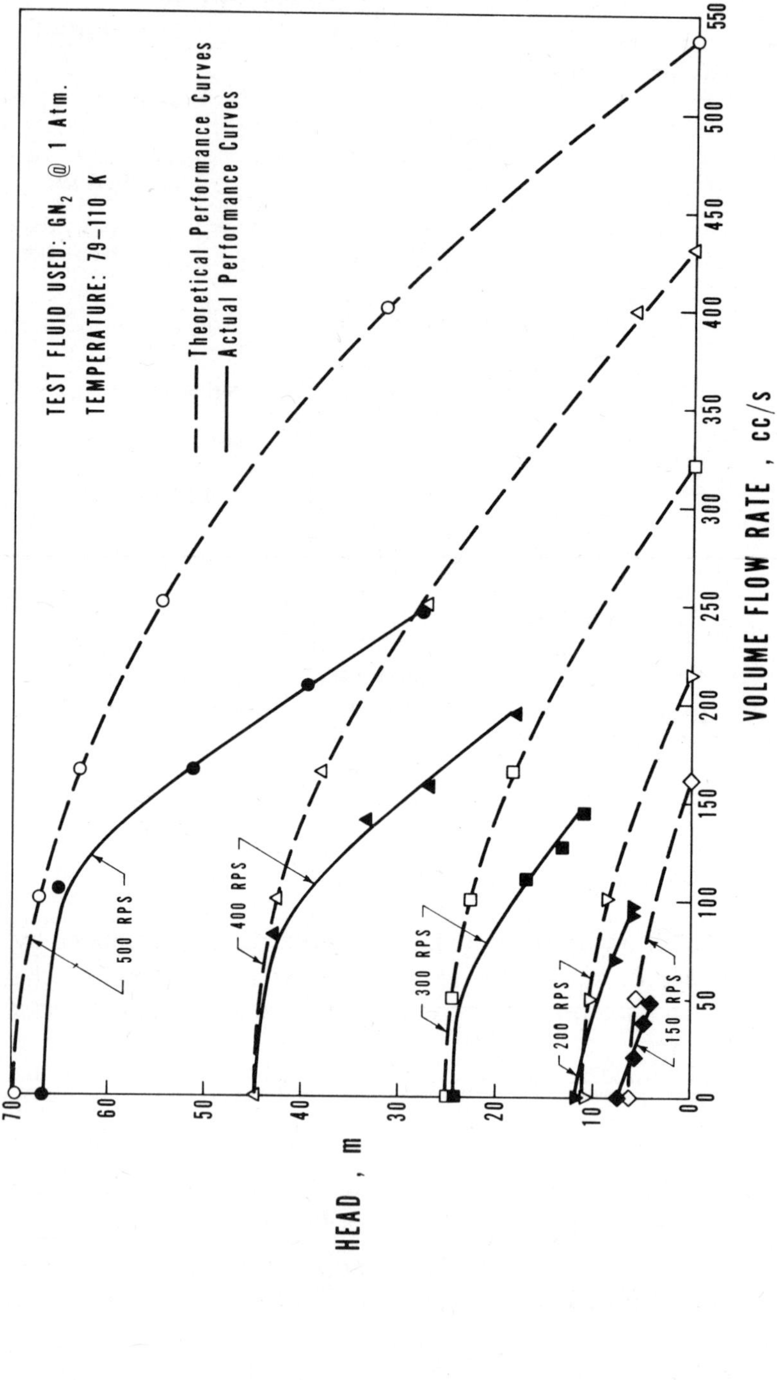

Fig. 4-3. Performance Characteristics of Helium Pump

Presentation of Experimental Results. The experimental data are presented graphically in Figure 4-4, where $(hD^{0.2}/G^{0.8})(T_w/T_b)^{0.716}$ is plotted versus T_b/T_{tc} for the range of pressures covered in the experimental measurements. T_{tc} is the transposed critical temperature, defined as the temperature at which the specific heat, C_p, is a maximum for the given pressure. T_{tc} as a function of pressure may be interpolated from the data of Figure 4-5. The ordinate for Figure 4-4 is suggested from the following correlation, which is developed in the next section.

$$\frac{hD^{0.2}}{G^{0.8}}\left(\frac{T_w}{T_b}\right)^{0.716} = 0.0259\,\frac{\kappa^{0.6}\,C_p^{0.4}}{\mu^{0.4}} \qquad (3)$$

The enhancement, or peak, in the properties parameter at T_{tc}, as shown in Figure 4-5, is reflected by the enhanced heat-transfer coefficient (for a given pressure, flow rate, diameter, and wall-to-bulk temperature ratio) at $T_b/T_{tc}=1$, as shown in Figure 4-4.

One data point reported by Johannes[17] is also shown in Figure 4-4. More data were obtained by Johannes but there is not sufficient information contained in this reference to calculate the ordinate of Figure 4-4 for more than the one data point shown. Johannes' data point is approximately 1.2 times higher than the 6 atm data of this work.

4-5. Comparison of Experimental Data with Predictive Correlations

A least-squares-fit technique[18] was used to determine the equation which best represents the data of this investigation. The correlations tested and the standard deviation (given in %) of the measured points from predicted values using each correlation are summarized in Table 4-1.

The first correlation tested is the classical Dittus-Boelter constant property turbulent heat-transfer correlation, expression (1)

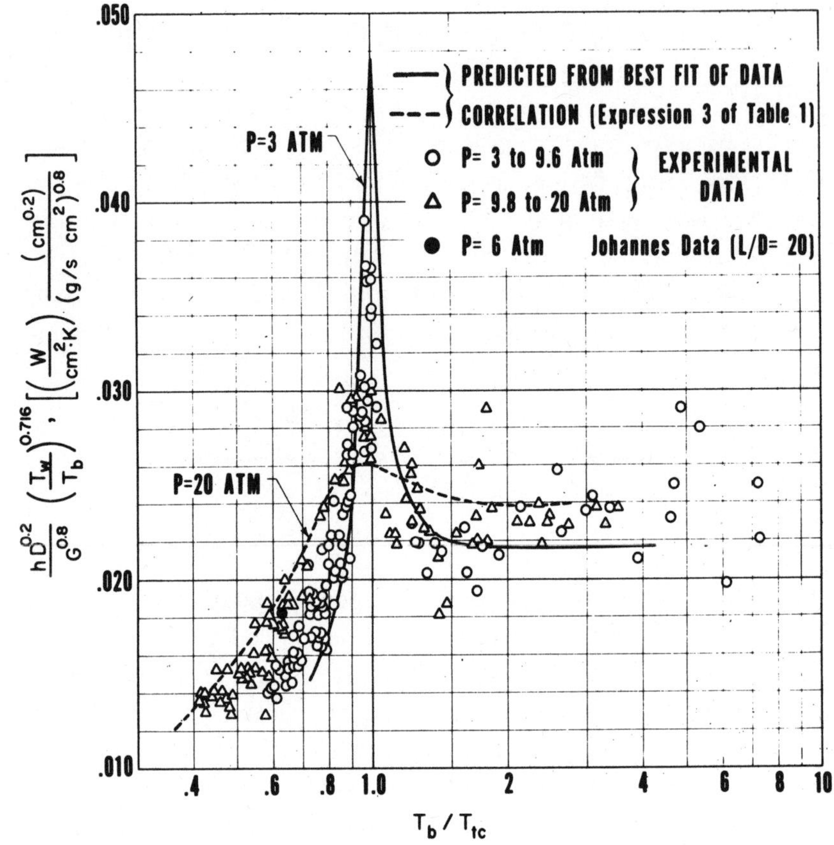

Fig. 4-4. Experimental and predicted heat transfer results using expression (3) of Table 4-1 (1 atm = .1 MN m^{-2}).

in Table 4-1. If the constant 0.023 is replaced by 0.022, the Dittus-Boelter expression can be used to predict the measured heat-transfer coefficient to within 14.8%. Brechna[19] reported that the constant as indicated by his data is 0.040.

A substantial improvement in % standard deviation (8.3%) is obtained with expression (2) of Table 4-1. This expression includes a wall-to-bulk temperature ratio parameter in an attempt to account for variation of fluid properties during heat transfer to near critical fluid. It is comforting to note that the exponents of

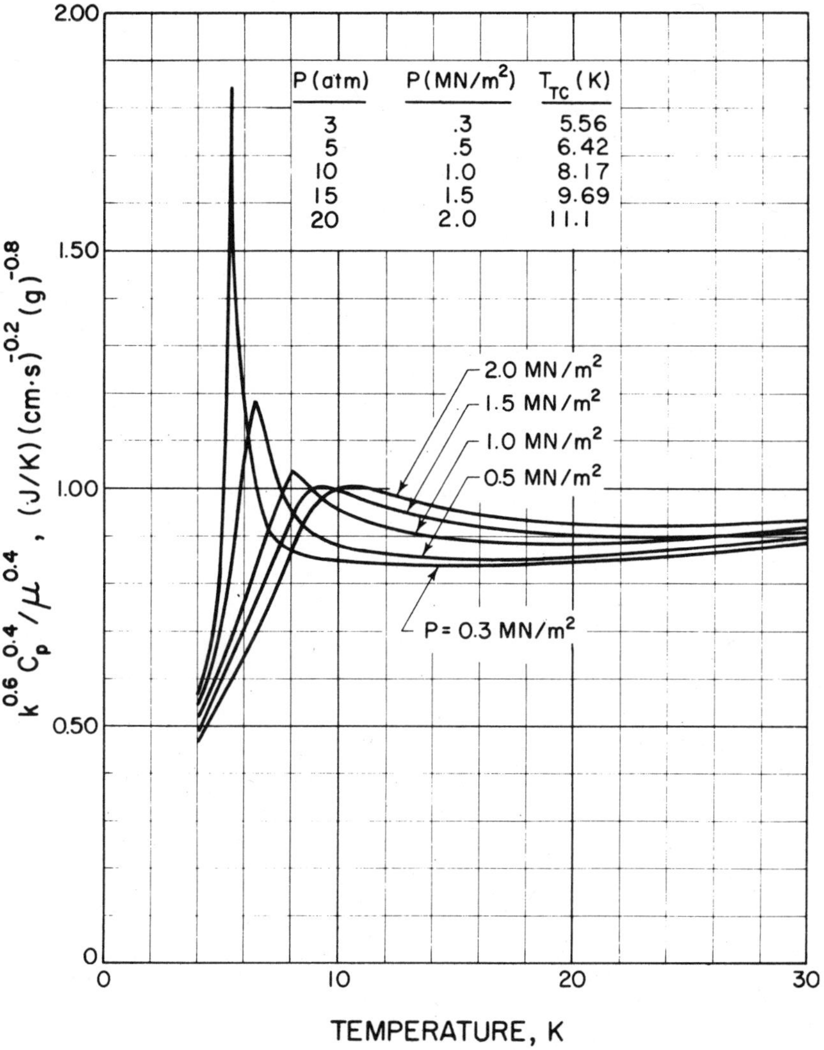

Fig. 4-5. Variation of properties parameter with temperature for constant pressure.

Table 4-1. Summary of Correlations Tested

Correlation		C_1	Standard Deviation for C_1	C_2	Standard Deviation for C_2	C_3	Standard Deviation for C_3	C_4	Standard Deviation for C_4	C_5	Standard Deviation for C_5	C_6	Standard Deviation for C_6	Standard Deviation for Correlation (percent)
$Nu = C_1 Re^{0.8} Pr^{0.4}$	(1)	.0219	.0002											14.8%
$Nu = C_1 Re^{C_2} Pr^{C_3} \left(\frac{T_w}{T_b}\right)^{C_4}$	(2)	.0258	.0053	.801	.0148	.461	.0186	-.711	.0532					8.3%
$Nu = C_1 Re^{.8} Pr^{.4} \left(\frac{T_w}{T_b}\right)^{C_4}$	(3)	.0259	.0002					-.716	.0268					8.5%
L/D = 20 $Nu = C_1 Re^{.8} Pr^{.4} \left(\frac{T_w}{T_b}\right)^{C_4}$	(4)	.0266	.0003					-.709	.0366					8.3%
L/D = 40 $Nu = C_1 Re^{.8} Pr^{.4} \left(\frac{T_w}{T_b}\right)^{C_4}$	(5)	.0260	.0003					-.822	.0433					8.3%
L/D = 20 $Nu = .023 Re^{.8} Pr^{.4} \left(\frac{T_w}{T_b}\right)^{C_4}$	(6)							-.373	.0378					13.1%
L/D = 40 $Nu = .023 Re^{.8} Pr^{.4} \left(\frac{T_w}{T_b}\right)^{C_4}$	(7)							-.495	.0372					11.5%
$Nu = .023 Re_f^{.8} Pr_f^{.4} \left(C_5 + C_6 \frac{\nu_w}{\nu_b}\right)$	(8)									1.028	.0202	-.00513	.0111	16.9%

the Re and Pr (0.801 and 0.461, respectively), as determined by the computer program, are in good agreement with the commonly used 0.8 and 0.4 numbers of the Dittus-Boelter expression.

Further, as seen in expression (3) of Table 4-1, forcing the exponents on Re and Pr to 0.8 and 0.4 and redetermination of the constant C_1 and exponent C_4 on the wall-to-bulk temperature ratio does not significantly increase the standard deviation. This correlation is similar to the one given by Taylor[15] for hydrogen, which is also valid in the entrance region:

$$Nu = 0.023\, Re^{0.8}\, Pr^{0.4} \left(\frac{T_w}{T_b}\right)^{-[0.57 - (1.59/L/D)]} \qquad (4)$$

The hydrogen data for obtaining this correlation are from ten different investigations covering a wide range of operating conditions. Taylor cautioned that the above equation does not accurately predict the heat-transfer coefficient in the near critical region where $0.76 \leq T_b/T_c \leq 1$ and $1 \leq P/P_c \leq 2.8$.

The data of this study, including the near critical region, were examined for an L/D effect on the wall-to-bulk temperature ratio exponent, as suggested by Taylor's correlation. The constants C_1 and C_4 of expression (3) were redetermined separately with the data from each of the two L/D positions along the test section (L/D=20 and L/D=40). The results are expressions (4) and (5) of Table 4-1. The constant C_1 does not change significantly in either case (about 2.3%), but the exponent C_4 on the temperature ratio does change by about 16% thus indicating an L/D effect on the exponent for the temperature ratio and a corresponding effect on the heat-transfer coefficient. This result implies that for some conditions the thermal entrance length may have been extended to at least 20 diameters of the heated test section. However for low wall-to-bulk temperature ratios, the effect is slight. As an example, expressions (4) and (5) in Table 4-1 show that

$$\text{if } \frac{T_w}{T_b} < 1.5, \text{ then } \frac{h_{L/D=20}}{h_{L/D=40}} < 1.05$$

The constant C_1 is forced to 0.023 in expressions (6) and (7) to compare the resulting C_4 temperature ratio exponent for each L/D with the exponents determined by Taylor for hydrogen data. The agreement with the exponents determined from these data is good at L/D=40 (-0.495 compared to Taylor's -0.530), but less satisfactory at L/D=20 (-0.373 compared to Taylor's -0.491).

Hess and Kunz[20] have developed a semi-empirical expression for forced convection heat transfer to supercritical hydrogen. This correlation, expression (8) of Table 4-1, was tested for the data of this study. As can be seen from the standard deviation for the coefficient of the wall-to-bulk kinematic viscosity ratio, this parameter does not appear to be significant, and the standard deviation for the fit, 16.9%, is the worst of all the correlations tested.

Summarizing, the single equation which best represents the supercritical helium data of this study and still retains the classical 0.8 and 0.4 exponents on Re and Pr, is

$$Nu = 0.0259\, Re^{0.8}\, Pr^{0.4} \left(\frac{T_w}{T_b}\right)^{-0.716} \quad (5)$$

It is useful to note that at a helium temperature of 4.2K, for pressures from 3 to 10 atm, the equation below fits the data

$$q\,(W/cm^2) \cong \frac{0.048}{D(cm)} \left(\frac{Re}{10^5}\right)^{0.8} \left(\frac{4.2}{T_w}\right)^{0.72} \left(T_w - 4.2\right) \quad (6)$$

within a few percent and may be used when convenient.

4-6. Comparison with Other Heat Transfer Modes

Various modes of helium heat transfer are compared in Figure 4-6. In considering a forced supercritical helium system versus a pool boiling or superfluid helium cooling system, one will note that the values of the heat-transfer coefficient are in the same general range. However, for forced flow of supercritical helium at a Re of 10^6 and conditions noted on the graph, the heat transfer attainable is somewhat more attractive. Further, the heat-transfer

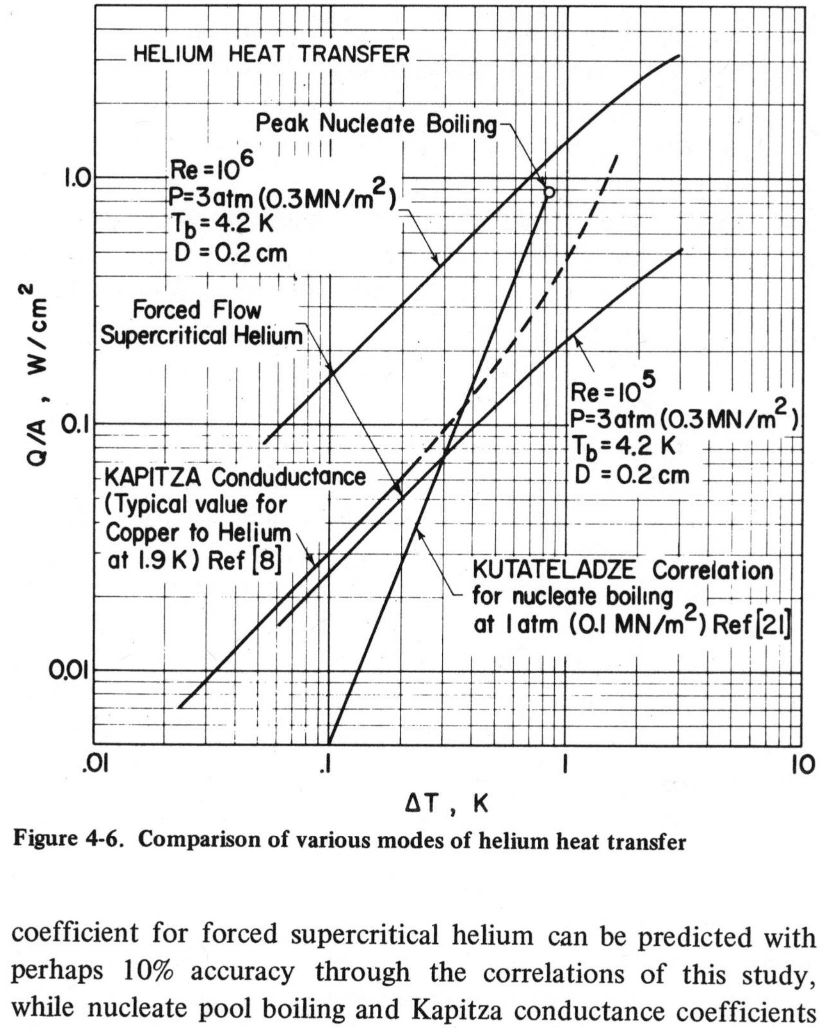

Figure 4-6. Comparison of various modes of helium heat transfer

coefficient for forced supercritical helium can be predicted with perhaps 10% accuracy through the correlations of this study, while nucleate pool boiling and Kapitza conductance coefficients often vary by factors of 2 or more, depending on the details of surface finish, etc. (as discussed by Smith[1] and Snyder[8]).

4-7. System Oscillations

A number of investigators have reported the appearance of oscillations during heat transfer to near critical fluids.[22-27] An apparent criterion for the onset of oscillations (judging from the above noted references) is the presence of vastly differing density states of the fluid within the system. This situation can be encountered in the near critical region where large changes in density occur with small changes in temperature, or in the region away from the critical under high heat flux or high wall-to-bulk temperature ratio conditions.

Low-frequency oscillations occasionally appeared during the helium heat-transfer studies of this work. The frequency of the oscillations varied from about 0.05 to 0.1 Hz with amplitude in temperature varying from about 0.01 to 1 K and amplitude in pressure generally less than 0.03 MN m^{-2}. The conditions necessary for the inception and maintenance of oscillations have not been well established here. The apparatus was not completely instrumented for an oscillations study; however, there is some evidence that oscillations were at least partially induced and sustained by instabilities in the heat exchanger cooling system and were not necessarily dependent upon the thermodynamic state of the fluid in the flow-loop (oscillations were observed at high pressures as well as near critical pressures). The oscillations could be eliminated by an increased flow rate through the heat exchanger. Further investigation is required to understand the nature and cause of the oscillations observed in this system.

4-8. Conclusions

For supercritical helium the heat-transfer coefficients predicted by the classical Dittus-Boelter correlation (if the 0.023 constant is replaced by 0.022) agree reasonably well with the experimentally determined coefficients over the range of operating conditions of this investigation. The standard deviation for this correlation is 14.8%. Modification of this correlation expression to include a wall-to-bulk temperature ratio, expression (3) of Table 4-1, substantially improves the predictive quality, resulting in a standard deviation of 8.5%. Comparison of the exponents C_4 in expressions (6) and (7) of Table 4-1, with the exponents required by Taylor's correlation for hydrogen, Equation (4) in the text, suggests that heat transfer to both these fluids in the supercritical state may be predicted by the same equation with reasonable accuracy, though text equation (5) does give a better fit of helium.

References

1. R. V. Smith, "Review of Heat Transfer to Helium I," *Cryogenics* **9**, No. 1, 11-19 (Feb. 1969) and Proceedings of the 1968 Summer Study on Superconducting Devices and Accelerations, Part K, BNL 50155 (C-55) 249-92, Brookhaven Natl. Lab., Upton, N. Y. (April 1969).
2. R. V. Smith, "The Influence of Surface Characteristics on the Boiling of Cryogenic Fluids," ASME *J. of Engr. for Industry* (Nov. 1969).
3. R. C. Hendricks, R. J. Simoneau, and R. V. Smith, "Survey of Heat Transfer to Near Critical Fluids," NASA Lewis Research Center, and National Bureau of Standards, Institute for Basic Standards, Boulder, Colorado, NASA TMX-5612 (1969), *Advances in Cryogenic Engineering,* **15**, 197-237, Plenum Press, N. Y. (1970) and NASA-TND-5886.
4. P. J. Giarratano, V. D. Arp, and R. V. Smith, "Forced Convection Heat Transfer to Supercritical Helium," *Cryogenics* **11**, 385 (1971).
5. V. Arp, "Forced Flow Single-Phase Helium Cooling Systems," *Advances in Cryogenic Engineering,* **17**, 342-351, Plenum Press, N. Y. (1972).
6. H. Sixsmith and P. J. Giarratano, "A Miniature Centrifugal Helium Pump," *Rev. of Sci. Inst.,* **41**, No. 11, 1570-73, (Nov. 1970).
7. J. E. Cruz and J. C. Jellison, "A Digital Technique for Generating Variable Frequency Multiple-Phase Waveforms," *Rev. of Sci. Inst.* **7**, 1099 (1970).
8. N. S. Snyder, "Thermal Conductance at the Interface of a Solid and Helium II," NBS Tech. Note 385 (1969), in a shorter form under the title, "Heat Transfer to Helium II: Kapitza Conductance," *Cryogenics,* **10**, No. 2, 89-95, (1970).
9. K. Mittag, "Kapitza Conductance and Thermal Conductivity of Copper, Niobium, and Aluminum in the Range from 1.3 to 2.1 K," to appear in *Cryogenics.*
10. V. D. Arp, "Heat Transport Through HeII," *Cryogenics* **10**, No. 2 (1970).
11. H. J. M. Hanley and G. E. Childs, "Interim Values for the Viscosity and Thermal Conductivity Coefficients of Fluid He^4 Between 2 and 50 K," *Cryogenics* **9**, No. 2, 106-111 (April 1969).
12. W. G. Steward and G. H. Wallace, "$Helium^4$ Viscosity Measurements– 4 to 20 K, 0 to 10 MN/m^2," forthcoming publication.
13. R. D. McCarty, "Thermophysical Properties of Helium-4 from 4 to 3000 R with Pressures to 15000 psia," NBS Tech. Note 622 (1972).
14. F. W. Dittus and L. M. Boelter, "Heat Transfer in Automobile Radiators of the Tubular Type," Univ. of Calif. Pubs. Eng. 2, 443 (1930).

15. M. F. Taylor, "Correlation of Local Heat Transfer Coefficients for Single-Phase Turbulent Flow of Hydrogen in Tubes with Temperature Ratios to 23," NASA TND-4332 (1968).
16. H. M. Roder, "Thermal Conductivity of Helium-4," published as subsection of Reference 13.
17. C. Johannes, "Studies of Forced Convection Heat Transfer to Helium I," *Advances in Cryogenic Engineering,* **17**, 352-360, Plenum Press, N. Y. (1952).
18. W. J. Hall, National Bureau of Standards (Private Communications).
19. H. Brechna, "Superconducting Magnets for High Energy Physics Applications," Stanford Linear Accelerator Center, Report No. SLAC-Pub 274 (1967).
20. H. L. Hess and H. R. Kunz, "A Study of Forced Convection Heat Transfer to Supercritical Hydrogen," *J. Heat Trans.* **87**, 41 (1965).
21. S. S. Kutateladze, State Sci. and Tech. Pub. of Lit. on Machinery Moscow (AEC Translation 3770, Tech. Info. Service, Oak Ridge, Tenn. (1949, 1952).
22. A. J. Cornelius, "An Investigation of Instabilities Encountered During Heat Transfer to a Supercritical Fluid," Argonne National Laboratory Report ANL-70392 (1965).
23. R. S. Thurston, "Probing Experiments on Pressure Oscillations in Two-Phase and Supercritical Hydrogen with Forced Convection Heat Transfer," *Adv. in Cryogenic Engineering,* **10**, 305-312 (Plenum Press N. Y., 1965).
24. K. Goldman, "Heat Transfer to Supercritical Water at 5000 psi Flowing at High Mass Flow Rates Through Round Tubes," Int. Developments in Heat Transfer, Part III, 561-568 (1961).
25. J. R. McCarty and H. Wolf, "The Heat Transfer Characteristics of Gaseous Hydrogen and Helium," Rocketdyne Research Report RR-60-12 (1960).
26. W. W. Hines and H. Wolf, "Pressure Oscillations Associated with Heat Transfer to Hydrocarbon Fluids at Supercritical Pressures and Temperatures," *ARS J.,* **32**, 361-366 (1962).
27. K. K. Knapp and R. H. Sabersky, "Free Convection Heat Transfer to Carbon Dioxide Near the Critical Point," *Int. J. Heat and Mass Trans.,* **9**, 41-51 (1966).

Acknowledgment

This work was sponsored by the U. S. Atomic Energy Commission.

Appendix

Tables of viscosity, thermal conductivity and Prandtl numbers for helium are given in the following pages.

Table 1A. Helium[4] Viscosity, μ, μg/cm-s (Ref. 12)

PRES MN/M²	0.010	0.051	0.101	0.152	0.182	0.193	0.203	0.213	0.223	0.233	0.243	0.253
ATM	0.100	0.500	1.000	1.500	1.800	1.900	2.000	2.100	2.200	2.300	2.400	2.500
TEMP,K												
3.0	7.8	37.4	38.6	39.7	40.3	40.6	40.8	41.0	41.2	41.4	41.6	41.8
3.5	9.1	35.2	36.4	37.5	38.2	38.4	38.6	38.9	39.1	39.3	39.5	39.7
4.0	10.3	10.9	33.4	34.6	35.4	35.6	35.8	36.1	36.3	36.5	36.7	36.9
4.2	10.8	11.4	31.9	33.3	34.1	34.3	34.5	34.8	35.0	35.2	35.5	35.7
4.4	11.3	11.8	12.8	31.8	32.6	32.9	33.1	33.4	33.7	33.9	34.1	34.4
4.6	11.7	12.3	13.1	29.9	31.0	31.3	31.6	31.9	32.1	32.4	32.7	32.9
4.8	12.2	12.7	13.5	14.8	28.9	29.3	29.7	30.0	30.4	30.7	31.0	31.3
5.0	12.6	13.1	13.9	15.0	16.0	16.7	26.9	27.5	28.1	28.5	29.0	29.4
5.1	12.8	13.3	14.1	15.1	16.0	16.4	17.1	25.4	26.3	27.0	27.6	28.1
5.2	13.1	13.6	14.3	15.2	16.0	16.4	16.8	17.4	18.5	24.4	25.6	26.5
5.3	13.3	13.8	14.5	15.4	16.1	16.4	16.8	17.2	17.7	18.6	20.8	24.1
5.4	13.5	14.0	14.7	15.5	16.2	16.5	16.8	17.1	17.5	18.0	18.7	19.9
5.5	13.7	14.2	14.9	15.7	16.3	16.6	16.8	17.1	17.5	17.9	18.3	18.9
5.6	13.9	14.4	15.1	15.9	16.5	16.7	16.9	17.2	17.5	17.8	18.2	18.6
5.7	14.1	14.6	15.2	16.0	16.6	16.8	17.0	17.3	17.5	17.8	18.1	18.5
5.8	14.3	14.8	15.4	16.2	16.7	16.9	17.1	17.4	17.6	17.8	18.1	18.4
5.9	14.5	15.0	15.6	16.4	16.9	17.1	17.3	17.5	17.7	17.9	18.2	18.4
6.0	14.7	15.2	15.8	16.5	17.0	17.2	17.4	17.6	17.8	18.0	18.2	18.5
6.2	15.1	15.6	16.2	16.9	17.3	17.5	17.7	17.8	18.0	18.2	18.4	18.6
6.4	15.5	16.0	16.5	17.2	17.6	17.8	17.9	18.1	18.3	18.4	18.6	18.8
6.6	15.9	16.3	16.9	17.5	18.0	18.1	18.2	18.4	18.5	18.7	18.9	19.0
6.8	16.3	16.7	17.3	17.9	18.3	18.4	18.5	18.7	18.8	19.0	19.1	19.3
7.0	16.7	17.1	17.6	18.2	18.6	18.7	18.8	19.0	19.1	19.3	19.4	19.5
7.5	17.6	18.0	18.5	19.0	19.4	19.5	19.6	19.7	19.9	20.0	20.1	20.2
8.0	18.5	18.9	19.4	19.9	20.2	20.3	20.4	20.5	20.6	20.7	20.8	20.9
8.5	19.4	19.7	20.2	20.7	21.0	21.1	21.2	21.3	21.4	21.5	21.6	21.7
9.0	20.2	20.6	21.0	21.5	21.7	21.8	21.9	22.0	22.1	22.2	22.3	22.4
9.5	21.1	21.4	21.8	22.2	22.5	22.6	22.7	22.7	22.8	22.9	23.0	23.1
10.0	21.9	22.2	22.6	23.0	23.2	23.3	23.4	23.5	23.6	23.6	23.7	23.8

11.0	23.5	23.8	24.1	24.5	24.7	24.8	24.8	24.9	25.0	25.1	25.1	25.2
12.0	25.0	25.2	25.6	25.9	26.1	26.2	26.2	26.3	26.4	26.4	26.5	26.6
13.0	26.4	26.7	27.0	27.3	27.5	27.5	27.6	27.7	27.7	27.8	27.9	27.9
14.0	27.9	28.1	28.4	28.6	28.8	28.9	28.9	29.0	29.0	29.1	29.2	29.2
15.0	29.2	29.4	29.7	30.0	30.1	30.2	30.2	30.3	30.3	30.4	30.4	30.5
16.0	30.5	30.7	31.0	31.2	31.4	31.4	31.5	31.5	31.6	31.6	31.7	31.7
17.0	31.8	32.0	32.2	32.5	32.6	32.7	32.7	32.8	32.8	32.8	32.9	32.9
18.0	33.1	33.2	33.5	33.7	33.8	33.9	33.9	34.0	34.0	34.0	34.1	34.1
19.0	34.3	34.4	34.7	34.9	35.0	35.0	35.1	35.1	35.2	35.2	35.2	35.3
20.0	35.5	35.6	35.8	36.0	36.1	36.2	36.2	36.3	36.3	36.3	36.4	36.4
25.0	41.0	41.1	41.3	41.5	41.6	41.6	41.6	41.7	41.7	41.7	41.8	41.8
30.0	46.1	46.2	46.3	46.5	46.6	46.6	46.6	46.7	46.7	46.7	46.7	46.8
35.0	50.8	50.9	51.0	51.1	51.2	51.2	51.3	51.3	51.3	51.3	51.4	51.4
40.0	55.2	55.3	55.4	55.5	55.6	55.6	55.6	55.7	55.7	55.7	55.7	55.8
45.0	59.4	59.5	59.6	59.7	59.8	59.8	59.8	59.8	59.9	59.9	59.9	59.9
50.0	63.4	63.5	63.6	63.7	63.8	63.8	63.8	63.8	63.8	63.9	63.9	63.9
60.0	71.0	71.1	71.2	71.3	71.3	71.3	71.3	71.4	71.4	71.4	71.4	71.4
80.0	84.9	85.0	85.0	85.1	85.2	85.2	85.2	85.2	85.2	85.2	85.2	85.3
100.0	97.7	97.7	97.8	97.8	97.9	97.9	97.9	97.9	97.9	97.9	98.0	98.0
120.0	109.7	109.7	109.8	109.8	109.9	109.9	109.9	109.9	109.9	109.9	109.9	110.0
140.0	121.2	121.2	121.3	121.3	121.3	121.3	121.4	121.4	121.4	121.4	121.4	121.4
160.0	132.2	132.3	132.3	132.4	132.4	132.4	132.4	132.4	132.4	132.5	132.5	132.5
180.0	143.0	143.1	143.1	143.2	143.2	143.2	143.2	143.2	143.2	143.2	143.2	143.2
200.0	153.6	153.6	153.7	153.7	153.7	153.7	153.7	153.8	153.8	153.8	153.8	153.8
220.0	164.0	164.0	164.0	164.1	164.1	164.1	164.1	164.1	164.1	164.1	164.1	164.1
240.0	174.2	174.2	174.2	174.3	174.3	174.3	174.3	174.3	174.3	174.3	174.3	174.3
260.0	184.3	184.3	184.3	184.4	184.4	184.4	184.4	184.4	184.4	184.4	184.4	184.4
280.0	194.3	194.3	194.3	194.4	194.4	194.4	194.4	194.4	194.4	194.4	194.4	194.4
300.0	204.2	204.2	204.3	204.3	204.3	204.3	204.3	204.3	204.3	204.3	204.3	204.3

Table 1A. Helium^4 Viscosity, μ, μg/cm-s (continued)

PRES MN/M²	0.263	0.274	0.284	0.294	0.304	0.355	0.405	0.456	0.507	0.557	0.608	0.659
ATM	2.600	2.700	2.800	2.900	3.000	3.500	4.000	4.500	5.000	5.500	6.000	6.500
TEMP,K												
3.0	42.1	42.3	42.5	42.7	42.9	43.9	44.9	45.9	46.9	47.9	48.9	49.9
3.5	39.9	40.1	40.3	40.5	40.7	41.7	42.7	43.7	44.6	45.6	46.5	47.4
4.0	37.2	37.4	37.6	37.8	38.0	39.0	40.0	40.9	41.9	42.8	43.7	44.5
4.2	35.9	36.1	36.4	36.6	36.8	37.8	38.8	39.8	40.7	41.6	42.5	43.3
4.4	34.6	34.8	35.1	35.3	35.5	36.6	37.6	38.6	39.5	40.4	41.3	42.1
4.6	33.2	33.5	33.7	33.9	34.2	35.3	36.3	37.3	38.3	39.2	40.1	40.9
4.8	31.6	31.9	32.2	32.4	32.7	33.9	35.0	36.1	37.0	38.0	38.9	39.7
5.0	29.7	30.1	30.4	30.7	31.0	32.4	33.6	34.7	35.8	36.7	37.6	38.5
5.1	28.6	29.0	29.4	29.8	30.1	31.6	32.9	34.0	35.1	36.1	37.0	37.9
5.2	27.1	27.7	28.2	28.6	29.0	30.7	32.1	33.3	34.4	35.4	36.4	37.3
5.3	25.4	26.2	26.9	27.5	28.0	29.9	31.4	32.7	33.9	34.9	35.9	36.8
5.4	22.1	24.1	25.2	26.1	26.8	29.1	30.7	32.1	33.3	34.3	35.3	36.3
5.5	19.8	21.1	22.7	24.1	25.2	28.1	30.0	31.4	32.7	33.8	34.8	35.7
5.6	19.2	19.8	20.7	21.9	23.1	27.1	29.2	30.8	32.1	33.2	34.3	35.2
5.7	18.9	19.4	20.0	20.7	21.5	25.9	28.4	30.1	31.5	32.7	33.7	34.7
5.8	18.8	19.2	19.6	20.1	20.7	24.5	27.4	29.4	30.8	32.1	33.2	34.2
5.9	18.7	19.1	19.4	19.8	20.3	23.3	26.5	28.6	30.2	31.5	32.7	33.7
6.0	18.7	19.0	19.3	19.7	20.0	22.5	25.5	27.8	29.5	30.9	32.1	33.2
6.2	18.8	19.1	19.3	19.6	19.9	21.6	23.9	26.2	28.2	29.8	31.1	32.2
6.4	19.0	19.2	19.4	19.6	19.9	21.2	23.0	24.9	26.9	28.6	30.0	31.2
6.6	19.2	19.4	19.6	19.8	20.0	21.1	22.5	24.1	25.8	27.5	29.0	30.3
6.8	19.4	19.6	19.8	20.0	20.1	21.1	22.3	23.6	25.1	26.6	28.0	29.3
7.0	19.7	19.9	20.0	20.2	20.3	21.2	22.2	23.4	24.6	25.9	27.2	28.5
7.5	20.4	20.5	20.6	20.8	20.9	21.6	22.4	23.3	24.2	25.1	26.1	27.1
8.0	21.1	21.2	21.3	21.4	21.5	22.2	22.8	23.5	24.3	25.0	25.8	26.6
8.5	21.8	21.9	22.0	22.1	22.2	22.7	23.3	23.9	24.6	25.2	25.9	26.6
9.0	22.5	22.6	22.7	22.8	22.9	23.4	23.9	24.4	25.0	25.6	26.1	26.7
9.5	23.2	23.3	23.4	23.5	23.6	24.0	24.5	25.0	25.5	26.0	26.5	27.0
10.0	23.9	24.0	24.1	24.1	24.2	24.7	25.1	25.5	26.0	26.5	27.0	27.4

11.0	25.3	25.4	25.4	25.5	25.6	26.0	26.4	26.7	27.1	27.5	28.0	28.4
12.0	26.6	26.7	26.8	26.9	26.9	27.3	27.6	28.0	28.3	28.7	29.0	29.4
13.0	28.0	28.0	28.1	28.2	28.2	28.5	28.9	29.2	29.5	29.8	30.1	30.5
14.0	29.3	29.3	29.4	29.4	29.5	29.8	30.1	30.4	30.7	31.0	31.3	31.6
15.0	30.5	30.6	30.6	30.7	30.8	31.0	31.3	31.6	31.8	32.1	32.4	32.6
16.0	31.8	31.8	31.9	31.9	32.0	32.2	32.5	32.7	33.0	33.2	33.5	33.7
17.0	33.0	33.0	33.1	33.1	33.2	33.4	33.7	33.9	34.1	34.4	34.6	34.8
18.0	34.2	34.2	34.3	34.3	34.4	34.6	34.8	35.0	35.2	35.5	35.7	35.9
19.0	35.3	35.4	35.4	35.5	35.5	35.7	35.9	36.1	36.3	36.6	36.8	37.0
20.0	36.5	36.5	36.5	36.6	36.6	36.8	37.0	37.2	37.4	37.6	37.8	38.0
25.0	41.8	41.9	41.9	41.9	42.0	42.1	42.3	42.5	42.6	42.8	43.0	43.1
30.0	46.8	46.8	46.8	46.9	46.9	47.0	47.2	47.3	47.5	47.6	47.7	47.9
35.0	51.4	51.4	51.5	51.5	51.5	51.6	51.8	51.9	52.0	52.1	52.3	52.4
40.0	55.8	55.8	55.8	55.9	55.9	56.0	56.1	56.2	56.3	56.4	56.5	56.7
45.0	59.9	60.0	60.0	60.0	60.0	60.1	60.2	60.3	60.4	60.5	60.6	60.7
50.0	63.9	63.9	64.0	64.0	64.0	64.1	64.2	64.3	64.4	64.5	64.6	64.7
60.0	71.4	71.5	71.5	71.5	71.5	71.6	71.7	71.8	71.9	71.9	72.0	72.1
80.0	85.3	85.3	85.3	85.3	85.3	85.4	85.5	85.5	85.6	85.7	85.8	85.8
100.0	98.0	98.0	98.0	98.0	98.0	98.1	98.2	98.2	98.3	98.4	98.4	98.5
120.0	110.0	110.0	110.0	110.0	110.0	110.1	110.1	110.2	110.2	110.3	110.4	110.4
140.0	121.4	121.4	121.4	121.5	121.5	121.5	121.6	121.6	121.7	121.7	121.8	121.8
160.0	132.5	132.5	132.5	132.5	132.5	132.6	132.6	132.7	132.7	132.8	132.8	132.9
180.0	143.3	143.3	143.3	143.3	143.3	143.3	143.4	143.4	143.5	143.5	143.5	143.6
200.0	153.8	153.8	153.8	153.8	153.8	153.9	153.9	153.9	154.0	154.0	154.0	154.1
220.0	164.1	164.1	164.2	164.2	164.2	164.2	164.2	164.3	164.3	164.3	164.4	164.4
240.0	174.3	174.3	174.4	174.4	174.4	174.4	174.4	174.5	174.5	174.5	174.5	174.6
260.0	184.4	184.4	184.4	184.4	184.4	184.5	184.5	184.5	184.5	184.6	184.6	184.6
280.0	194.4	194.4	194.4	194.4	194.4	194.4	194.5	194.5	194.5	194.5	194.6	194.6
300.0	204.3	204.3	204.3	204.3	204.3	204.3	204.4	204.4	204.4	204.4	204.4	204.5

Table 1A. Helium^4 Viscosity, μ, μg/cm-s (continued)

PRES MN/M	0.608	0.709	0.811	0.912	1.013	1.520	2.026	2.533	3.040	3.546	4.053	4.560
ATM	6.000	7.000	8.000	9.000	10.000	15.000	20.000	25.000	30.000	35.000	40.000	45.000
TEMP,K												
3.0	48.9	50.8	52.8	54.7	56.6	66.1	75.9	86.1	96.8	108.3	120.5	133.6
3.5	46.5	48.3	50.1	51.9	53.6	62.3	71.0	79.9	89.2	98.9	109.2	120.0
4.0	43.7	45.4	47.1	48.7	50.4	58.3	66.1	73.9	82.0	90.4	99.1	108.2
4.2	42.5	44.2	45.8	47.5	49.0	56.7	64.2	71.7	79.4	87.3	95.5	104.0
4.4	41.3	43.0	44.6	46.2	47.8	55.2	62.4	69.6	76.9	84.4	92.2	100.2
4.6	40.1	41.8	43.4	45.0	46.5	53.7	60.7	67.6	74.6	81.7	89.0	96.6
4.8	38.9	40.6	42.2	43.7	45.3	52.3	59.0	65.7	72.4	79.2	86.1	93.3
5.0	37.6	39.4	41.0	42.5	44.0	51.0	57.5	63.9	70.3	76.8	83.4	90.2
5.1	37.0	38.8	40.4	41.9	43.4	50.3	56.7	63.0	69.3	75.7	82.1	88.8
5.2	36.4	38.2	39.8	41.4	42.8	49.7	56.0	62.2	68.3	74.6	80.9	87.4
5.3	35.9	37.7	39.3	40.9	42.3	49.1	55.4	61.4	67.5	73.6	79.8	86.1
5.4	35.3	37.1	38.8	40.4	41.8	48.6	54.7	60.7	66.6	72.6	78.7	84.9
5.5	34.8	36.6	38.3	39.9	41.4	48.0	54.1	60.0	65.8	71.7	77.6	83.7
5.6	34.3	36.1	37.8	39.4	40.9	47.5	53.5	59.3	65.1	70.8	76.6	82.6
5.7	33.7	35.7	37.4	38.9	40.4	47.0	53.0	58.7	64.3	70.0	75.7	81.5
5.8	33.2	35.2	36.9	38.5	40.0	46.5	52.4	58.1	63.6	69.2	74.8	80.4
5.9	32.7	34.7	36.4	38.0	39.5	46.1	51.9	57.5	62.9	68.4	73.9	79.4
6.0	32.1	34.2	36.0	37.6	39.1	45.6	51.4	56.9	62.3	67.6	73.0	78.5
6.2	31.1	33.3	35.1	36.7	38.3	44.7	50.4	55.8	61.0	66.2	71.4	76.6
6.4	30.0	32.3	34.2	35.9	37.5	43.9	49.5	54.7	59.8	64.8	69.9	75.0
6.6	29.0	31.4	33.4	35.1	36.7	43.1	48.6	53.8	58.7	63.6	68.5	73.4
6.8	28.0	30.5	32.6	34.4	36.0	42.4	47.8	52.9	57.7	62.4	67.2	71.9
7.0	27.2	29.7	31.8	33.6	35.3	41.7	47.1	52.0	56.7	61.4	65.9	70.6
7.5	26.1	28.2	30.2	32.0	33.7	40.2	45.4	50.1	54.6	58.9	63.2	67.5
8.0	25.8	27.5	29.1	30.8	32.4	38.8	44.0	48.5	52.8	56.9	61.0	65.0
8.5	25.9	27.3	28.6	30.1	31.5	37.7	42.7	47.1	51.2	55.2	59.0	62.8
9.0	26.1	27.3	28.5	29.7	31.0	36.8	41.7	46.0	49.9	53.7	57.3	60.9
9.5	26.5	27.6	28.6	29.7	30.8	36.1	40.8	45.0	48.8	52.4	55.9	59.3
10.0	27.0	27.9	28.9	29.9	30.8	35.6	40.1	44.2	47.8	51.3	54.7	58.0

11.0	28.0	28.8	29.6	30.4	31.2	35.2	39.2	42.9	46.3	49.6	52.7	55.7
12.0	29.0	29.7	30.5	31.2	31.9	35.4	38.8	42.2	45.3	48.4	51.3	54.1
13.0	30.1	30.8	31.4	32.1	32.7	35.9	38.9	41.9	44.8	47.6	50.3	53.0
14.0	31.3	31.8	32.4	33.0	33.6	36.5	39.3	42.0	44.6	47.2	49.7	52.2
15.0	32.4	32.9	33.5	34.0	34.5	37.2	39.8	42.3	44.7	47.1	49.4	51.7
16.0	33.5	34.0	34.5	35.0	35.5	38.0	40.4	42.7	45.0	47.2	49.4	51.5
17.0	34.6	35.1	35.5	36.0	36.5	38.8	41.0	43.2	45.3	47.4	49.5	51.5
18.0	35.7	36.1	36.6	37.0	37.5	39.6	41.8	43.8	45.8	47.6	49.7	51.6
19.0	36.8	37.2	37.6	38.0	38.4	40.5	42.5	44.5	46.4	48.2	50.1	51.9
20.0	37.8	38.2	38.6	39.0	39.4	41.4	43.3	45.2	47.0	48.8	50.5	52.2
25.0	43.0	43.3	43.6	43.9	44.2	45.8	47.4	48.9	50.4	51.9	53.3	54.6
30.0	47.7	48.0	48.3	48.6	48.8	50.2	51.6	52.9	54.2	55.5	56.7	57.9
35.0	52.3	52.5	52.6	53.0	53.2	54.4	55.6	56.8	58.0	59.1	60.2	61.4
40.0	56.5	56.8	57.0	57.2	57.4	58.5	59.6	60.7	61.7	62.8	63.8	64.8
45.0	60.6	60.8	61.1	61.3	61.5	62.5	63.5	64.5	65.4	66.4	67.4	68.3
50.0	64.6	64.8	65.0	65.1	65.3	66.3	67.2	68.1	69.1	70.0	70.8	71.7
60.0	72.0	72.2	72.4	72.5	72.7	73.6	74.4	75.2	76.0	76.9	77.7	78.4
80.0	85.8	85.9	86.1	86.2	86.3	87.1	87.8	88.5	89.2	89.9	90.5	91.2
100.0	98.4	98.6	98.7	98.8	98.9	99.6	100.2	100.8	101.4	102.0	102.6	103.2
120.0	110.4	110.5	110.6	110.7	110.8	111.4	111.9	112.5	113.1	113.6	114.1	114.7
140.0	121.8	121.9	122.0	122.1	122.2	122.7	123.2	123.7	124.2	124.7	125.2	125.7
160.0	132.8	132.9	133.0	133.1	133.2	133.6	134.1	134.5	135.0	135.4	135.9	136.3
180.0	143.5	143.6	143.7	143.8	143.9	144.3	144.7	145.1	145.5	145.9	146.3	146.7
200.0	154.0	154.1	154.2	154.3	154.3	154.7	155.1	155.4	155.8	156.2	156.5	156.9
220.0	164.4	164.4	164.5	164.6	164.6	165.0	165.3	165.6	165.9	166.2	166.6	166.9
240.0	174.5	174.6	174.7	174.7	174.8	175.1	175.3	175.6	175.9	176.2	176.5	176.7
260.0	184.6	184.6	184.7	184.7	184.8	185.0	185.3	185.5	185.8	186.0	186.3	186.5
280.0	194.6	194.6	194.6	194.7	194.7	194.9	195.1	195.4	195.6	195.8	196.0	196.2
300.0	204.4	204.5	204.5	204.5	204.6	204.8	204.9	205.1	205.3	205.4	205.6	205.8

Table 1A. Helium^4 Viscosity, μ, μg/cm-s (concluded)

PRES MN/M	5.066	5.573	6.079	6.586	7.093	7.599	8.106	8.613	9.119	9.626	10.133	10.639
ATM	50.000	55.000	60.000	65.000	70.000	75.000	80.000	85.000	90.000	95.000	100.000	105.000
TEMP,K												
3.0	147.8	163.0	179.6	197.5	217.1	238.4						
3.5	131.5	143.7	156.6	170.5	185.2	201.0	217.9	236.0	255.5	276.3	298.7	
4.0	117.7	127.8	138.3	149.4	161.2	173.5	186.6	200.5	215.1	230.6	247.1	264.5
4.2	113.0	122.3	132.1	142.4	153.2	164.5	176.5	189.1	202.4	216.4	231.2	246.8
4.4	108.6	117.3	126.4	135.9	145.9	156.4	167.4	179.0	191.1	203.9	217.3	231.4
4.6	104.5	112.7	121.2	130.1	139.4	149.1	159.3	169.9	181.1	192.8	205.0	217.9
4.8	100.7	108.4	116.4	124.8	133.4	142.5	151.9	161.8	172.1	182.9	194.1	205.9
5.0	97.3	104.5	112.1	119.9	128.0	136.5	145.3	154.5	164.0	174.0	184.4	195.3
5.1	95.6	102.7	110.0	117.6	125.5	133.7	142.2	151.1	160.3	169.9	179.9	190.4
5.2	94.1	101.0	108.1	115.4	123.1	131.0	139.3	147.8	156.7	166.0	175.7	185.7
5.3	92.6	99.3	106.3	113.4	120.9	128.6	136.5	144.8	153.5	162.4	171.7	181.4
5.4	91.2	97.8	104.5	111.5	118.7	126.2	134.0	142.0	150.3	159.0	168.0	177.4
5.5	89.9	96.3	102.9	109.7	116.7	124.0	131.5	139.3	147.4	155.8	164.5	173.5
5.6	88.6	94.9	101.3	107.9	114.7	121.8	129.1	136.7	144.5	152.7	161.1	169.9
5.7	87.4	93.5	99.8	106.2	112.9	119.8	126.9	134.2	141.8	149.7	157.9	166.4
5.8	86.3	92.2	98.3	104.6	111.1	117.8	124.7	131.9	139.3	146.9	154.9	163.1
5.9	85.1	91.0	96.9	103.1	109.4	115.9	122.7	129.6	136.8	144.3	152.0	159.9
6.0	84.0	89.7	95.6	101.6	107.8	114.1	120.7	127.5	134.5	141.7	149.2	156.9
6.2	82.0	87.5	93.1	98.8	104.7	110.8	117.0	123.4	130.1	136.9	144.0	151.3
6.4	80.1	85.4	90.7	96.2	101.9	107.7	113.6	119.7	126.0	132.6	139.3	146.2
6.6	78.4	83.4	88.6	93.8	99.2	104.8	110.5	116.3	122.3	128.5	134.9	141.5
6.8	76.7	81.6	86.6	91.6	96.8	102.1	107.6	113.2	118.9	124.8	130.9	137.1
7.0	75.2	79.9	84.7	89.6	94.6	99.7	104.9	110.3	115.8	121.4	127.2	133.2
7.5	71.8	76.2	80.6	85.1	89.6	94.3	99.0	103.9	108.8	113.9	119.1	124.4
8.0	69.0	73.0	77.1	81.3	85.5	89.7	94.1	98.5	103.0	107.6	112.4	117.2
8.5	66.6	70.4	74.2	78.0	81.9	85.9	89.9	94.0	98.1	102.3	106.7	111.1
9.0	64.5	68.1	71.7	75.3	78.9	82.6	86.3	90.1	93.9	97.8	101.8	105.9
9.5	62.7	66.1	69.5	72.9	76.3	79.7	83.2	86.7	90.3	93.9	97.6	101.4
10.0	61.2	64.4	67.6	70.8	74.0	77.3	80.5	83.9	87.2	90.6	94.0	97.5

11.0	56.7	61.6	64.5	67.4	70.3	73.3	76.2	79.1	82.1	85.1	88.1	91.2
12.0	56.9	59.6	62.3	64.9	67.6	70.2	72.9	75.6	78.2	80.9	83.6	86.4
13.0	55.5	58.1	60.6	63.1	65.5	68.0	70.4	72.8	75.3	77.7	80.2	82.7
14.0	54.6	57.0	59.3	61.7	64.0	66.2	68.5	70.8	73.0	75.3	77.6	79.8
15.0	54.0	56.3	58.5	60.6	62.8	64.9	67.1	69.2	71.3	73.4	75.5	77.6
16.0	53.7	55.8	57.8	59.9	61.9	63.9	65.9	67.9	69.9	71.9	73.9	75.8
17.0	53.5	55.5	57.5	59.4	61.3	63.2	65.1	67.0	68.9	70.7	72.6	74.4
18.0	53.5	55.4	57.3	59.1	60.9	62.7	64.5	66.3	68.1	69.8	71.6	73.3
19.0	53.7	55.5	57.2	59.0	60.7	62.4	64.1	65.8	67.5	69.2	70.8	72.5
20.0	53.9	55.6	57.3	59.0	60.6	62.3	63.9	65.5	67.1	68.7	70.3	71.9
25.0	56.2	57.6	58.9	60.3	61.6	63.0	64.3	65.6	66.9	68.3	69.6	70.8
30.0	59.2	60.4	61.6	62.8	63.9	65.1	66.2	67.4	68.5	69.6	70.7	71.9
35.0	62.5	63.5	64.6	65.7	66.7	67.8	68.8	69.8	70.8	71.8	72.8	73.8
40.0	65.8	66.8	67.8	68.8	69.7	70.7	71.6	72.6	73.5	74.4	75.3	76.2
45.0	69.2	70.2	71.1	72.0	72.9	73.8	74.6	75.5	76.4	77.2	78.1	78.9
50.0	72.6	73.5	74.3	75.2	76.0	76.8	77.7	78.5	79.3	80.1	80.9	81.7
60.0	79.2	80.0	80.8	81.5	82.3	83.0	83.8	84.5	85.2	86.0	86.7	87.4
80.0	91.9	92.5	93.2	93.9	94.5	95.1	95.8	96.4	97.0	97.6	98.2	98.8
100.0	103.8	104.4	105.0	105.6	106.1	106.7	107.2	107.8	108.3	108.9	109.4	110.0
120.0	115.2	115.7	116.2	116.8	117.3	117.8	118.3	118.8	119.2	119.7	120.2	120.7
140.0	126.1	126.6	127.1	127.5	128.0	128.5	128.9	129.4	129.8	130.2	130.7	131.1
160.0	136.7	137.2	137.6	138.0	138.4	138.8	139.2	139.6	140.0	140.4	140.8	141.2
180.0	147.1	147.5	147.8	148.2	148.6	149.0	149.3	149.7	150.1	150.4	150.8	151.1
200.0	157.2	157.6	157.9	158.2	158.6	158.9	159.2	159.6	159.9	160.2	160.5	160.9
220.0	167.2	167.5	167.8	168.1	168.4	168.7	169.0	169.3	169.6	169.9	170.1	170.4
240.0	177.0	177.3	177.5	177.8	178.1	178.3	178.6	178.9	179.1	179.4	179.6	179.9
260.0	186.7	187.0	187.2	187.4	187.7	187.9	188.1	188.3	188.6	188.8	189.0	189.2
280.0	196.4	196.6	196.8	197.0	197.2	197.4	197.6	197.7	197.9	198.1	198.3	198.5
300.0	205.9	206.1	206.3	206.4	206.6	206.8	206.9	207.1	207.2	207.4	207.6	207.7

Table 2A. Helium[4] Thermal Conductivity, κ, mW/cm-K (Ref. 16)

PRES MN/M²	0.010	0.051	0.101	0.152	0.182	0.193	0.203	0.213	0.223	0.233	0.243	0.253
ATM	0.100	0.500	1.000	1.500	1.800	1.900	2.000	2.100	2.200	2.300	2.400	2.500
TEMP,K												
3.0	0.065	0.179	0.182	0.184	0.185	0.185	0.185	0.186	0.186	0.187	0.187	0.187
3.5	0.075	0.188	0.191	0.194	0.195	0.196	0.196	0.196	0.197	0.197	0.198	0.198
4.0	0.085	0.092	0.196	0.200	0.202	0.203	0.204	0.204	0.205	0.205	0.206	0.207
4.2	0.088	0.094	0.197	0.201	0.204	0.204	0.205	0.206	0.207	0.207	0.208	0.209
4.4	0.092	0.097	0.113	0.201	0.204	0.205	0.206	0.207	0.207	0.208	0.209	0.210
4.6	0.096	0.100	0.112	0.204	0.202	0.203	0.204	0.206	0.207	0.208	0.209	0.210
4.8	0.099	0.103	0.112	0.141	0.212	0.211	0.210	0.210	0.210	0.210	0.210	0.211
5.0	0.103	0.106	0.114	0.131	0.169	0.224	0.239	0.228	0.222	0.219	0.217	0.216
5.1	0.104	0.108	0.115	0.129	0.153	0.172	0.217	0.316	0.254	0.235	0.227	0.223
5.2	0.106	0.109	0.116	0.128	0.145	0.156	0.173	0.210	0.366	0.348	0.259	0.238
5.3	0.107	0.111	0.117	0.128	0.141	0.148	0.158	0.174	0.203	0.277	0.696	0.340
5.4	0.109	0.112	0.118	0.128	0.139	0.144	0.151	0.160	0.174	0.197	0.241	0.346
5.5	0.111	0.114	0.119	0.128	0.137	0.141	0.146	0.153	0.162	0.174	0.292	0.222
5.6	0.112	0.115	0.121	0.129	0.136	0.140	0.144	0.149	0.155	0.163	0.174	0.188
5.7	0.114	0.117	0.122	0.129	0.136	0.139	0.142	0.146	0.151	0.157	0.164	0.173
5.8	0.115	0.118	0.123	0.130	0.136	0.138	0.141	0.145	0.148	0.153	0.158	0.165
5.9	0.117	0.120	0.124	0.131	0.136	0.138	0.141	0.144	0.147	0.150	0.155	0.160
6.0	0.119	0.121	0.126	0.132	0.137	0.139	0.141	0.143	0.146	0.149	0.152	0.156
6.2	0.122	0.124	0.128	0.134	0.138	0.139	0.141	0.143	0.145	0.148	0.150	0.153
6.4	0.125	0.127	0.131	0.136	0.139	0.141	0.142	0.144	0.146	0.147	0.149	0.152
6.6	0.128	0.130	0.134	0.138	0.141	0.142	0.144	0.145	0.147	0.148	0.150	0.152
6.8	0.130	0.133	0.136	0.140	0.143	0.144	0.145	0.147	0.148	0.149	0.151	0.152
7.0	0.133	0.135	0.139	0.143	0.145	0.146	0.147	0.148	0.150	0.151	0.152	0.153
7.5	0.140	0.142	0.145	0.148	0.151	0.152	0.152	0.153	0.154	0.155	0.156	0.157
8.0	0.147	0.149	0.151	0.154	0.156	0.157	0.158	0.159	0.159	0.160	0.161	0.162
8.5	0.153	0.155	0.157	0.160	0.162	0.163	0.163	0.164	0.165	0.165	0.166	0.167
9.0	0.159	0.161	0.163	0.166	0.168	0.168	0.169	0.169	0.170	0.171	0.171	0.172
9.5	0.165	0.167	0.169	0.172	0.173	0.174	0.174	0.175	0.175	0.176	0.176	0.177
10.0	0.171	0.173	0.175	0.177	0.178	0.179	0.179	0.180	0.180	0.181	0.182	0.182

11.0	0.182	0.184	0.186	0.188	0.189	0.189	0.190	0.190	0.191	0.191	0.192	0.192
12.0	0.193	0.194	0.196	0.198	0.199	0.200	0.200	0.200	0.201	0.201	0.202	0.202
13.0	0.203	0.204	0.206	0.208	0.209	0.209	0.210	0.210	0.211	0.211	0.211	0.212
14.0	0.212	0.214	0.216	0.217	0.218	0.219	0.219	0.220	0.220	0.220	0.221	0.221
15.0	0.222	0.223	0.225	0.227	0.228	0.228	0.228	0.229	0.229	0.229	0.230	0.230
16.0	0.231	0.232	0.234	0.235	0.236	0.237	0.237	0.237	0.238	0.238	0.239	0.239
17.0	0.240	0.241	0.242	0.244	0.245	0.245	0.246	0.246	0.246	0.247	0.247	0.247
18.0	0.248	0.249	0.251	0.253	0.253	0.254	0.254	0.254	0.255	0.255	0.255	0.256
19.0	0.257	0.258	0.259	0.261	0.262	0.262	0.262	0.263	0.263	0.263	0.264	0.264
20.0	0.265	0.266	0.267	0.269	0.270	0.270	0.270	0.271	0.271	0.271	0.272	0.272
25.0	0.303	0.304	0.306	0.307	0.308	0.308	0.308	0.309	0.309	0.309	0.309	0.310
30.0	0.339	0.340	0.341	0.342	0.343	0.343	0.343	0.344	0.344	0.344	0.344	0.345
35.0	0.373	0.374	0.375	0.376	0.376	0.377	0.377	0.377	0.377	0.377	0.378	0.378
40.0	0.405	0.406	0.407	0.408	0.408	0.409	0.409	0.409	0.409	0.409	0.410	0.410
45.0	0.436	0.437	0.438	0.439	0.439	0.439	0.440	0.440	0.440	0.440	0.440	0.440
50.0	0.466	0.467	0.468	0.469	0.469	0.469	0.469	0.470	0.470	0.470	0.470	0.470
60.0	0.524	0.525	0.525	0.526	0.527	0.527	0.527	0.527	0.527	0.527	0.528	0.528
80.0	0.633	0.633	0.634	0.634	0.635	0.635	0.635	0.635	0.635	0.636	0.636	0.636
100.0	0.735	0.735	0.736	0.736	0.736	0.737	0.737	0.737	0.737	0.737	0.737	0.737
120.0	0.831	0.832	0.832	0.833	0.833	0.833	0.833	0.833	0.833	0.833	0.834	0.834
140.0	0.923	0.924	0.924	0.925	0.925	0.925	0.925	0.925	0.925	0.926	0.926	0.926
160.0	1.012	1.012	1.013	1.013	1.013	1.014	1.014	1.014	1.014	1.014	1.014	1.014
180.0	1.097	1.097	1.098	1.098	1.098	1.098	1.099	1.099	1.099	1.099	1.099	1.099
200.0	1.179	1.179	1.180	1.180	1.180	1.180	1.180	1.180	1.180	1.180	1.181	1.181
220.0	1.258	1.258	1.258	1.259	1.259	1.259	1.259	1.259	1.259	1.259	1.259	1.259
240.0	1.334	1.334	1.334	1.335	1.335	1.335	1.335	1.335	1.335	1.335	1.335	1.335
260.0	1.407	1.408	1.408	1.408	1.408	1.408	1.408	1.409	1.409	1.409	1.409	1.409
280.0	1.478	1.479	1.479	1.479	1.479	1.479	1.479	1.479	1.479	1.480	1.480	1.480
300.0	1.547	1.547	1.547	1.548	1.548	1.548	1.548	1.548	1.548	1.548	1.548	1.548

Table 2A. Helium^4 Thermal Conductivity, κ, mW/cm-K (continued)

PRES MN/M^2	0.263	0.274	0.284	0.294	0.304	0.355	0.405	0.456	0.507	0.557	0.608	0.659
ATM	2.600	2.700	2.800	2.900	3.000	3.500	4.000	4.500	5.000	5.500	6.000	6.500
TEMP,K												
3.0	0.188	0.188	0.188	0.189	0.189	0.191	0.192	0.194	0.195	0.197	0.198	0.200
3.5	0.199	0.199	0.200	0.200	0.200	0.202	0.204	0.206	0.208	0.210	0.211	0.213
4.0	0.207	0.208	0.208	0.209	0.209	0.212	0.215	0.217	0.219	0.221	0.223	0.225
4.2	0.209	0.210	0.211	0.211	0.212	0.215	0.218	0.220	0.223	0.225	0.227	0.229
4.4	0.211	0.211	0.212	0.213	0.214	0.217	0.220	0.223	0.226	0.228	0.231	0.233
4.6	0.211	0.212	0.213	0.213	0.214	0.218	0.222	0.225	0.228	0.231	0.234	0.236
4.8	0.209	0.210	0.211	0.212	0.213	0.218	0.222	0.226	0.230	0.233	0.236	0.239
5.0	0.215	0.215	0.215	0.215	0.216	0.216	0.222	0.226	0.230	0.234	0.237	0.240
5.1	0.220	0.219	0.218	0.217	0.217	0.219	0.221	0.226	0.230	0.234	0.238	0.241
5.2	0.228	0.223	0.221	0.219	0.219	0.219	0.222	0.225	0.229	0.234	0.238	0.241
5.3	0.264	0.241	0.231	0.226	0.223	0.220	0.222	0.225	0.229	0.234	0.238	0.242
5.4	0.460	0.321	0.266	0.245	0.234	0.222	0.223	0.226	0.228	0.233	0.238	0.242
5.5	0.272	0.338	0.354	0.299	0.264	0.226	0.224	0.226	0.230	0.233	0.238	0.242
5.6	0.210	0.240	0.277	0.306	0.306	0.233	0.226	0.227	0.230	0.234	0.237	0.242
5.7	0.185	0.201	0.222	0.246	0.267	0.245	0.228	0.228	0.231	0.234	0.238	0.241
5.8	0.173	0.183	0.196	0.211	0.227	0.257	0.232	0.229	0.231	0.234	0.238	0.241
5.9	0.166	0.173	0.181	0.191	0.203	0.252	0.237	0.231	0.231	0.234	0.238	0.242
6.0	0.161	0.166	0.173	0.180	0.188	0.234	0.241	0.232	0.232	0.235	0.238	0.242
6.2	0.156	0.160	0.164	0.168	0.173	0.204	0.230	0.235	0.233	0.235	0.238	0.242
6.4	0.154	0.157	0.159	0.163	0.166	0.187	0.211	0.228	0.233	0.235	0.238	0.241
6.6	0.153	0.156	0.158	0.160	0.163	0.178	0.197	0.215	0.227	0.233	0.237	0.240
6.8	0.154	0.155	0.157	0.159	0.161	0.173	0.188	0.204	0.218	0.228	0.234	0.239
7.0	0.155	0.156	0.158	0.159	0.161	0.171	0.183	0.196	0.209	0.220	0.229	0.236
7.5	0.158	0.159	0.160	0.161	0.163	0.169	0.177	0.186	0.196	0.206	0.215	0.224
8.0	0.163	0.163	0.164	0.165	0.166	0.171	0.177	0.184	0.191	0.199	0.207	0.214
8.5	0.167	0.168	0.169	0.170	0.170	0.175	0.179	0.185	0.190	0.197	0.203	0.209
9.0	0.172	0.173	0.174	0.174	0.175	0.179	0.183	0.187	0.192	0.197	0.202	0.208
9.5	0.177	0.178	0.179	0.179	0.180	0.183	0.187	0.190	0.195	0.199	0.203	0.208
10.0	0.183	0.183	0.184	0.184	0.185	0.188	0.191	0.194	0.198	0.202	0.206	0.210

11.0	0.193	0.193	0.194	0.194	0.195	0.197	0.200	0.203	0.206	0.209	0.212	0.215
12.0	0.203	0.203	0.203	0.204	0.204	0.207	0.209	0.211	0.214	0.217	0.219	0.222
13.0	0.212	0.213	0.213	0.213	0.214	0.216	0.218	0.220	0.223	0.225	0.228	0.230
14.0	0.221	0.222	0.222	0.223	0.223	0.225	0.227	0.229	0.231	0.234	0.236	0.238
15.0	0.230	0.231	0.231	0.232	0.232	0.234	0.236	0.238	0.240	0.242	0.244	0.246
16.0	0.239	0.240	0.240	0.240	0.241	0.242	0.244	0.246	0.248	0.250	0.252	0.254
17.0	0.248	0.248	0.248	0.249	0.249	0.251	0.253	0.254	0.256	0.258	0.260	0.262
18.0	0.256	0.256	0.257	0.257	0.257	0.259	0.261	0.262	0.264	0.266	0.268	0.270
19.0	0.264	0.265	0.265	0.265	0.265	0.267	0.269	0.270	0.272	0.274	0.275	0.277
20.0	0.272	0.272	0.273	0.273	0.273	0.275	0.276	0.278	0.280	0.281	0.283	0.285
25.0	0.310	0.310	0.310	0.311	0.311	0.312	0.314	0.315	0.316	0.318	0.319	0.320
30.0	0.345	0.345	0.345	0.346	0.346	0.347	0.348	0.349	0.351	0.352	0.353	0.354
35.0	0.378	0.378	0.379	0.379	0.379	0.380	0.381	0.382	0.383	0.384	0.385	0.387
40.0	0.410	0.410	0.410	0.411	0.411	0.412	0.413	0.414	0.415	0.416	0.417	0.418
45.0	0.441	0.441	0.441	0.441	0.441	0.442	0.443	0.444	0.445	0.446	0.447	0.448
50.0	0.470	0.471	0.471	0.471	0.471	0.472	0.473	0.474	0.475	0.475	0.476	0.477
60.0	0.528	0.528	0.528	0.528	0.528	0.529	0.530	0.531	0.532	0.532	0.533	0.534
80.0	0.636	0.636	0.636	0.636	0.636	0.637	0.638	0.638	0.639	0.640	0.640	0.641
100.0	0.737	0.738	0.738	0.738	0.738	0.738	0.739	0.740	0.740	0.741	0.741	0.742
120.0	0.834	0.834	0.834	0.834	0.834	0.835	0.835	0.836	0.836	0.837	0.837	0.838
140.0	0.926	0.926	0.926	0.926	0.926	0.927	0.927	0.928	0.928	0.929	0.929	0.930
160.0	1.014	1.014	1.014	1.014	1.014	1.015	1.015	1.016	1.016	1.017	1.017	1.018
180.0	1.099	1.099	1.099	1.099	1.099	1.100	1.100	1.101	1.101	1.101	1.102	1.102
200.0	1.181	1.181	1.181	1.181	1.181	1.181	1.182	1.182	1.183	1.183	1.183	1.184
220.0	1.259	1.260	1.260	1.260	1.260	1.260	1.260	1.261	1.261	1.261	1.262	1.262
240.0	1.335	1.336	1.336	1.336	1.336	1.336	1.336	1.337	1.337	1.337	1.338	1.338
260.0	1.409	1.409	1.409	1.409	1.409	1.409	1.410	1.410	1.410	1.411	1.411	1.411
280.0	1.480	1.480	1.480	1.480	1.480	1.480	1.480	1.481	1.481	1.481	1.482	1.482
300.0	1.548	1.548	1.548	1.548	1.548	1.549	1.549	1.549	1.549	1.550	1.550	1.550

Table 2A. Helium^4 Thermal Conductivity, κ, mW/cm-K (continued)

PRES MN/M	0.608	0.709	0.811	0.912	1.013	1.520	2.026	2.533	3.040	3.546	4.053	4.560
ATM	6.000	7.000	8.000	9.000	10.000	15.000	20.000	25.000	30.000	35.000	40.000	45.000
TEMP,K												
3.0	0.198	0.201	0.203	0.206	0.208	0.219	0.229	0.238	0.246	0.254	0.261	0.269
3.5	0.211	0.215	0.218	0.221	0.223	0.236	0.246	0.256	0.265	0.274	0.282	0.289
4.0	0.223	0.227	0.231	0.234	0.237	0.251	0.263	0.274	0.284	0.293	0.302	0.311
4.2	0.227	0.231	0.235	0.239	0.242	0.257	0.270	0.281	0.292	0.301	0.310	0.319
4.4	0.231	0.235	0.239	0.243	0.247	0.263	0.276	0.288	0.299	0.309	0.319	0.328
4.6	0.234	0.239	0.243	0.247	0.251	0.268	0.282	0.295	0.306	0.317	0.327	0.336
4.8	0.236	0.241	0.246	0.251	0.255	0.273	0.288	0.301	0.313	0.324	0.334	0.344
5.0	0.237	0.243	0.249	0.254	0.258	0.278	0.294	0.307	0.320	0.331	0.342	0.352
5.1	0.238	0.244	0.250	0.255	0.260	0.280	0.296	0.310	0.323	0.334	0.345	0.356
5.2	0.238	0.245	0.251	0.256	0.261	0.282	0.299	0.313	0.326	0.338	0.349	0.359
5.3	0.238	0.245	0.252	0.257	0.262	0.284	0.301	0.316	0.329	0.341	0.352	0.363
5.4	0.238	0.246	0.252	0.258	0.264	0.286	0.304	0.319	0.332	0.344	0.356	0.367
5.5	0.238	0.246	0.253	0.259	0.265	0.288	0.306	0.321	0.335	0.348	0.359	0.370
5.6	0.237	0.246	0.253	0.260	0.266	0.290	0.308	0.324	0.338	0.351	0.363	0.374
5.7	0.238	0.246	0.254	0.260	0.267	0.291	0.310	0.326	0.341	0.354	0.366	0.377
5.8	0.238	0.245	0.254	0.261	0.267	0.293	0.312	0.329	0.343	0.357	0.369	0.381
5.9	0.238	0.245	0.254	0.261	0.268	0.294	0.314	0.331	0.346	0.360	0.372	0.384
6.0	0.238	0.245	0.253	0.261	0.268	0.295	0.316	0.333	0.348	0.362	0.375	0.387
6.2	0.238	0.245	0.253	0.261	0.269	0.298	0.319	0.337	0.353	0.368	0.381	0.393
6.4	0.238	0.245	0.253	0.260	0.269	0.300	0.322	0.341	0.358	0.373	0.386	0.399
6.6	0.237	0.244	0.253	0.260	0.268	0.301	0.325	0.345	0.362	0.377	0.391	0.405
6.8	0.234	0.243	0.252	0.260	0.268	0.302	0.327	0.348	0.365	0.381	0.396	0.410
7.0	0.229	0.241	0.250	0.259	0.267	0.303	0.329	0.351	0.369	0.385	0.400	0.414
7.5	0.215	0.232	0.245	0.255	0.265	0.303	0.333	0.356	0.376	0.394	0.410	0.425
8.0	0.207	0.222	0.236	0.249	0.260	0.302	0.334	0.360	0.381	0.400	0.418	0.434
8.5	0.203	0.216	0.229	0.242	0.254	0.300	0.334	0.362	0.385	0.405	0.423	0.440
9.0	0.202	0.213	0.225	0.236	0.248	0.297	0.333	0.362	0.387	0.409	0.428	0.446
9.5	0.203	0.213	0.223	0.234	0.244	0.293	0.331	0.362	0.388	0.411	0.431	0.450
10.0	0.206	0.214	0.223	0.232	0.242	0.289	0.328	0.360	0.388	0.412	0.433	0.452

11.0	0.212	0.219	0.226	0.234	0.241	0.282	0.322	0.356	0.385	0.411	0.434	0.455
12.0	0.219	0.225	0.231	0.238	0.244	0.280	0.316	0.350	0.381	0.406	0.432	0.454
13.0	0.228	0.233	0.238	0.244	0.250	0.282	0.314	0.347	0.377	0.404	0.429	0.452
14.0	0.236	0.241	0.246	0.251	0.256	0.285	0.315	0.345	0.373	0.400	0.426	0.449
15.0	0.244	0.248	0.253	0.258	0.263	0.289	0.317	0.345	0.372	0.398	0.423	0.446
16.0	0.252	0.256	0.260	0.265	0.269	0.293	0.319	0.345	0.371	0.396	0.420	0.443
17.0	0.260	0.264	0.268	0.272	0.276	0.298	0.322	0.347	0.371	0.395	0.418	0.441
18.0	0.268	0.271	0.275	0.279	0.283	0.304	0.326	0.349	0.372	0.395	0.417	0.439
19.0	0.275	0.279	0.282	0.286	0.290	0.309	0.330	0.352	0.374	0.396	0.417	0.438
20.0	0.263	0.286	0.290	0.293	0.297	0.315	0.335	0.355	0.376	0.397	0.417	0.437
25.0	0.319	0.322	0.325	0.327	0.330	0.345	0.361	0.377	0.393	0.410	0.427	0.443
30.0	0.353	0.355	0.358	0.360	0.363	0.376	0.389	0.402	0.416	0.430	0.444	0.458
35.0	0.385	0.388	0.390	0.392	0.394	0.406	0.417	0.429	0.441	0.453	0.465	0.477
40.0	0.417	0.419	0.421	0.423	0.425	0.435	0.445	0.456	0.467	0.477	0.488	0.499
45.0	0.447	0.449	0.451	0.453	0.454	0.464	0.473	0.483	0.493	0.502	0.512	0.522
50.0	0.476	0.478	0.480	0.482	0.483	0.492	0.501	0.510	0.519	0.528	0.537	0.546
60.0	0.533	0.535	0.536	0.538	0.539	0.547	0.555	0.563	0.571	0.579	0.587	0.595
80.0	0.640	0.642	0.643	0.644	0.646	0.652	0.659	0.665	0.672	0.679	0.685	0.692
100.0	0.741	0.743	0.744	0.745	0.746	0.752	0.758	0.763	0.769	0.775	0.781	0.786
120.0	0.837	0.838	0.839	0.840	0.842	0.847	0.852	0.857	0.862	0.868	0.873	0.878
140.0	0.929	0.930	0.931	0.932	0.933	0.938	0.942	0.947	0.952	0.957	0.961	0.966
160.0	1.017	1.018	1.019	1.020	1.021	1.025	1.029	1.034	1.038	1.043	1.047	1.051
180.0	1.102	1.103	1.103	1.104	1.105	1.109	1.113	1.117	1.121	1.125	1.129	1.133
200.0	1.183	1.184	1.185	1.186	1.186	1.190	1.194	1.198	1.201	1.205	1.209	1.212
220.0	1.262	1.263	1.263	1.264	1.265	1.268	1.272	1.275	1.279	1.282	1.285	1.289
240.0	1.338	1.338	1.339	1.340	1.340	1.343	1.347	1.350	1.353	1.356	1.359	1.363
260.0	1.411	1.411	1.412	1.413	1.413	1.416	1.419	1.422	1.425	1.428	1.431	1.434
280.0	1.482	1.482	1.483	1.483	1.484	1.487	1.489	1.492	1.495	1.497	1.500	1.503
300.0	1.550	1.550	1.551	1.551	1.552	1.554	1.557	1.560	1.562	1.565	1.567	1.570

Table 2A. Helium^4 Thermal Conductivity, κ, mW/cm-K (concluded)

PRES MN/M	5.066	5.573	6.079	6.586	7.093	7.599	8.106	8.613	9.119	9.626	10.133	10.639
ATM	50.000	55.000	60.000	65.000	70.000	75.000	80.000	85.000	90.000	95.000	100.000	105.000
TEMP,K												
3.0	0.276	0.283	0.290	0.297	0.303	0.310						
3.5	0.297	0.304	0.311	0.319	0.325	0.332	0.339	0.346	0.352	0.359	0.366	
4.0	0.319	0.326	0.334	0.342	0.349	0.356	0.363	0.370	0.377	0.384	0.390	0.397
4.2	0.327	0.336	0.343	0.351	0.358	0.366	0.373	0.380	0.387	0.394	0.401	0.408
4.4	0.336	0.345	0.353	0.361	0.368	0.376	0.383	0.390	0.398	0.405	0.412	0.419
4.6	0.345	0.353	0.362	0.370	0.378	0.386	0.393	0.401	0.408	0.415	0.423	0.430
4.8	0.353	0.362	0.371	0.379	0.387	0.395	0.403	0.411	0.419	0.426	0.433	0.441
5.0	0.361	0.371	0.380	0.388	0.397	0.405	0.413	0.421	0.429	0.436	0.444	0.451
5.1	0.365	0.375	0.384	0.393	0.401	0.410	0.418	0.426	0.434	0.442	0.449	0.457
5.2	0.369	0.379	0.388	0.397	0.406	0.414	0.423	0.431	0.439	0.447	0.455	0.462
5.3	0.373	0.383	0.392	0.402	0.410	0.419	0.427	0.436	0.444	0.452	0.460	0.468
5.4	0.377	0.387	0.397	0.406	0.415	0.424	0.432	0.441	0.449	0.457	0.465	0.473
5.5	0.381	0.391	0.401	0.410	0.419	0.428	0.437	0.445	0.454	0.462	0.470	0.478
5.6	0.385	0.395	0.405	0.414	0.424	0.433	0.441	0.450	0.459	0.467	0.475	0.483
5.7	0.388	0.399	0.409	0.418	0.428	0.437	0.446	0.455	0.463	0.472	0.480	0.488
5.8	0.392	0.402	0.413	0.422	0.432	0.441	0.450	0.459	0.468	0.477	0.485	0.493
5.9	0.395	0.406	0.416	0.426	0.436	0.445	0.455	0.464	0.473	0.481	0.490	0.498
6.0	0.399	0.410	0.420	0.430	0.440	0.450	0.459	0.468	0.477	0.486	0.494	0.503
6.2	0.405	0.416	0.427	0.438	0.448	0.458	0.467	0.477	0.486	0.495	0.504	0.512
6.4	0.411	0.423	0.434	0.445	0.455	0.465	0.475	0.485	0.494	0.503	0.512	0.521
6.6	0.417	0.429	0.440	0.451	0.462	0.473	0.483	0.492	0.502	0.512	0.521	0.530
6.8	0.423	0.435	0.447	0.458	0.469	0.479	0.490	0.500	0.510	0.519	0.529	0.538
7.0	0.428	0.440	0.452	0.464	0.475	0.486	0.497	0.507	0.517	0.527	0.537	0.546
7.5	0.439	0.452	0.465	0.478	0.489	0.501	0.512	0.523	0.534	0.544	0.554	0.564
8.0	0.448	0.463	0.476	0.489	0.502	0.514	0.525	0.537	0.548	0.559	0.570	0.580
8.5	0.456	0.471	0.485	0.499	0.512	0.525	0.537	0.549	0.560	0.572	0.583	0.594
9.0	0.462	0.478	0.493	0.507	0.521	0.534	0.547	0.559	0.571	0.583	0.594	0.606
9.5	0.467	0.483	0.499	0.513	0.528	0.541	0.554	0.567	0.580	0.592	0.604	0.616
10.0	0.470	0.487	0.503	0.519	0.533	0.547	0.561	0.574	0.587	0.600	0.612	0.624

11.0	0.474	0.492	0.509	0.525	0.541	0.556	0.570	0.584	0.598	0.611	0.624	0.636
12.0	0.475	0.494	0.512	0.529	0.545	0.561	0.576	0.590	0.604	0.618	0.631	0.645
13.0	0.473	0.493	0.512	0.530	0.547	0.563	0.576	0.593	0.608	0.622	0.636	0.650
14.0	0.471	0.491	0.511	0.529	0.546	0.563	0.579	0.595	0.610	0.624	0.638	0.652
15.0	0.468	0.489	0.508	0.527	0.545	0.562	0.578	0.594	0.609	0.624	0.639	0.653
16.0	0.465	0.486	0.505	0.524	0.542	0.560	0.576	0.592	0.608	0.623	0.638	0.652
17.0	0.462	0.483	0.502	0.521	0.539	0.557	0.573	0.590	0.605	0.621	0.635	0.650
18.0	0.460	0.480	0.499	0.518	0.536	0.554	0.570	0.587	0.602	0.618	0.633	0.647
19.0	0.458	0.478	0.497	0.515	0.533	0.551	0.567	0.584	0.599	0.615	0.630	0.644
20.0	0.457	0.476	0.495	0.513	0.531	0.548	0.564	0.581	0.596	0.611	0.626	0.641
25.0	0.460	0.476	0.492	0.508	0.524	0.539	0.555	0.569	0.584	0.598	0.613	0.626
30.0	0.472	0.486	0.500	0.514	0.527	0.541	0.554	0.568	0.581	0.594	0.607	0.619
35.0	0.489	0.502	0.514	0.526	0.538	0.550	0.562	0.574	0.586	0.598	0.609	0.621
40.0	0.510	0.521	0.532	0.543	0.553	0.564	0.575	0.586	0.596	0.607	0.618	0.628
45.0	0.532	0.542	0.552	0.562	0.572	0.581	0.591	0.601	0.611	0.620	0.630	0.640
50.0	0.555	0.564	0.573	0.582	0.591	0.601	0.610	0.619	0.628	0.637	0.646	0.654
60.0	0.603	0.610	0.618	0.626	0.634	0.642	0.650	0.658	0.666	0.674	0.682	0.690
80.0	0.698	0.705	0.711	0.718	0.725	0.731	0.738	0.744	0.751	0.757	0.764	0.770
100.0	0.792	0.798	0.804	0.809	0.815	0.821	0.827	0.832	0.838	0.844	0.849	0.855
120.0	0.883	0.888	0.894	0.899	0.904	0.909	0.914	0.919	0.924	0.929	0.934	0.940
140.0	0.971	0.976	0.980	0.985	0.990	0.994	0.999	1.004	1.008	1.013	1.018	1.022
160.0	1.056	1.060	1.064	1.069	1.073	1.077	1.082	1.086	1.090	1.094	1.099	1.103
180.0	1.137	1.141	1.145	1.149	1.153	1.157	1.161	1.165	1.169	1.173	1.177	1.181
200.0	1.216	1.220	1.224	1.227	1.231	1.235	1.238	1.242	1.246	1.249	1.253	1.257
220.0	1.292	1.296	1.299	1.303	1.306	1.309	1.313	1.316	1.320	1.323	1.327	1.330
240.0	1.366	1.369	1.372	1.375	1.379	1.382	1.385	1.388	1.391	1.394	1.398	1.401
260.0	1.437	1.440	1.443	1.446	1.449	1.452	1.455	1.458	1.460	1.463	1.466	1.469
280.0	1.506	1.508	1.511	1.514	1.517	1.519	1.522	1.525	1.527	1.530	1.533	1.536
300.0	1.572	1.575	1.577	1.580	1.582	1.585	1.587	1.590	1.592	1.595	1.597	1.600

Table 3A. Helium[4] Prandtl Number, $C_p\mu/\kappa$ (Ref. 12, 13, 16)

PRES MN/M²	0.010	0.051	0.101	0.152	0.182	0.193	0.203	0.213	0.223	0.233	0.243	0.253
ATM	0.100	0.500	1.000	1.500	1.800	1.900	2.000	2.100	2.200	2.300	2.400	2.500
TEMP,K												
3.0	0.667	0.500	0.500	0.501	0.501	0.502	0.502	0.502	0.502	0.502	0.503	0.503
3.5	0.659	0.579	0.571	0.565	0.562	0.562	0.561	0.560	0.550	0.559	0.558	0.558
4.0	0.654	0.754	0.692	0.664	0.652	0.648	0.645	0.641	0.638	0.636	0.633	0.630
4.2	0.652	0.748	0.788	0.733	0.710	0.704	0.698	0.692	0.687	0.682	0.678	0.674
4.4	0.652	0.736	0.961	0.843	0.798	0.786	0.775	0.765	0.756	0.747	0.740	0.733
4.6	0.651	0.726	0.900	1.033	0.952	0.925	0.902	0.881	0.853	0.847	0.832	0.819
4.8	0.651	0.718	0.859	1.238	1.230	1.163	1.110	1.067	1.031	1.000	0.973	0.949
5.0	0.651	0.711	0.830	1.079	1.512	1.929	1.793	1.571	1.427	1.323	1.245	1.182
5.1	0.651	0.708	0.818	1.031	1.328	1.527	1.894	2.602	2.000	1.707	1.526	1.401
5.2	0.651	0.706	0.808	0.992	1.218	1.344	1.530	1.844	2.584	2.786	2.082	1.760
5.3	0.651	0.703	0.798	0.961	1.141	1.233	1.354	1.524	1.738	2.267	3.331	2.730
5.4	0.651	0.701	0.790	0.936	1.085	1.155	1.243	1.357	1.511	1.731	2.072	2.610
5.5	0.651	0.699	0.783	0.914	1.040	1.097	1.166	1.249	1.355	1.492	1.676	1.931
5.6	0.651	0.698	0.776	0.895	1.005	1.052	1.108	1.173	1.252	1.349	1.470	1.624
5.7	0.651	0.696	0.770	0.879	0.975	1.016	1.062	1.115	1.178	1.251	1.339	1.446
5.8	0.651	0.694	0.765	0.865	0.951	0.986	1.025	1.070	1.121	1.179	1.248	1.327
5.9	0.652	0.693	0.760	0.853	0.929	0.960	0.994	1.033	1.076	1.124	1.179	1.242
6.0	0.652	0.692	0.755	0.841	0.911	0.938	0.969	1.002	1.039	1.080	1.125	1.177
6.2	0.653	0.690	0.747	0.823	0.881	0.903	0.927	0.953	0.982	1.013	1.046	1.083
6.4	0.653	0.688	0.740	0.807	0.857	0.876	0.896	0.917	0.940	0.964	0.991	1.019
6.6	0.654	0.686	0.735	0.794	0.836	0.854	0.871	0.889	0.907	0.928	0.949	0.971
6.8	0.655	0.685	0.730	0.784	0.822	0.836	0.850	0.866	0.882	0.899	0.916	0.935
7.0	0.655	0.684	0.725	0.774	0.808	0.821	0.834	0.847	0.851	0.875	0.891	0.906
7.5	0.657	0.682	0.717	0.757	0.783	0.793	0.802	0.812	0.822	0.833	0.844	0.855
8.0	0.660	0.681	0.711	0.744	0.766	0.773	0.781	0.789	0.797	0.805	0.813	0.822
8.5	0.662	0.681	0.707	0.735	0.753	0.759	0.765	0.772	0.778	0.785	0.791	0.798
9.0	0.664	0.681	0.704	0.728	0.743	0.748	0.754	0.759	0.754	0.770	0.775	0.781
9.5	0.666	0.682	0.702	0.723	0.736	0.740	0.745	0.750	0.754	0.759	0.763	0.768
10.0	0.668	0.682	0.700	0.719	0.730	0.734	0.738	0.742	0.746	0.750	0.754	0.758

11.0	0.672	0.684	0.699	0.714	0.723	0.726	0.729	0.732	0.735	0.739	0.742	0.745
12.0	0.676	0.686	0.698	0.710	0.718	0.721	0.723	0.726	0.728	0.731	0.733	0.736
13.0	0.679	0.688	0.698	0.708	0.715	0.717	0.719	0.721	0.723	0.725	0.727	0.730
14.0	0.683	0.690	0.698	0.707	0.713	0.714	0.716	0.718	0.720	0.722	0.723	0.725
15.0	0.685	0.691	0.699	0.707	0.711	0.713	0.714	0.716	0.717	0.719	0.720	0.722
16.0	0.688	0.693	0.700	0.706	0.710	0.712	0.713	0.714	0.716	0.717	0.718	0.719
17.0	0.690	0.695	0.701	0.706	0.710	0.711	0.712	0.713	0.714	0.715	0.717	0.718
18.0	0.693	0.697	0.702	0.707	0.709	0.710	0.711	0.712	0.713	0.714	0.715	0.716
19.0	0.695	0.698	0.702	0.707	0.709	0.710	0.711	0.712	0.713	0.714	0.715	0.715
20.0	0.696	0.699	0.703	0.707	0.709	0.710	0.711	0.712	0.712	0.713	0.714	0.715
25.0	0.703	0.704	0.707	0.709	0.710	0.711	0.711	0.712	0.712	0.712	0.713	0.713
30.0	0.706	0.707	0.709	0.710	0.711	0.711	0.711	0.712	0.712	0.712	0.713	0.713
35.0	0.708	0.709	0.709	0.710	0.711	0.711	0.711	0.712	0.712	0.712	0.712	0.712
40.0	0.708	0.709	0.709	0.710	0.710	0.711	0.711	0.711	0.711	0.711	0.711	0.711
45.0	0.708	0.708	0.709	0.709	0.709	0.709	0.709	0.710	0.710	0.710	0.710	0.710
50.0	0.707	0.707	0.707	0.708	0.708	0.708	0.708	0.708	0.708	0.708	0.708	0.708
60.0	0.704	0.704	0.704	0.704	0.705	0.705	0.705	0.705	0.705	0.705	0.705	0.705
80.0	0.697	0.697	0.697	0.697	0.697	0.697	0.697	0.697	0.697	0.697	0.697	0.697
100.0	0.691	0.691	0.691	0.691	0.691	0.691	0.691	0.691	0.691	0.691	0.691	0.691
120.0	0.685	0.685	0.685	0.685	0.685	0.685	0.685	0.685	0.685	0.685	0.685	0.685
140.0	0.681	0.681	0.681	0.681	0.681	0.681	0.681	0.681	0.681	0.681	0.681	0.681
160.0	0.679	0.679	0.679	0.679	0.679	0.679	0.679	0.679	0.679	0.679	0.679	0.679
180.0	0.677	0.677	0.677	0.677	0.677	0.677	0.677	0.677	0.677	0.677	0.677	0.677
200.0	0.677	0.677	0.677	0.677	0.677	0.677	0.677	0.677	0.677	0.677	0.676	0.676
220.0	0.677	0.677	0.677	0.677	0.677	0.677	0.677	0.677	0.677	0.677	0.677	0.677
240.0	0.678	0.678	0.678	0.678	0.678	0.678	0.678	0.678	0.678	0.678	0.678	0.678
260.0	0.680	0.680	0.680	0.680	0.680	0.680	0.680	0.680	0.680	0.680	0.680	0.680
280.0	0.683	0.683	0.682	0.682	0.682	0.682	0.682	0.682	0.682	0.682	0.682	0.682
300.0	0.686	0.686	0.686	0.685	0.685	0.685	0.685	0.685	0.685	0.685	0.685	0.685

Table 3A. Helium[4] Prandtl Number, $C_p\mu/\kappa$ (concluded)

PRES MN/M²	0.263	0.274	0.284	0.294	0.304	0.355	0.405	0.456	0.507	0.557	0.608	0.659
ATM	2.600	2.700	2.800	2.900	3.000	3.500	4.000	4.500	5.000	5.500	6.000	6.500
TEMP,K												
3.0	0.503	0.503	0.504	0.504	0.504	0.506	0.507	0.509	0.511	0.512	0.514	0.516
3.5	0.557	0.557	0.557	0.556	0.556	0.554	0.553	0.552	0.552	0.552	0.552	0.552
4.0	0.628	0.626	0.623	0.621	0.619	0.611	0.604	0.599	0.594	0.591	0.588	0.585
4.2	0.670	0.666	0.662	0.659	0.656	0.642	0.632	0.623	0.616	0.610	0.606	0.602
4.4	0.726	0.720	0.714	0.708	0.703	0.682	0.665	0.653	0.642	0.634	0.627	0.621
4.6	0.807	0.736	0.786	0.777	0.768	0.734	0.708	0.689	0.674	0.661	0.651	0.643
4.8	0.939	0.917	0.898	0.880	0.865	0.805	0.764	0.735	0.712	0.695	0.680	0.669
5.0	1.132	1.030	1.054	1.023	0.996	0.908	0.839	0.793	0.759	0.734	0.714	0.698
5.1	1.309	1.237	1.179	1.132	1.092	0.958	0.885	0.826	0.786	0.756	0.732	0.714
5.2	1.567	1.434	1.337	1.261	1.200	1.015	0.916	0.856	0.807	0.772	0.745	0.724
5.3	2.126	1.810	1.612	1.474	1.371	1.092	0.962	0.884	0.831	0.789	0.757	0.733
5.4	3.085	2.609	2.142	1.850	1.654	1.203	1.025	0.925	0.865	0.814	0.777	0.748
5.5	2.273	2.632	2.758	2.456	2.116	1.356	1.104	0.975	0.896	0.843	0.799	0.766
5.6	1.821	2.062	2.313	2.493	2.505	1.565	1.204	1.035	0.937	0.872	0.825	0.786
5.7	1.576	1.732	1.912	2.096	2.248	1.831	1.331	1.109	0.986	0.907	0.852	0.810
5.8	1.421	1.531	1.658	1.797	1.938	2.054	1.486	1.197	1.043	0.947	0.882	0.837
5.9	1.314	1.396	1.490	1.594	1.706	2.060	1.653	1.301	1.109	0.994	0.916	0.861
6.0	1.234	1.299	1.372	1.452	1.539	1.918	1.787	1.416	1.185	1.046	0.955	0.891
6.2	1.124	1.168	1.216	1.268	1.325	1.631	1.780	1.615	1.355	1.170	1.046	0.961
6.4	1.049	1.082	1.117	1.154	1.195	1.422	1.606	1.641	1.493	1.302	1.152	1.044
6.6	0.996	1.021	1.048	1.077	1.107	1.279	1.447	1.537	1.523	1.402	1.256	1.133
6.8	0.955	0.976	0.997	1.020	1.044	1.178	1.320	1.423	1.451	1.428	1.330	1.216
7.0	0.923	0.940	0.958	0.977	0.997	1.105	1.222	1.322	1.379	1.391	1.352	1.272
7.5	0.867	0.879	0.891	0.904	0.917	0.987	1.064	1.139	1.201	1.241	1.259	1.258
8.0	0.830	0.839	0.848	0.858	0.867	0.918	0.972	1.027	1.077	1.119	1.148	1.164
8.5	0.805	0.812	0.819	0.827	0.834	0.873	0.913	0.955	0.995	1.030	1.060	1.081
9.0	0.787	0.793	0.798	0.804	0.810	0.841	0.873	0.905	0.937	0.967	0.994	1.015
9.5	0.773	0.778	0.783	0.788	0.793	0.818	0.844	0.870	0.896	0.921	0.944	0.964
10.0	0.762	0.766	0.771	0.775	0.779	0.800	0.822	0.844	0.856	0.887	0.906	0.924

11.0	0.748	0.751	0.754	0.758	0.761	0.777	0.793	0.810	0.826	0.841	0.856	0.869
12.0	0.739	0.741	0.744	0.746	0.749	0.762	0.775	0.787	0.800	0.812	0.824	0.834
13.0	0.732	0.734	0.736	0.738	0.740	0.751	0.761	0.771	0.782	0.791	0.801	0.809
14.0	0.727	0.729	0.730	0.732	0.734	0.743	0.752	0.760	0.769	0.777	0.784	0.792
15.0	0.724	0.725	0.727	0.728	0.730	0.737	0.745	0.752	0.759	0.766	0.773	0.779
16.0	0.721	0.722	0.723	0.725	0.726	0.732	0.739	0.745	0.751	0.757	0.762	0.768
17.0	0.719	0.720	0.721	0.722	0.723	0.729	0.734	0.739	0.745	0.750	0.755	0.759
18.0	0.717	0.718	0.719	0.720	0.721	0.726	0.731	0.735	0.740	0.744	0.748	0.753
19.0	0.716	0.717	0.718	0.719	0.720	0.724	0.728	0.732	0.736	0.740	0.743	0.747
20.0	0.715	0.716	0.717	0.718	0.718	0.722	0.726	0.729	0.733	0.736	0.739	0.743
25.0	0.714	0.714	0.715	0.715	0.715	0.718	0.720	0.722	0.724	0.726	0.728	0.729
30.0	0.713	0.713	0.714	0.714	0.714	0.716	0.717	0.718	0.719	0.721	0.722	0.723
35.0	0.712	0.713	0.713	0.713	0.713	0.714	0.715	0.716	0.717	0.717	0.718	0.719
40.0	0.711	0.712	0.712	0.712	0.712	0.712	0.713	0.714	0.714	0.715	0.715	0.716
45.0	0.710	0.710	0.710	0.710	0.710	0.711	0.711	0.712	0.712	0.713	0.713	0.713
50.0	0.708	0.709	0.709	0.709	0.709	0.709	0.709	0.710	0.710	0.710	0.711	0.711
60.0	0.705	0.705	0.705	0.705	0.705	0.705	0.705	0.706	0.706	0.706	0.706	0.706
80.0	0.697	0.697	0.697	0.697	0.697	0.697	0.698	0.698	0.698	0.698	0.698	0.698
100.0	0.691	0.691	0.691	0.691	0.691	0.691	0.691	0.691	0.691	0.691	0.691	0.691
120.0	0.685	0.685	0.685	0.685	0.685	0.685	0.685	0.685	0.685	0.685	0.685	0.685
140.0	0.681	0.681	0.681	0.681	0.681	0.681	0.681	0.681	0.681	0.681	0.681	0.681
160.0	0.679	0.679	0.679	0.679	0.679	0.679	0.679	0.678	0.678	0.678	0.678	0.678
180.0	0.677	0.677	0.677	0.677	0.677	0.677	0.677	0.677	0.677	0.677	0.677	0.677
200.0	0.676	0.676	0.676	0.676	0.676	0.676	0.676	0.676	0.676	0.676	0.676	0.676
220.0	0.677	0.677	0.677	0.677	0.677	0.677	0.677	0.677	0.677	0.677	0.677	0.677
240.0	0.678	0.678	0.678	0.678	0.678	0.678	0.678	0.678	0.678	0.678	0.678	0.678
260.0	0.680	0.680	0.680	0.680	0.680	0.680	0.680	0.680	0.680	0.680	0.680	0.679
280.0	0.682	0.682	0.682	0.682	0.682	0.682	0.682	0.682	0.682	0.682	0.682	0.682
300.0	0.685	0.685	0.685	0.685	0.685	0.685	0.685	0.685	0.685	0.685	0.685	0.685

5

LARGE HELIUM SYSTEMS FOR SUPERCONDUCTING POWER APPLICATIONS AND HIGH ENERGY PHYSICS

Jack E. Jensen

The emergence of superconductivity, from the laboratory, as a practical means of transporting large electrical currents promises to have a profound effect on liquid helium temature refrigeration systems. Some of these systems will require the distribution of the refrigeration over ten's of kilometers of length while others will require ten's of kilowatts of refrigeration in relatively small areas. Examples of such systems are described, bringing out some of the heat transfer and hydrodynamic requirements. The effects of the transport properties of helium on the system requirements in several pressure-temperature regions are presented. Estimates of the quantity of helium required are included.

5-1. Introduction

In any prediction of the near future of a technology, it is well to look critically at the past. The growth in use and size of helium temperature refrigeration is and has been due to another technology: the development and application of superconductivity. The discovery of this quirk of nature over 60 years ago by Kamerlingh Onnes did not cause a revolution in science and industry, but superconductivity remained an interesting scientific curiosity relegated to the laboratories of the pure scientists and theorists. It was not until the work of Kunsler in 1961 showed that various compounds and alloys of niobium were not only superconducting, but could support large currents in the presence of high magnetic fields, that this technology began to blossom. From 1961 to the present the application of this exciting tool has still been extremely slow and even painful. Recently, Henry Kolm of the National Magnet Laboratory said "superconductivity has been going through a very long adolescence." At the recent Fourth International Magnet Conference at Brookhaven, it was pointed out that most early workers in the superconductivity field took the attitude that "here is a solution, now what is the problem?" More recently the order has been reversed; many studies and proposals now invoke superconductivity to solve real problems in all fields of science and engineering. This is a much more orderly approach. There is every reason to believe that applications in this field will bear fruit within the next decade, and that a revolution similar to the invention of the wheel is in the future.

The foregoing brings us to the topic of this paper. Two areas of applied superconductivity are considered: the transmission of large blocks of electrical power in the industrial domain, and the construction of high magnetic field magnets to be used to extend the energy of particle accelerators in the scientific domain. Both of these applications will require helium temperature refrigerators, as well as rather large quantities of helium in inventory. Examples of several systems for each application will be given, pointing out the reasons for selection and the advantages or disadvantages. Included is a discussion of helium refrigerators and the effects of the transport properties of helium on efficiencies of such machines in several pressure-temperature regions.

5-2. Refrigeration for Power Transmission

The use of superconductors for transmission lines has been studied by a number of groups. Most of these studies have come to the conclusion that it is both feasible and necessary.[1] Such studies, followed by research proposals, have been made for both dc and ac applications. Although superconductors are only truly lossless when carrying dc currents, modern electric power is generated as ac, and it is in this area that the largest number of workers are engaged. When transporting ac currents, superconductors have magnetic losses associated with the self-field produced by a conductor carrying a current. As the current is increased the magnetic field is increased and consequently the losses also increase. If the field is increased sufficiently the superconductor becomes normal, and the losses become I^2R or purely resistive. The other parameter that enters into consideration is the temperature. Each superconducting material has a critical temperature, at a given critical temperature, at a given critical field and current, above which it will no longer carry supercurrents.

The two superconducting materials most considered for power transmission at the present time are pure niobium (Nb) and Nb_3Sn. Nb is typified by very low losses, but has a very low (<3kG at 4.2 K) critical field, and a critical temperature of 9 K at zero field. The upper critical field of Nb_3Sn is > 150 kG at 4.2 K and the critical temperature is 16 K; however, this material does have higher magnetic losses.

The magnetic losses are in effect sources of heat to the helium temperature region, as are the dielectric losses because of the high voltages (69 to 270 kV) and the thermal heat inleak from ambient. The total low temperature load can be calculated and is shown in Table 5-1 for a sample 3000 MVA line using parallel coaxial cables, 5 per phase, 3 phases. It is interesting to note (and also a subject of much controversy) that although the Nb line has lower losses, the Nb_3Sn requires less ambient horsepower to cool it.

The line consists of 15 coaxial cables operating at a linear current density of 320 A/cm (rms). It has a nominal capacity of 3000 MVA. Characteristics of the line are given in Table 5-1. Although the magnetic losses increase faster for Nb with increasing temperature, the losses including shields are only 0.135 watts/meter as compared with 1.018 W/m for Nb_3Sn, each taken at its average temperature. The dielectric losses are the same for both lines (0.166 W/m); and other losses are 0.031 W/m. The total electrical losses are then equal to 0.332 and 1.215 W/m for the Nb and Nb_3Sn versions, respectively. The heat leak and support losses are the same for each line (0.093 W/m).

Table 5-1. Tentative Characteristics of a 3000-MVA Line Using Parallel Coaxial Cables

No. of cables per phase		5
Total No. of cables		15
Rated Voltage, line-to-line (E), kV		132
Rated line current (I), kA		13.8
Rated power (P_r), MVA		3161
Critical length (l_c) miles		≈ 240
Per unit series inductive impedance per mile		2.16×10^{-3}
Per unit shunt capacitive impedance per mile		240
Per unit surge impedance		0.72
Charging power, MVAR/mile		13.2
Dewar i.d., cm		30
Overall pipe diameter, cm		42
Dewar heat influx (with LN_2 shield), W/mile		150
Refrigerator station spacing,* km		10
Refrigerator station spacing,* miles		6.2
Superconducting material	*Nb*	*Nb_3Sn*
Temp. range, inlet to outlet, K	4.4 to 5.0	6.2 to 8.2
Magnetic losses, W/mile	217	1638
Dielectric losses, W/mile	267	267
Other losses,† W/mile	50	50
Total refrigerator line load	684	2105
Power input to refrigerator,* kW	3981	2108
Duration of 1.5 times rated current,≠ hr	5	12

* Spacing is 10 km, feeding 5 km of line in either direction.
† Other losses include the possibility of losses at splices, etc.
≠ Followed by suitable recovery period.

The incoming coolant passes through the central hole in each coaxial cable to remove the electrical heat. The return flow passes through an annular passage surrounding, and insulated from, the pipe containing the coaxial cables. This return stream intercepts the insulation and support heat influx. It was decided to use the whole temperature range in each case to remove the electrical load and allow the return stream to exceed the upper temperatures of

5 and 8.2 K by the amount necessary to take care of the heat leakage. Because the heat capacity of helium in the 4.4 to 5.0 K range decreases with increasing pressure, it is desirable to keep the peak inlet pressure as low as possible. However, if the pressure loss along the length of the line allows the pressure to fall below the critical pressure of helium (2.245 atm), a two-phase (gas and liquid) mixture will form that is unstable in a hydrodynamic sense. A cooling length of 5 km is found to be close to optimum for this design. This allows refrigerator stations to be placed 10 km apart.

The cooling circuit for the Nb line is assumed to be a closed loop with a cold pump used to force the circulation. The helium fluid returning from the line passes through the pump, which raises the pressure so that the desired line inlet pressure is maintained after the fluid has passed through a heat exchanger and the temperature has been reduced to 4.4 K. In the heat exchanger the heat in the cooling circuit is rejected to the main refrigerator. The pump adds to the other loads already mentioned, since the work of compression appears as heat in the circulating fluid. In this case the pump efficiency is assumed to be 50%. A pump is not required for the circuit operating in the 6.2 to 8.2 K range.

With all the parameters fixed, only one variable is left, the inlet pressure of the cooling fluid. In the case of the Nb line the minimum inlet pressure is found to be 4.4 atm. For the Nb_3Sn line it is desirable to find an optimum pressure in the region of 6 atm or less.

Table 5-2 is a tabulation of the results from computer output for the two versions of the line. The large ambient work for the Nb line is due almost entirely to the pump work required. Using the fixed parameters of length and flow area places a severe penalty on this line, as shown by the high flow rate and the large pressure loss. If a length of 2.5 km is used the pumping losses fall to $<50\%$ of the thermal load. Although the refrigerator stations are then closer together, the ambient work required per unit length would be about equal to that of the Nb_3Sn line. This shows quite clearly that the stage length of a line is affected not only by the heat load but also by the temperature region within which it is

Table 5-2. Results of Cryogenic Study of 3000-MVA Parallel Coaxial Cable Line*

Temperature range of conductor, K	*Superconducting material*	*Total ΔT, go and return, K*	*Inlet pressure atm*	*Total pressure drop, atm*	*Mass flow rate, g/sec*	*Line heat load, W*	*Pump heat load, W*	*Total heat load, W*	*Hot work at 25% Carnot, kW*
4.4-5.0	Nb	0.61	4.4	2.15	713	2125	5796	7921	1991
6.2-8.2	Nb_3Sn	2.07	6.0	0.83	279	6540	–	6540	1054
6.2-7.2	Nb_3Sn	0.81	6.0	1.64	447	6540	–	6540	1156

*5-km line length.

operated. Table 5-2 also shows the hot-work penalty of having a 1 K temperature differential. The refrigerator input is increased by about 10%.

If one neglects problems associated with critical point phenomena, the computations to date indicate no fundamental problems in cooling relatively long sections of superconducting power transmission cables. This presupposes that some control is maintained over flow areas and consequently the mass velocity of the fluid, and also that the heat flux (W/cm^2) is kept as small as possible, of the order of 20 $\mu W/cm^2$. These two variables are interdependent in the sense that the mass flow is directly dependent on the heat flux. The effect of larger heat fluxes is an increase in pressure loss for a fixed temperature rise and flow area. This then decreases the stage length between refrigerators or enlarges the cross-sectional area to permit lower mass flow velocities. In the case of the coaxial design the central flow channels are fixed by optimizing the electrical properties, which may not be optimum from a cooling standpoint.

Because of the low mass-flow velocity (< 1 m/sec), the cooling system time constants in lines of the length discussed are of the order of 8 to 24 hr. Therefore the effect of a fault, even if the line does not go normal, will exist for 1 to 2 time constants of the system. Cool-down time of the line is also affected by the time constant, although a detailed examination of the problem was not made for this study.

In the cable configuration investigated here, the higher magnetic losses of Nb_3Sn, compared with Nb, are offset by marked improvements in the refrigeration system at the higher allowable operating temperatures. It is likely that optimum designs of lines using either of the superconducting materials will have comparable refrigeration system power inputs.

The quantity of helium in such a system will depend on the operating pressure and temperature, but for the Nb_3Sn line in this example the equivalent of 42,000 liquid liters/mile will be needed.

Figure 5-1 shows schematically the two cold end systems assumed for the line designs. System (a) is intended for use at

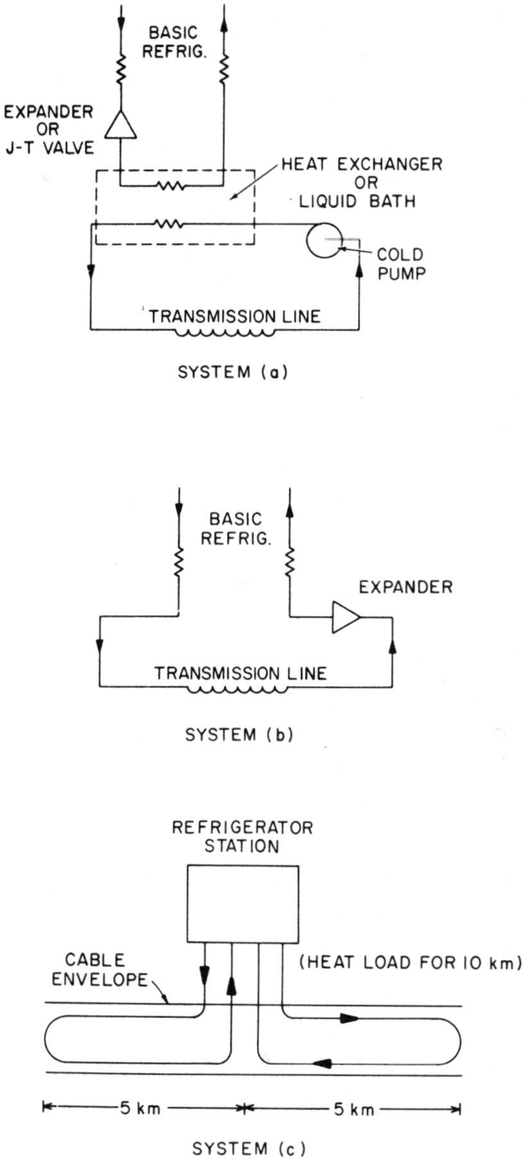

Fig. 5-1. Cooling systems for superconducting cables. (a) System for Nb cable (4.4 to 5 K); (b) system for Nb_3Sn cable (6.2 to 8.2 K); (c) configuration used in cooling calculations.

temperatures below the critical temperature of helium and is suitable for a niobium line system, while (b) is for niobium-tin cables operating above this temperature. These systems are capable of producing refrigeration at an efficiency >22% Carnot in the temperature regions for which they have been specified. It must be remembered that system (a) must produce an amount of refrigeration equal to the line losses plus the heat added by the cold pump. System (b) has pumping losses, but these are rejected at ambient temperature rather than the 5 K level. Pumping lengths and refrigerator spacing used in cooling calculations for practical lines are shown in system (c).

The regions of pressure and temperature mandated by the choice of superconductor may also be examined for other important helium properties. Figure 5-2 (from Ref. 2) shows contours of constant Joule-Thomson (J-T) coefficient (ψ) and the transposed critical line. The region associated with niobium lies in the area of negative J-T coefficient; i.e., pressure drop along the pipe causes further heating, which must be removed by the refrigerator. The niobium-tin region corresponds to a positive coefficient of about 0.2, which improves the refrigeration efficiency.

A typical system for the 4.4 to 5 K range is shown in Figure 5-1 (a). As noted, the need for a cold pump in this system decreases the overall efficiency. There are probably no problems associated with fluid property oscillations if the maximum temperature does not exceed 5 K. However, a major drawback is that the J-T coefficient is negative in much of the operating region of this system.

If the inlet pressure is increased, the warming effect is increased. This effect was noted in the calculations made for the niobium coaxial line. As the inlet pressure was increased, the mass rate of flow had to be increased to hold the temperature increment constant at 0.6 K. In so doing, the velocity of the flowing stream increased, the pressure loss increased because of the velocity increase, and the heat capacity decreased. All these factors resulted in an increase in the pumping loss and a consequent higher total heat load.

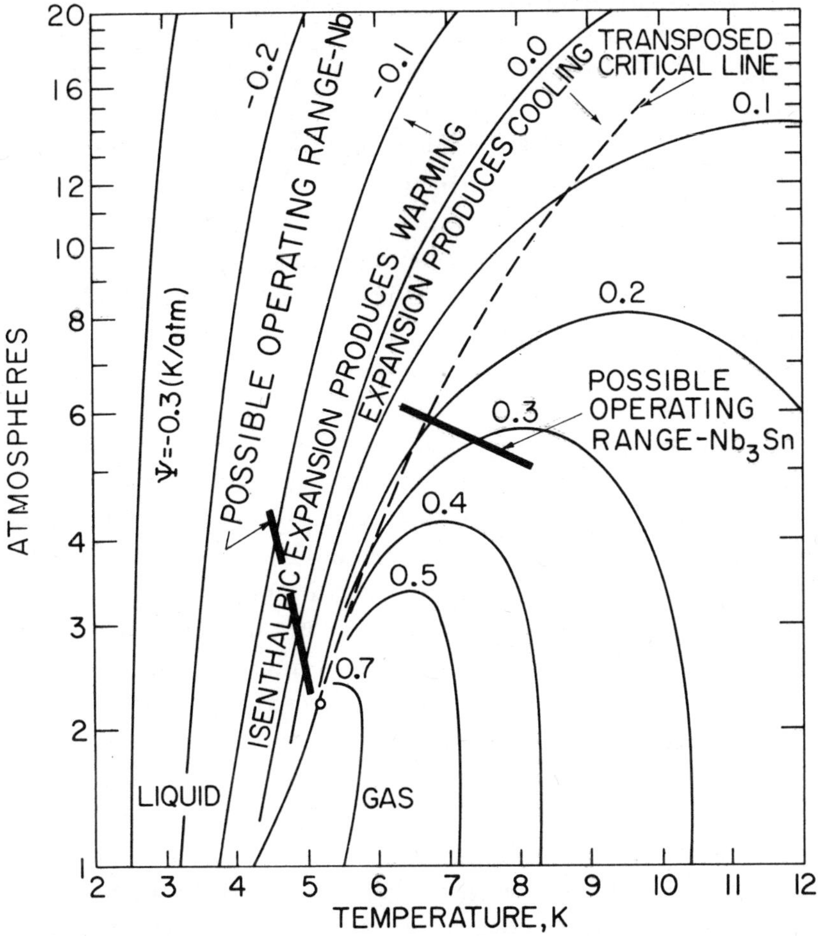

Fig. 5-2. Locus of the transposed critical line, and contours of constant Joule-Thomson coefficient, as a function of pressure and temperature.

Decreasing the inlet temperature makes matters worse. In creasing the allowable temperature interval and thus allowing the exit temperature to be higher appears to be the only possibility for improvement, but this may move a portion of the cable into the region of critical-point oscillation. If inlet pressures are below 5 atm, where little is known of supercritical helium behavior, total breakdown of the cooling system is possible!

System (b) in Figure 5-1 is not penalized by the J-T coefficient. In fact, for the temperature and pressure ranges specified for most of the calculations, it is a positive number and therefore enhances the refrigerator performance. The operating range is shown in Figure 5-2. The problem of critical-point phenomena is obviously present, however. In the various line designs it was decided to maximize the amount of heat removed per unit mass flow and thus minimize the mass flow rate. This procedure places the system on the transposed critical line to obtain the highest available specific heat for helium. Since there is no analytical method of predicting the oscillations or their magnitude, the effect of possible oscillations was ignored in calculating the performance.

Until more is learned about these oscillations, the regions in which they occur, their effect on flowing systems, and methods of damping or stopping them, a practical approach may be to avoid the transposed critical line. Measurements to date of dielectric strength in helium gas show a density effect. At lower pressures and consequently lower densities the dielectric strength is also lower. This factor may mean that only the highest pressure region is available if a helium-impregnated dielectric construction is considered. If this is the case, the use of the region in which maximum use of a favorable J-T coefficient can be made will be precluded. Avoiding oscillations may also mean operating in a region where the refrigeration per unit mass flow is not optimum, or at a temperature at which the superconductor losses are not at a minimum. These factors could affect the line size and stage length between refrigerators and thus have an important effect on the economics of the complete system.

5-3. Refrigeration for High Energy Accelerators

There are a number of installations being considered at the present time; three of these will be used to demonstrate the systems here. All of these have similarities, but operation of the accelerators dictates the optimum cycle parameters. Two of these machines are intended to extend the energy in a pulsed synchrotron to 1000 GeV, while the other is a double storage ring facility with an increase in energy from 30 GeV to 200 GeV. As pointed out previously, superconductors are essentially lossless when operated in a dc mode. However, in all of these accelerators the magnets will start at a low field and then be raised to a higher field as the particles are accelerated. During this increase in field the superconductors will have losses which will be a function of the rate of change of the magnetic field. The total heat load at 4.5 K will vary depending on the cyclic rate of the machine. At Brookhaven National Laboratory (BNL), where the storage ring facility (ISABELLE) is proposed, the cycle will be once per day. At the National Accelerator Laboratory (NAL) and at CERN II, the cyclic rate is proposed to be several times per minute. It is not surprising, therefore, that the heat load due to pulsing the magnets is larger for the two latter systems.

The magnets for all of these machines will use a niobium-titanium (NbTi) alloy as the superconductor in a fine filament configuration surrounded by a normal metal. NbTi is a stable material up to magnetic fields of 100 kG at 4.2 K. It does have a rather low critical temperature of ≈ 9 K. Magnets in the region of 60 kG have been operated successfully in a bath of liquid helium, but a number of magnets of NbTi have not reached design fields. Failures have been traced to a number of causes including coil motion and poor cooling. The fields planned for these machines will all be ≥ 40 kG, and since critical current increases with decreasing temperature, the lowest practical operating temperature is desired.

A study of various methods to cool ISABELLE was made at BNL,[3,4] from which the decision to cool the magnets in boiling liquid helium was indicated. At the same time, Dean[5,6] at Rutherford High Energy Laboratory in England was also looking at this problem. His conclusion was also to cool with boiling helium, but in addition there might be an economic advantage in running the refrigerator as cold as 3.5 K. More recently Vander Arend[7] of Cryogenic Consultants, Inc. indicated NAL was considering a supercritical, pump-circulated system.

The CERN II system will require 97 kW of refrigeration at a temperature near 3.5 K, driven by 42 MW of electrical power.

Approximately 20% of the total capacity is required to distribute the refrigerant around a ring 7 km in circumference. The preliminary numbers for the NAL system are ≈30 kW at 4.5 to 5 K distributed around a ring 6 km in circumference. The ISABELLE system requires 20 kW at 4.5 K distributed around a ring 2 km in circumference. This translates to an ambient temperature compressor power of 4.8 MW.

The cold end cooling system for the NAL machine is essentially the same as that shown in Figure 5-1 (a), repeated some 24 times, breaking the system into units of 1260 watts capacity each. The systems proposed for CERN II and ISABELLE are essentially the same as shown in Figure 5-3. In both of these systems the refrigerant is distributed as a supercritical fluid and then expanded to the operating temperature equilibrium pressure in the magnet cryostats. The return line in the CERN II system operates below 1 atm while in the ISABELLE system it is slightly above 1 atm. The proposed liquid storage capacity of each of these systems is as follows:

NAL	160,000 liters
CERN II	77,000 liters
ISABELLE	80,000 liters

The amount of liquid required in each case is determined by the size of the piping in the distribution system and the free volume in the magnet cryostats. In the ISABELLE system there are two 2 km rings, one on top of the other, which accounts for the large volume required.

5-4. Concluding Remarks

There are many problems yet to be solved for the applications discussed. It is obvious that helium refrigerators of the size required are within present day capability, and it is also probable that the reliability will be satisfactory. The problems of distributing cold helium long distances need consideration since the distribution losses always make up a large percentage of the total load. Perhaps the two most pressing problems are those associated with transient conditions in the system and cool-down of the

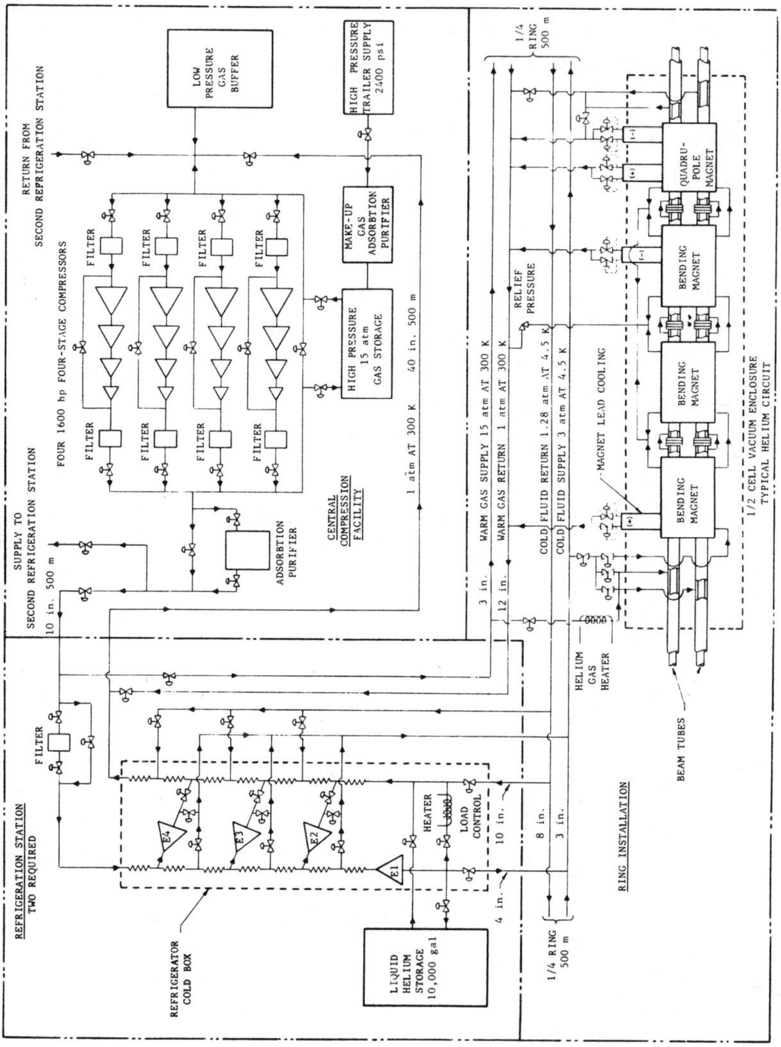

Fig. 5-3. Refrigeration Cycle and Cooling System

devices. Helium is subject to oscillatory behavior when operating near the critical point. What happens to the coolant stream when a transient heat spike is imposed on a transmission line? If the fault is removed will the line recover and operate or will the power have to be shut off for a period of hours? The latter is totally unacceptable to the power industry. During cool-down the region of instability in helium must be crossed. Will there be a smooth crossing or will there be oscillations somewhere in the long lines that essentially block the system? In the preceding discussion, the refrigeration capacities given were for steady-state operation and did not include allowances for excess capacity during a fault or cool-down.

The time scale for all of the projects discussed is within the coming decade. There are proposals for demonstration transmission lines of 1 km length to be completed within 5 years. Most of the accelerator builders would hope to have operating accelerators in that same period.

Assuming these projects are pursued to completion, it is obvious that we are on the brink of a rapid expansion in the use of helium. It may well be true, as Nicholas Kurti has suggested, that the world should go on a "helium standard."

References

1. "Underground Power Transmission by Superconducting Cable," Brookhaven Power Transmission Group, Edited by E. B. Forsyth, Rept. BNL 50325, March 1972.
2. V. Arp et al., "Helium Heat Transfer," Nat'l. Bur Stds., U.S. Rept. No. 10703, July 1, 1971.
3. J. E. Jensen, "Refrigeration Systems for Large Magnet Dewar Systems," BNL 15977, July 1971.
4. "200-GeV Intersecting Storage Accelerators 'ISABELLE,' A Preliminary Design Study," BNL 16716, May 1972.
5. J. W. Dean, "The Cryogenic Aspects of the Proposed European 1000-GeV Superconducting Synchrotron," Rutherford High Energy Laboratory, RHEL/M/A22, December 1971.
6. J. W. Dean, "Refrigeration Systems for DC and Pulsed Superconducting Magnets in High Energy Physics," Proc. Fourth Intern. Magnet Conf.

on Magnet Technology, Brookhaven National Laboratory, Paper 7-4, Sept. 1972.
7. P. C. Vander Arend, "Magnet Cooling and Energizing Systems," *Ibid.*, Paper 7-1.

Acknowledgment

Work performed under the auspices of the U.S. Atomic Energy Commission and the National Science Foundation.

The author would like to acknowledge the efforts of the ISABELLE Design Study Group under the direction of F. E. Mills and the Brookhaven Power Transmission Group under the direction of E. B. Forsyth. The requirements of the systems and equipment proposed by these groups was the stimulus for the studies which have resulted in the material presented here.

6

GAS PERMEATION FOR HELIUM EXTRACTION

Lawrence M. Litz and George E. Smith

The use of organic polymeric membranes to obtain an enriched helium product by selective permeation from helium containing natural gas streams has been studied by a number of investigators. This paper discusses the development of a commercial size permeator unit using a special high-flux membrane and septum type construction. The design allows operation with over 1000 pounds per square inch differential across the membrane and very low pressure drop on both the feed and permeator side. Details of the permeation system are presented together with operating experience obtained over a two-year period in a pilot plant facility producing gas containing several thousand cubic feet of helium per month.

6-1. Introduction

This paper presents the highlights of a cooperative field test program between Union Carbide's Linde Division and the UCC Corporate Research Department to examine the characteristics of a new helium separation capability, developed by the latter group, based on gas permeation. Among the areas to be discussed are the nature of the permeation process, the particular permselective material used, the general design of the permeator module and the pilot plant system, the mode of operation, and some of the results obtained.

6-1. Permeation Theory

The fundamental concept of the type of gas permeator under consideration is that certain gases will pass through a thin layer of a polymer film diffusively under the driving force of a partial pressure gradient across the film. For gases with essentially ideal behavior such as helium, methane and nitrogen (which are of primary concern in processing natural gas for helium extraction), the quantity of permeate Q is proportional to the partial pressure difference across the membrane Δp and the area of the membrane A and inversely proportional to the thickness of the membrane t. Thus, the throughput for any given gas can be expressed by an equation of the form

$$Q = P\,A\,\Delta p/t \qquad (1)$$

where P is the permeability constant of the membrane for the specific gas. Its units are conventionally cc-cm/cm^2-sec-cmHg. In the absence of materials which would affect the polymer's morphology, such as plasticizing or swelling agents, the permeability toward one of the above gases is not affected by the presence of the others. That is, in the ideal case, the separation factor, given as the ratio of the permeabilities involved, such as P_{He}/P_{CH_4}, will be the same for mixtures as when measured with pure gases. Discrepancies from this ideal condition will be discussed later when actual permeator performance is analyzed.

6-3. Membrane Considerations

The realization that mixtures of gases may be separated by taking advantage of the fact that different gases will permeate a polymeric membrane at different rates has been understood and recognized for over 100 years.[1] Surveys and discussions of the problems and capabilities have been reviewed in a number of articles by Stern,[2] Kammermeyer,[3] Friedlander and Litz,[4] etc. The ability to make such separations on an economically practical basis has been limited by the need, with the membrane materials generally available, for extremely large areas and therefore extremely high capital investment. The basis for the work discussed in this paper is associated with the idea that an ultra-thin membrane can be created which has as a consequence of its thinness a very high throughput under normal operating pressures. Because the formation of very thin films has normally led to defects and, therefore, intolerable leak rates in most systems considered in early permeator devices, the polymer film would be no less than about 0.001 inch to 0.002 inch thick.

In recent years the asymmetric form of membrane has been extensively developed in the water desalination field to provide an extremely thin layer of dense film, nominally about 0.00001 inch thick, on top of a thicker, low-flow resistance, porous support layer which was of the order of 0.003 inch thick. Such a thin, dense layer would have a throughput of the order of 100 times that of a fully dense film 0.001 inch thick. The ability to use such asymmetric membranes for gas permeator devices has made a major change in the economics.

The polymer chosen for use in the program under discussion was cellulose acetate. It has a reasonably high permeability for helium, has fairly good selectivity between helium and the other components of natural gas, and, most of all, is readily available in the thin skin asymmetric form, since it is the prime reverse osmosis water desalting membrane. All of the units built for this study were made of Eastman Chemical Company's R0-89 or KP-98 cellulose acetate desalination membrane. A special processing and drying technique was developed to obtain the film characteristics suitable for the helium extraction pilot plant. Typically, the permeation constant for the membrane in the initial dry state was between 2 and 5 x 10^{-9} cc-cm/cm^2-sec-cmHg. The separation factor, P_{He}/P_{CH_4} might vary between 30 and 100, depending upon the particular processing procedure.

6-4. Permeator Design

The important design factors for a commercial type permeator unit are dictated by the process considerations. Gas at high pressure is to be fed over a substantial area of the semipermeable membrane to permit the extraction from the feed stream of an economically useful fraction of the desired gas. In the helium-natural gas system the low concentration of helium, from 0.2% to perhaps as high as 10% in workable feed stocks, requires fairly high feed pressures to supply a reasonable partial pressure gradient high feed pressures to supply a reasonable partial pressure gradient of helium across the membrane. A pressure of 1000 psig is deemed quite practical. Further, to maximize the pressure differential, the pressure on the downstream side of the membrane should be kept as low as practical.

As an illustration, if the feed contains the fairly high level of 5% helium at a total pressure of 1000 psia, the inlet helium partial pressure is only 50 psi while that of the other components is 950 psi. If the downstream side was under an absolute vacuum, these would represent the effective driving forces. Then, for a membrane with a reasonably good separation factor P_{He}/P_X of 50, the relative rates of permeation would be

$$\frac{He}{X} = \frac{50 \text{ psi x } 50}{950 \text{ psi x } 1} = \frac{2500}{950} = 2.63 \tag{2}$$

and the helium concentration would be given by

$$\%He = \frac{2500}{2500 + 950} = 72.5\%. \tag{3}$$

However, this applies only at the inlet to the permeator. As the feed gas traverses the permeator unit, helium is extracted. If for example, one half of the contained helium is removed before the tail gas is discharged, and if no pressure drop occurs as the feed moves down the permeator path, the helium content on the exit end of the unit would be

$$\%He = \frac{1250}{975 + 1250} = 56.2\%. \tag{4}$$

Should the feed contain only 0.5% helium, under the same conditions, the helium content of the permeating gas at the inlet would be 20.1% and at the exit 11.1%.

If the resistance to flow over the membrane surface is high, the available pressure on the feed side would be reduced by whatever feed side pressure drop is developed. If the permeate side pressure drop is high, once again a loss in driving force as well as product purity would occur. Therefore, a good permeator design will have adequate feed channels and an adequate permeate collection system. The latter is particularly important since the high helium content in the permeate can lead to a very low difference in pressure across the membrane if the permeate pressure drop results in high permeate pressure in the further reaches of the permeate collection system.

The permeator design chosen for this program was based on the septum concept developed by Stern and his coworkers.[5,6] It provides not only the desired low pressure drop on the feed and permeate side of the membrane but also has a reasonable high membrane area per unit pressure vessel volume. Of major importance, it provides adequate support for the typically thin, somewhat fragile membrane material. The membrane element configuration is evident from the sketch of Figure 6-1. The permeable membrane is the outer envelope of a sandwich structure having as a membrane support a paper layer in contact with each face of the envelope with a wire screen layer in the center. This wire screen layer provides a low resistance path for the permeate gas to travel from where it permeates through the membrane and paper layer to the collection points, which are at the tie-bolt holes indicated as permeate outlets in the sketch.

These septa or sandwiches are assembled in a stack as depicted in Figure 6-2. The clamp paltes above and below the stack are pulled together with tie-bolts which go through the indicated holes. The tie-bolt hole area is sealed from the high pressure side of the system by use of O-rings between each face of adjoining sandwiches. Permeate flows in the plane of the screen to the tie-bolt location and then passes to the external manifold through slots in the tie-bolts. The assembled stack is sealed along the long edges and the exit end so that the feed flowing into the pressure vessel must pass into the stack at one end and exit through the flange plate at the other end of the stack. A picture of a module containing a 200-sq. ft. active area membrane stack, such as used in this program, is shown in Figure 6-3. Each of the sandwich

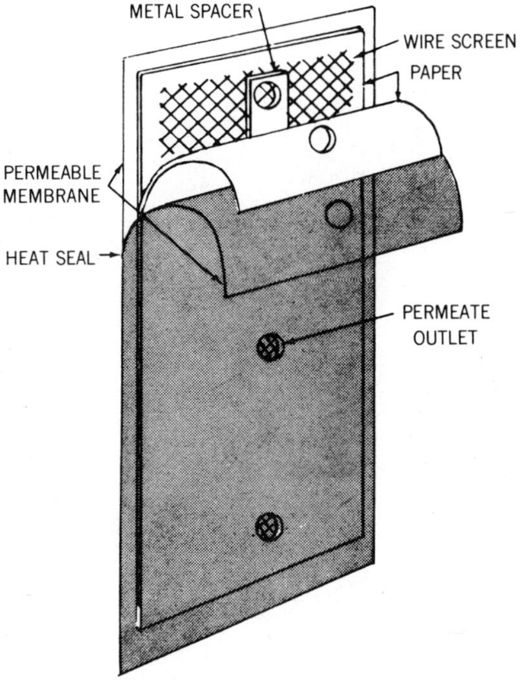

Fig. 6-1. Structure of a Typical Permeation Septum

elements in this stack is approximately 7 inches wide and 4 feet long to provide a total working area on the two sides of each sandwich of 4 sq. ft. Each of these elements is approximately 1/16 inch thick and the interelement spacing is approximately 0.060 inch. The 50-elements shown in Figure 6-3 are clamped between a pair of metal plates approximately 6¼ inches apart to give a roughly square array. The cartridge is fastened to the flange plate by a suitable means so that the gas flow path is limited as described above. The pressure vessel into which the cartridge fits is a piece of 10-inch, schedule 40, carbon steel pipe with a 1½ inch NPT coupling welded into the bell on the gas inlet end; it is flanged on the exit end. In the flange plate carrying the cartridge are the penetrations to carry the tail gas fitting and the permeate manifold line. The assembly was light enough, and so designed, that the vessel could be opened, the cartridge removed and a new one replaced on the flange, and the module put back on stream in less than one-half hour.

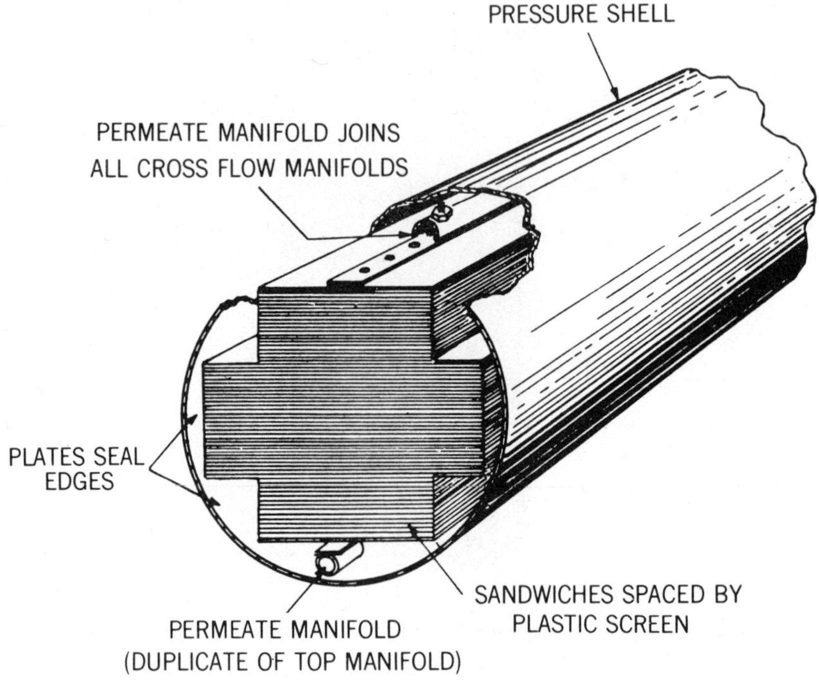

Fig. 6-2. Module Configuration

Laboratory studies with permeator modules of this design showed them to be mechanically sound, to have very low pressure drop across the feed space and to allow the support of the dried cellulose acetate membranes at pressures in excess of 1000 psi differential. Techniques for fabricating the structures on a routine basis were developed and procedures for processing the membrane on a large scale were generated. Preliminary economic analysis indicated that a helium extraction system based on the properties of the dry cellulose acetate membrane could be economically competitive with existing cryogenic systems.

6-5. Pilot Plant Installation

Having determined that practical permeator membranes and module design were available, we initiated the design and construction of a pilot plant to provide an engineering field test. The site

Fig. 6-3 Permeator Cartridge and Shell

chosen for this test was the Navajo Helium Facility in Shiprock, N. M. This facility contained a cryogenic helium separation plant in 1943 for the Bureau of Mines. In the late 1960's, it was deeded to the Navajo Tribe, who leased it at various times to different operators. In June of 1970, the Linde Division of Union Carbide leased the site for the permeator pilot study.

The pilot plant was installed to take maximum advantage of existing equipment. This included the interstage and product compressors, oil traps, knock-out tower, some analytical equipment and miscellaneous valves, pipe, etc. The site also had existing tie-ins to what was hoped to be an adequate natural gas well producing a rather rich 5.3 percent helium content gas. Unfortunately, operation was interrupted for several periods as one well after another became inoperative, presumably because of intrusion of water into the gas strata. An analysis of the typical feed gas is given in Table 6-1. All of the gas was purchased from the Eastern Petroleum Co.

Table 6-1. Feed Gas Analysis
(Bureau of Mines, Nov/5/71)

Gas		*Volume Percent*
Helium		5.8
Nitrogen		87.4
Carbon Dioxide		0.4
Argon		0.7
Methane		3.7
Ethane		0.5
Propane		0.4
Butanes		0.7
Pentanes		0.2
Hexanes plus		0.2
	Total	100.0%

Gross BTU/cu. ft., dry at 60 °F., 30 in. Hg – 94

Overall, the system design requirements were relatively simple. Flow and pressure measuring and control devices were conventional. The factors of greatest concern were the relative humidity and cleanliness of the gas passing over the membrane. Heavy oils and gross particulate matter were undesirable from the standpoint of fouling or damaging the membrane surface. These were easily eliminated with the usual knock-out drums and coarse filter beds. Moisture in the gas is a problem because in the presence of high humidity the cellulose acetate membrane tends to lose a significant fraction of its permeability. Accordingly, the design specifications called for a drier bed to be installed which would maintain the feed gas dew point below 0°C (32°F).

The initial plans called for a two-stage system with up to a total of ten 200 ft^2 permeators. Figure 6-4 shows the flow diagram of the system as it finally evolved. The feed gas was brought into the plant from one of the four different wells tapped during the plant's operation. It passed through a recording orifice meter which measured its flow rate and then into a knock-out tower where any residual liquid was removed. A knock-out drum at the well-head removed the bulk of the entrained liquid hydrocarbons and water. A glass-wool trap removed any fine mist not separated earlier. The indicated booster compressor was installed in the last months to compensate for erratic and low well pressure. It permitted smooth operation at the typical 850 psig on the first stage. To remove oil mist from this compressor, another glass-wool trap followed it after which the gas was passed through a silica gel absorber system to reduce the water vapor pressure. One-half of the silica gel towers were regenerated while the other half was on stream. The tail gas from the first stage permeators was expanded, passed through a molten salt preheater and then the gel bed to raise the bed's temperature to about 300°F. The bed was cooled to ambient temperature before going back on service. Regeneration was typically on a 24-hour cycle.

The dried feed was passed through a heat-exchanger to bring it to the desired operating temperature prior to admission to the first stage permeators. Firstly, the cellulose acetate has an appreciable

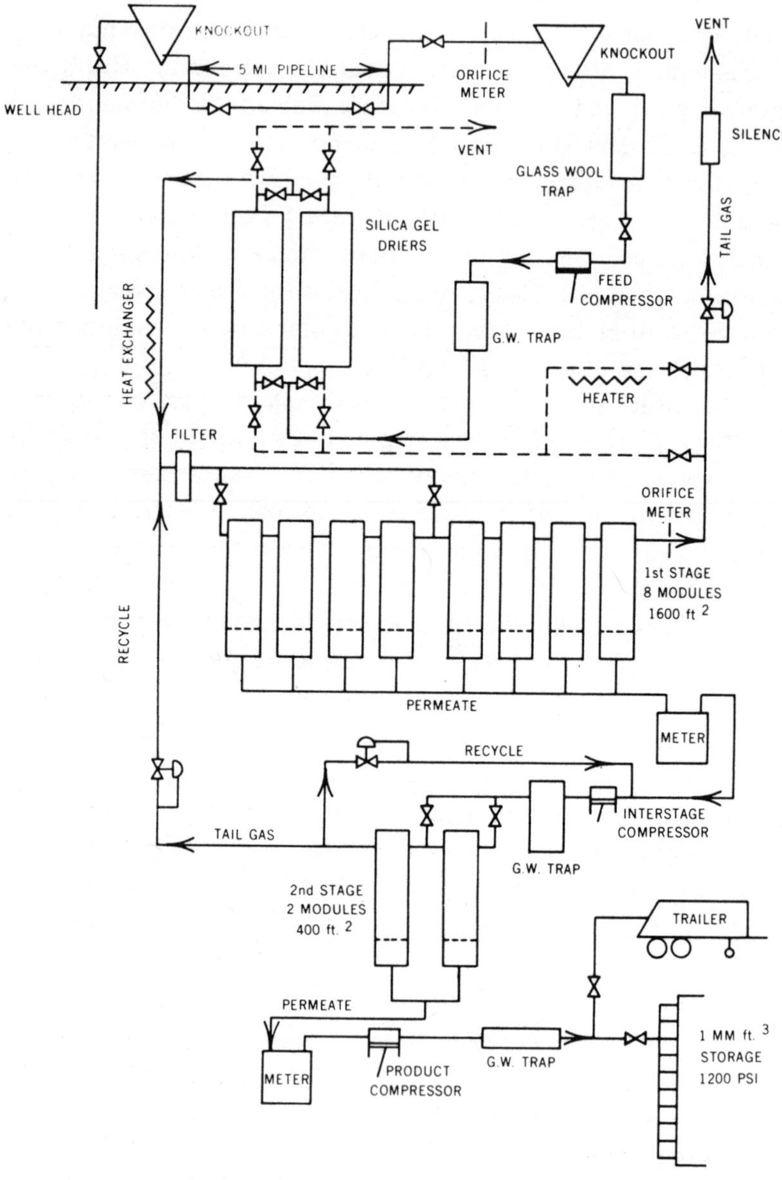

Fig. 6-4. Helium Extraction System Schematic–Pilot Facility, Shiprock, N.M.

temperature coefficient. Operation at elevated temperatures increased the unit's productivity but decreased the operating life, as will be discussed later. Also, as the temperature increased, the selectivity and product purity decreased. Examination of these parameters was desirable and therefore required installation of the heat-exchanger in the line immediately before the permeators.

Another important reason to control the temperature became evident in the early days of operation. Apparently as a result of the fairly large swings in temperature from day to night, several gallons of light gasoline condensed in the feed lines, found its way into the first-stage permeator units, dissolved some of the sealing materials and deposited them on the membrane surface. Needless to say, a significant loss in performance resulted. By keeping the operating temperature above the ambient, the problem was avoided. Fixing the temperature to eliminate the normal swings in ambient also simplified analysis of the operating data. Conventional blanket-type insulation helped minimize temperature loss in the winter.

The permeator modules in the first stage were in series on the feed side and in parallel on the permeate side. Figure 6-5 is a photograph of the eight modules of this first-stage bank. These were arranged at a convenient level on a rack with unions in the piping lines so that a single operator could easily remove and exchange the permeator cartridges as desired. Valves on the permeate line from each module permitted the rapid switching of the permeate from collection manifold to the analytical system for measurement of the permeation rate and composition. Sample points on the high pressure system allowed analyses to be made of the feed and tail gas in both the first and second stages as well as on the gas coming from the fourth module and feeding the fifth module in line. Thermocouples on the various module shells, under the insulation, gave the temperature readings on each.

In the early stages of operation, a downstream pressure regulator before the heat exchanger provided pressure control, and a throttling valve on the last permeator exit line gave flow regulation. When the booster compressor was installed on the feed line,

Fig. 6-5. First Stage Permeators

its speed was regulated to adjust the feed rate, and a back-pressure regulator on the tail gas set the system pressure.

Permeate from the first stage, controlled at 10" w.c. by a regulator on the interstage compressor intake, was compressed to 950 psig, typically, and fed through the second-stage permeators. Second-stage tail gas was fed back into the first-stage feed. Provisions were made to allow some recycle of the second-stage tail gas back through the second-stage permeator so as to give increased gas velocity through this unit for reasons discussed below. Second-stage permeate was compressed by the product compressor and delivered to either a tube trailer truck for transport to the final user or to the large storage bank at the site. This bank, the size of a football field, consisted of one hundred 10-inch diameter pressure vessels, 100 feet long each, with a capacity of one million SCF at 1200 psig.

The most critical components of the system from the operational analysis point of view were the metering and analytical devices. In order to evaluate the performance of individual permeators, it was necessary to accurately measure the permeation rate and the permeate composition of each one and to know the feed or tail gas rate and composition. Early experience showed that rotameters on the permeate stream were unsatisfactory because of the complication of the variable correction for viscosity from one unit to another arising from varying feed gas concentrations. A positive displacement dry gas meter resolved this difficulty. Analysis of the several gas streams, varying from as low as 1.2% He in the tail gas to as high as 92% in the second-stage permeate, also proved less simple than expected. A Gow-Mac thermal conductivity cell–Wheatstone Bridge unit–was satisfactory between 10% to 90% He. A Leonco gas chromatograph system was used for concentrations under 10% and to measure the true helium concentration in the gases used to calibrate the Gow-Mac analyzer. To minimize the influence of the components other than helium in the gas mixtures, these calibration gases were actual process gas samples.

6-6. System Operation and Results

The first permeator units were put on stream on June 20, 1970, and the system was run 24 hours a day, seven days a week, almost continuously until August 25, 1972. Over most of this period it was operated without on-site engineering personnel. Five on-the-job trained people provided the three shift, seven day, one man per shift operating crew under the general, long distance supervision of the Linde Division's Amarillo, Texas Plant Manager. Most operating variations were instituted via telephone orders from UCC's Tarrytown, New York site. Data collected at Shiprock was mailed to Tarrytown for reduction and analysis.

No serious problems were encountered with the permeator modules even though a number of sudden shut-downs were induced by local power failures. A number of planned shut-downs occurred to remove or replace a membrane cartridge for repair or to examine a modified construction or material. The operating nature of the system was such that these changes could be made with the plant down for less than one hour. The ease of changing cartridges permitted shifting a given cartridge from one location in the first stage to another, or to the second stage, without moving the pressure shells. This allowed a direct probing of the performance under the variations of feed composition and velocity provided by different points in the system.

A complete data record was taken every two hours by the shift operator to provide tracking information. Reproductions of actual log sheets for the first and second stages are shown in Figures 6-6 and 6-7. The second-stage sheet contains also entries for the ambient conditions, the pipeline gas data, the silica gel trap condition and the trailer and/or storage pressure and temperature.

A computer program was written to calculate the individual module performance based on the raw data obtained from Shiprock. Figure 6-8 shows the computer output based on the date of Figures 6-6 and 6-7 and the membrane area in the various modules. As indicated on the line marked "Sq. Ft.," on this particular date, only three of the nine modules on stream at this time had full cartridges; the balance had half cartridges or less. This illustrates one of the advantages of this particular design, in that if for one reason or another, less than a full stack is desired, it can be readily accommodated in the system.

The computer output provided a material balance check on the analytical and measuring devices by giving the computed as well as the measured plant input and tail gas rates as well as the computed and measured helium concentration of the 'E' position module. Also calculated was the percent recovery of the helium entering the plant, the linear velocity through the feed passages in each module, the permeability of both helium and the total of the other components, reported as P(25)He and P(25)N_2, as well as the ratio of these, P(25)He/P(25)N_2, reported as A(25)He. P(T)He and A(T)He are the respective data at the operating temperature of the module. For purpose of convenience, the permeability values are multiplied by 10^{+12} to give a whole number. The eighth and ninth lines above the bottom, P(I) and A(I)He give the permeability and separation ratio, normalized to 25°C, of the modules as measured with pure gases in Tarrytown before the

FIRST STAGE LOG SHEET

DATE 11-11-71 / TIME	12 AM	2 AM	4 AM	6 AM	8 AM	10 AM	12 PM	2 PM	4 PM	6 PM	8 PM	10 PM
AMBIENT TEMP. °F	50°	50°	48°	46°	46°	58°	65°	70°	71°	64°	54°	56°
BAROMETRIC PRESSURE IN. Hg.	25.1	25.1	25.1	25.1	25.1	25.1	25.1	25.0	24.9	24.9	24.9	24.9
FEED — PRESSURE PSIG	850	850	850	850	850	850	850	850	850	850	850	850
FEED — TEMPERATURE °F	122°	122°	120°	118°	116°	120°	120°	116°	114°	118°	117°	120°
FEED — HELIUM %	6.40	6.45	6.35	6.30	6.38	6.43	6.33	6.40	6.43	6.42	6.62	6.60
FIRST STAGE MODULE DATA — A FLOW CFH	325	325	318		317	326	322	335	331	335	324	325
HELIUM %	39.0	40.0	40.0	40.0	40.0	40.0	39.0	39.5	39.0	39.5	39.0	39.5
SHELL TEMP °F	111°	111°	111°	109°	108°	111°	108°	108°	109°	110°	109°	110°
B FLOW CFH	234	230	228		226	231	226	239	239	231	237	234
HELIUM %	45.0	46.0	46.0	46.0	46.0	46.0	45.5	46.0	45.5	45.5	45.5	45.5
SHELL TEMP °F	108°	110°	110°	108°	107°	109°	108°	107°	108°	109°	109°	109°
C FLOW CFH	325	324	318		318	318	324	330	333	327	333	327
HELIUM %	44.0	45.0	44.5	44.5	44.5	45.0	44.5	44.5	44.5	44.0	44.5	44.5
SHELL TEMP °F	102°	101°	101°	100°	99°	102°	101°	101°	102°	103°	100°	102°
D FLOW CFH	736	720	720		730	735	735	760	750	744	743	745
HELIUM %	16.5	17.0	17.0	16.5	16.5	16.5	16.0	16.0	16.0	16.0	16.0	16.0
SHELL TEMP °F	106°	107°	107°	105°	103°	106°	106°	103°	108°	106°	106°	105°
D TO E STREAM HELIUM %	4.35	4.40	4.30	4.33	4.43	4.43	4.40	4.42	4.90	4.80	4.77	4.74
E FLOW CFH	589	577	575		581	585	600	595	595	595	600	590
HELIUM %	14.0	15.0	15.0	15.0	14.5	14.5	14.5	14.5	14.5	14.5	14.5	14.0
SHELL TEMP °F	103°	104°	104°	102°	101°	104°	104°	102°	105°	103°	104°	103°
F FLOW CFH	310	307	302		302	306	311	318	319	318	315	314
HELIUM %	25.0	26.5	26.5	26.5	26.5	26.5	26.0	26.0	26.0	26.0	26.0	26.0
SHELL TEMP °F	106°	107°	106°	103°	103°	106°	106°	103°	107°	105°	105°	104°
G FLOW CFH	221	219	216		214	214	220	220	225	223	221	220
HELIUM %	43.0	44.5	44.0	44.0	44.0	43.5	43.0	44.0	44.0	43.5	43.0	43.0
SHELL TEMP °F	100°	101°	101°	99°	98°	101°	101°	100°	102°	102°	101°	100°
H FLOW CFH	645	660	643		637	650	650	655	660	655	650	713
HELIUM %	29.0	29.5	29.5	29.5	29.5	29.0	29.0	28.5	29.0	28.5	28.0	28.0
SHELL TEMP °F						100	100	99		1		
TAIL GAS He %	2.28	2.30	2.20		2.38	2.28	2.26	2.28	2.25	2.27	2.25	2.30
PRODUCT — METER/MANIFOLD PRES. IN W.G.	10"	10"	10"		8"	10"	9"	10"	9"	8"	11"	9"
#1 METER TEMP °F	56"	56"	55°		51°	63°	71°	77°	81°	74°	66°	67°
# COMPRESSOR He%	24.0	25.0	24.5	25.0	24.5	24.5	24.0	24.5	24.0	24.5	24.5	24.5
TOTAL FLOW CFH					3323	3373	3388	3444	3452	3428	3428	3468
DATA BY: INITIALS	R.R.	R.R.	R.R.	R.R.	[illegible]	[illegible]	[illegible]	[illegible]	[illegible]	[illegible]	[illegible]	[illegible]

ENTER ADDITIONAL INFORMATION ON REVERSE SIDE

Fig. 6-6. First Stage Log Sheet

cartridges were sent to Shiprock, while the last two lines give the comparative performance on the date the data was taken.

The correction for temperature was made in the computer utilizing the formulas of Figure 6-9. Laboratory data on a number of different cellulose acetate membranes showed that for gases like helium, hydrogen, methane, nitrogen and carbon monoxide,

PLANT LOG SHEET

DATE 11-11-71 ITEM / TIME	1 AM	3 AM	5 AM	7 AM	9 AM	11 AM	1 PM	3 PM	5 PM	7 PM	9 PM	11 PM
AMBIENT TEMP. °F	50°	49°	47°	45	50°	62°	67°	70°	68	60°	57°	56
BAROMETRIC PRESSURE IN. Hg	25.1	25.1	25.1	25.1	25.1	25.1	25.1	25.0	24.9	24.4	24.9	24.9
PIPELINE PRESSURE PSIG	530	565	565	565	560	560	560	550	540	530	520	540
PIPELINE TEMPERATURE °F	40°	40	40°	40°	41°	44°	46°	47°	46°	44°	42°	42°
PIPELINE FLOWMETER ΔP IN. W.G.	75"	71"	72"	72"	69"	66"	68"	73"	72"	69"	71"	73"
PIPELINE MOISTURE	1.1	1.1	1.0	0.9	0.8	1.3	1.8	1.8	1.8	1.4	2.0	1.3
GEL TRAP IN SERVICE	R	R	R	R	R	R	R	R	R	L	L	L
REGEN. TRAP TEMP. °F	306°	306°	306°	306°	298°	245°	203°	166°	145°	70°	214°	280°
MOISTURE	0	0	0	0	0	0	0	0	0	0	0	0
PIPELINE HELIUM %	5.80	5.90	5.80	5.75	5.70	5.65	5.75	5.73	5.73	5.80	5.85	5.88
CHROMATOGRAPH SPAN He %	5.53	5.53	5.53	5.53	5.53	5.53	5.53	5.53	5.53	5.53	5.53	5.53
FIRST STAGE – TAIL GAS – PRESSURE PSIG	845	845	845	840	840	840	840	840	845	845	840	840
FIRST STAGE – TAIL GAS – TEMPERATURE °F	98°	99°	97°	97°	96°	99°	99°	98°	101°	99°	98°	100°
FIRST STAGE – TAIL GAS – FLOWMETER ΔP IN W.G.	58	55	56"	56"	52"	50"	55"	52"	55"	48"	48"	52"
FIRST STAGE – TAIL GAS – HELIUM %	2.30	2.20	2.23	2.30	2.31	2.30	2.33	2.31	2.28	2.23	2.25	2.31
SECOND STAGE PRODUCT – FEED – PRESSURE PSIG	950	950	950	950	950	950	950	950	950	950	950	950
SECOND STAGE PRODUCT – FEED – TEMPERATURE °F	77°	80°	80°	78°	78°	81°	76°	78°	81°	78°	76°	78°
SECOND STAGE PRODUCT – FEED – HELIUM %	24.0	25.0	25.0	25.0	25.0	24.0	25.0	24.5	24.5	24.0	25.0	25.0
SECOND STAGE PRODUCT – MODULE – I. FLOW CFH	823	830	819	813	817	810	823	815	820	812	813	815
SECOND STAGE PRODUCT – MODULE – HELIUM %	82.0	82.0	82.5	82.5	82.5	82.5	82.5	82.5	82.5	82.5	82.0	82.5
SECOND STAGE PRODUCT – MODULE – SHELL TEMP. °F	72°	75°	73°	72°	73°	76°	74°	75°	80°	77°	73°	74°
SECOND STAGE PRODUCT – MODULE – METER PRESSURE IN W.G.	7"	8"	7"	7	5"	8"	7"	10"	10"	10	6"	7"
SECOND STAGE PRODUCT – MODULE – METER TEMP. °F	67°	67°	67°	65°	66°	73°	74°	75°	75°	70°	68°	68°
SECOND STAGE PRODUCT – MODULE – METER TOTAL / TIME	418900	420500	422200	4238	425600	427200	428800	430500	432100	433800	435500	436900
SECOND STAGE PRODUCT – TAIL GAS – PRESSURE PSIG	950	950	950	950	950	950	950	950	950	950	950	950
SECOND STAGE PRODUCT – TAIL GAS – TEMPERATURE °F	77°	80°	80°	78°	78°	81°	76°	78°	81°	78°	76°	78°
SECOND STAGE PRODUCT – TAIL GAS – HELIUM %	11.0	11.0	11.5	11.0	11.0	11.0	12.0	11.5	10.5	10.5	11.0	11.0
SECOND STAGE PRODUCT – TAIL GAS – FLOWMETER ΔP PSIG	14.0	14.0	14.1	14.3	13.7	13.0	12.8	12.5	12.5	13.4	13.9	13.7
TRAILER PRESSURE PSIG	1610	1620	1630	1650	1670	1740	1810	1870	1900	1900	1900	1900
TRAILER TEMP. °F	41°	37°	34°	32	34°	38°	49°	59°	63°	61°	56°	51°
# STORAGE BANK PRESS. PSIG												
STORAGE BANK TEMP. °F												
DATA BY: INITIALS	K.R.	R.R.	K.R.	R.R.	DN	DN	DN	DN	J.J.	L.J.	L.J.	J.J.

Fig. 6-7. Second Stage Log Sheet

PERFORMANCE OF HELIUM RECOVERY UNIT AT SHIPROCK, N.M.

TIME 2PM — TEST DATE 11/11/71

PLANT INPUT: 16.26 MSCFH(MEASURED); 5.7 % HELIUM(MEASURED)
14.97 MSCFH(COMPUTED)
TAIL GAS, 1ST STAGE: 14.15 MSCFH(MEASURED); 2.28 % HELIUM(MEASURED)
14.32 MSCFH(COMPUTED)
2ND STAGE: 2876. SCFH(COMPUTED): 11.50 % HELIUM(MEASURED)
PRODUCT: 644.65 SCFH(MEASURED); 82.5 % HELIUM(MEASURED)
RECOVERY 62.0 % (BASED ON MEASURED PRODUCT VOL. AND COMPUTED FEED VOL.)
FEED TEMPERATURE: 1ST STAGE 116.F ; 2ND STAGE 78.F

POSITION	A	B	C	D	E	F	G	H	I
MODULE	C10L	C6L	C9	C5	C8U	C1U	C1L	C11	C12
TEMP.(F)	108.	107.	101.	103.	102.	103.	100.	99.	75.
SQ.FT.	104.	112.	208.	208.	100.	104.	100.	196.	64.
LIN.VEL.F/S	0.66	0.61	0.32	0.31	0.63	0.60	0.61	0.30	0.15
FEED:									
MSCFH	17.04	16.78	16.59	16.33	15.73	15.26	15.01	14.84	3.521
PSIA.	863.	863.	863.	863.	853.	853.	853.	853.	963.
% HE(COMP)	6.40	5.88	5.42	4.80	4.38	4.06	3.70	3.23	24.50
% HE (MEASURED)					4.42				
PRODUCT:									
SCFH	264.0	188.3	260.1	598.9	468.9	250.6	173.4	516.2	644.6
PSIA.	12.9	12.9	12.9	12.9	12.9	12.9	12.9	12.9	12.9
% HE	39.50	46.00	44.50	16.00	14.50	26.00	44.00	29.50	82.50
P(I)HE*	1330.0	1800.0	900.0	925.0	459.0	666.0	1180.0	3200.0	2961.0
A(I)HE*	66.00	40.13	55.30	28.75	25.30	17.25	23.12	40.00	46.00
P(25)HE	638.2	555.6	472.2	392.3	643.0	673.0	1051.5	1310.9	2162.5
P(25)N2	50.0	29.8	24.7	83.5	141.5	61.7	34.8	67.2	95.7
A(25)HE	12.76	18.64	19.14	4.70	4.54	10.90	30.19	19.50	22.61
P(T)HE	871.0	751.2	603.0	510.7	829.0	876.1	1330.0	1642.0	2124.9
A(T)HE	10.94	16.05	16.95	4.12	4.01	9.57	26.87	17.44	22.81
P(25)/P(I)	0.480	0.309	0.525	0.424	1.401	1.011	0.891	0.410	0.730
A(25)/A(I)	0.193	0.464	0.346	0.163	0.180	0.632	1.306	0.487	0.491

* P = PERMEABILITY, P,CC-CM X 10^{12}/CM^2 SEC-CM Hg

* A = SELECTIVITY RATIO - P_{He}/P_{N_2}, etc.

01/31/72

Figure 6-8. Performance Data Analysis – Computer Readout

the temperature dependence of permeability was accurately represented by an Arrhenius equation of the form

$$P = P_o e^{-\Delta E/RT} \tag{5}$$

where

- P = permeability, cc-cm/cm^2-sec-cmHg.
- P_o = specific gas membrane parameter (temperature) independent).
- ΔE = activation energy of permeability, kcal/g mole.
- R = gas constant, kcal/g mole, °C.
- T = absolute temperature, °K.

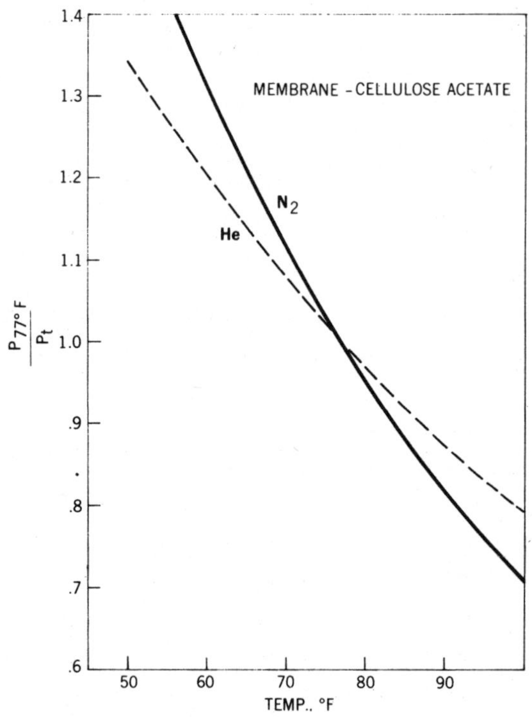

Fig. 6-9. Temperature Dependence

The equations shown on Figure 6-9 were derived from equation (5) using the activation energies ΔE_{He} = -3.34 kcal/gmole, ΔE_{N_2}, CH_4 = -5.0 kcal/gmole.

The graph of Figure 6-9 illustrates the significant temperature effect on permeability relative to that at the chosen standard temperature of 25°C (77°F). For example, a 10°C increase in temperature from 25°C to 35°C (98°F) results in a permeation rate for helium which is multiplied by 0.805 and one for nitrogen (or methane) which is multiplied by 0.725 to obtain the respective rates at 25°C. Not only is there almost a 25% increase in helium permeability in this 10°C range, but there is approximately a 10% loss in selectivity. Obviously, plant design and operation must take these opposing factors into account.

It was observed that at low feed velocities, in the laminar flow range, a significant loss in permeability and selectivity was obtained relative to the permeability and selectivity measured with pure helium and methane or nitrogen. When the feed flow was in the turbulent range, the performance of a given module corresponded reasonably well with that predicted by the pure gas data. One approach taken to maintain higher feed velocities in the second stage was to recycle a portion of the tail gas back to the interstate compressor suction. Although this resulted in a dilution of the feed to the second stage, an optimum recycle ratio could be determined for a given set of operating conditions which maximized the productivity without seriously lowering the product purity.

Another option available to provide increased flow velocity was to reduce the number of elements in a given module, thereby reducing the parallel flow cross-section, and to put more modules in series. As is apparent from Figure 6-8, on the date shown, five of the eight modules in the first-stage positions A through H were half stacks (about 100 ft^2 area). The second-stage module, position I, only had 16 elements in it. Unfortunately, because of the insufficient gas supply problem, it was not usually practical to substantially increase the feed to the plant. Also, doing so would reduce the helium recovery ratio, which on Figure 6-8 was 62%.

The specific helium production over most of the last year of operation of the system, expressed as the number of thousands of cubic feet of helium in the gas produced in the pilot plant per thousand square feet of membrane in service, on a daily basis averaged over monthly periods, is shown in Figure 6-10 for the first- and second-stage units. The approximately 10-fold higher specific output of the second stage reflects the higher helium content of its feed. Interestingly, despite the decline in performance of the various modules with time, the specific output of the system as a whole did not change very drastically. To a large extent, this reflects the occasional replacement of low productivity modules with better ones, changes in operating conditions, such as increasing the flow velocity and, most of all, the fact that the performance of almost all the modules had leveled off because of their age. Thus, this data could be anticipated to represent a permeator

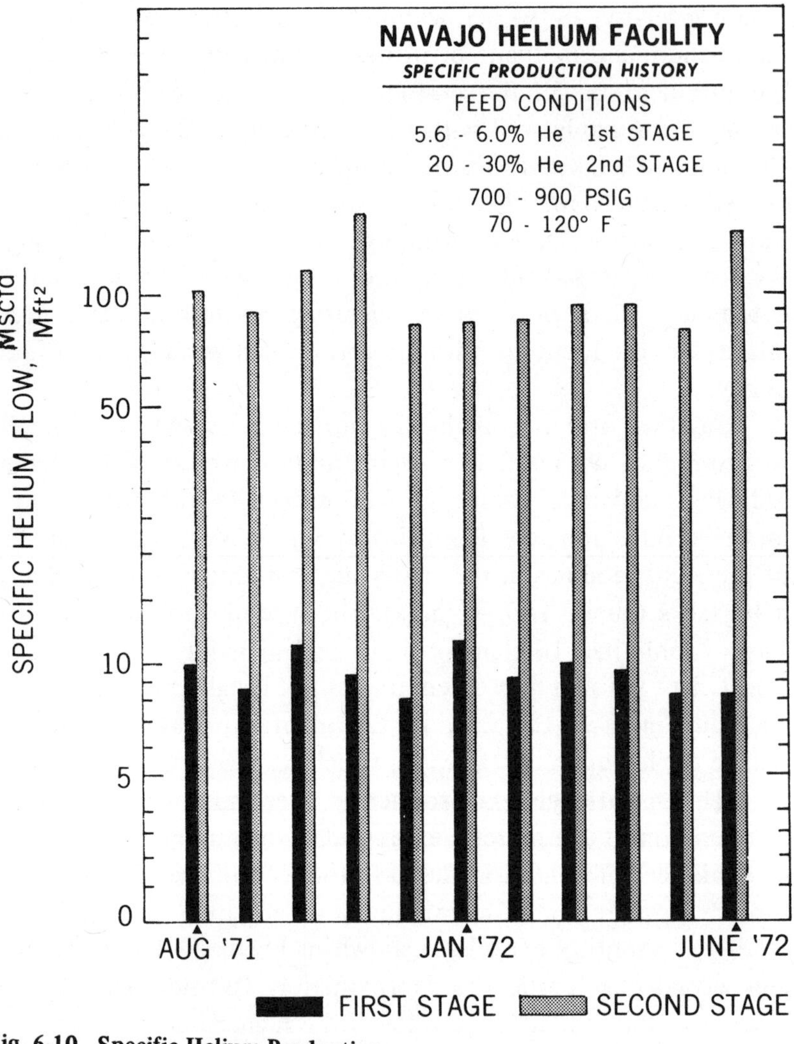

Fig. 6-10. Specific Helium Production

system in its typical steady-state condition. Therefore, it is the sort of data which could form a design base. With the type of feed used these values are about 8500 scfd of helium per 1000 ft^2 of membrane in the first stage and about 90,000 scfd of helium for 1000 ft^2 of membrane in the second stage. For a three million ft^3/month plant, this would translate into about 12,000 ft^2 of this type of membrane in the first stage and about 1100 ft^2 in the second stage. These quantities are expected to be reduced by as much as a factor of two or more, based on the performance of the newest modules installed before the test was terminated. Improvements in membrane properties and fabrication and operating procedures promise to give performance characteristics like those of module C-12 in the data of Figure 6-8 with decidedly lower time-flux loss coefficients than most of the units tested in this program.

6-7. Conclusions

On the basis of this pilot plant study, it may be concluded that:

1. High productivity gas permeation plants can be built to provide enriched helium from natural gas feeds using commercially-available, high flux, asymmetric membranes of cellulose acetate.
2. Plant operation requires minimum attention of relatively unskilled operators.
3. Improved membrane performance would be desirable and is potentially available based on the most recent permeator modules.
4. Neither scheduled nor unscheduled shut-downs had a deleterious effect on the permeators.

References

1. T. Graham, *Phil. Mag.* **32**, 401 (1866).
2. S. A. Stern, "Industrial Processing with Membranes," R. E. Lacey and S. Loeb, Ed., *Interscience*, 279-339, Wiley, New York, 1972.
3. K. Kammermeyer, "Progress in Separation and Purification," E. S. Perry, Ed., Vol. 1, *Interscience*, 335-72, Wiley, New York, 1968.
4. H. Z. Friedlander and L. M. Litz, *Membrane Procession in Industry and Biomedicine*, M. Bier, Ed., 73-99, Plenum Press, 1971.
5. S. A. Stern, T. F. Sinclair, P. J. Gareis, N. P. Vahldieck, and P. H. Mohr, *Ind. Eng. Chem.* **57**, 49, 1965.
6. S. A. Stern, U.S. Pat. 3,332,216, July 25, 1967.

Acknowledgment

The authors would like particularly to acknowledge the support and consultation provided by Mr. Arthur W. Francis, UCC, Linde, New York, the very excellent construction and operational supervision of Mr. W. J. C. Murray of Linde, Amarillo, and the excellent day-to-day on-site contributions made by Mr. Richard Redhouse, Shift Foreman for most of this program, in addition to the many other individuals who made it possible to carry out this effort. We would also like to acknowledge the cooperation of the Navajo Tribal Council for the use of the facilities and those of the Navajo Tribe who worked on the project.

7

RECOVERY AND LIQUEFACTION OF HELIUM FROM NATURAL GAS IN POLAND

J. K. Jones and J. M. Stacey

Poland is currently expanding its utilisation of natural gas. Several of the gas fields contain up to 45% nitrogen, which will be rejected prior to distribution, and 0.4% helium, which will be recovered. This paper is concerned with the process design considerations of extracting helium from lean natural gas and the subsequent purification, liquefaction and storage.

The particularly interesting feature of the design of the plant was the combination of separation of large quantities of methane from the inert nitrogen with extraction of very much smaller volumes of helium. The separation of the methane-nitrogen fraction is achieved in a double distillation column which also produces a small purge stream containing almost all the helium.

The Polish plant, with an annual capability of 150 MMSCF (4 million cubic metres), will make available to Europe a large quantity of helium, which in the past was nearly all imported from the Western Hemisphere.

7-1. Introduction

In the past few years Poland, in common with many other energy conscious countries, has been carrying out an exploration and drilling programme. This programme is designed to determine the presence and extent of any hydrocarbon fuels which could be used to meet the country's growing need for energy. The search has located several natural gas fields of significantly different composition.

The gases found in Poland fall into two broad composition categories. The first category contains gases rich in methane and other hydrocarbons and can be used directly as fuel. The second category includes gases which have a high nitrogen content varying from 38 mol. % up to 75 mol. %. Additionally, some of these gases also contain helium.[1] It is the purpose of this paper to deal with the treatment of these so-called "lean" gases and to indicate how the two process requirements, namely the rejection of the inert nitrogen and the recovery of the valuable helium, are combined.

Although this paper concentrates on the recovery and liquefaction of helium it also deals briefly with the overall picture and relates how the economic exploitation of a lean gas results in a bonus. This bonus is the concentration of the helium from the lean gas into a single stream which can be readily purified, liquefied and marketed. The availability of large quantities of helium in Europe must substantially change the present supply and usage situation. Why this particular helium plant has such significance for Europe is indicated in the next section.

7-2. European Helium Situation

Helium is a material in relatively short supply in Europe at the present time. The best estimate of helium consumption for 1971 in Western Europe is about 65 MMSCFA. Traditionally all this helium has been imported; mostly from the United States of America with smaller quantities from Canada. In recent years two plants have been started up in France. These are expected to recover 15 MMSCFA of gaseous helium from Dutch Gas.[2] This still leaves a substantial shortfall, particular as helium consumption is continually increasing.

There were hopes that the natural gases found in the British sector of North Sea would contain recoverably quantities of helium, but of those gases at present being utilised the maximum helium concentration is only 0.12 mol.%, while the average helium content is only 0.05 mol.%.[3] Since these gases are rich in hydrocarbons they require no cryogenic treatment and offer little possibility of economic helium recovery.

Faced with the situation of increasing helium consumption and an almost total dependence on supplies from the Western Hemisphere, the emergence of a new indigenous source of helium has been widely welcomed in Europe. The decision of Poland to go ahead with the exploitation of their lean natural gas fields will make available a large quantity of helium. It is anticipated that the ready availability of this scarce but versatile element will stimulate more helium-related research and development, and that the European usage of helium will increase substantially in the latter part of this decade.

7-3. The Economic Exploitation of Lean Natural Gas

The choice facing the gas distributor who has access to lean gases is whether to find a use for them in the well-head condition, or to remove the inert components by a suitable process and then to distribute the enriched gas in the conventional manner.

The first course of action would not require a nitrogen removal step. In this situation it is unlikely that helium recovery would be considered unless a policy of helium conservation was in force. The second course is subject to certain economic considerations but opens up the possibility of adapting the methane upgrading process to recover any contained helium.

The basic problem in the utilisation of lean natural gas is determination of the point at which the costs of installing and operating a methane upgrading plant become significantly lower than those of using the lean gas directly from the well head. The disadvantages of distributing the lean gas are that the compressor stations and pipelines have to be increased in size to cope with the additional inert gas volume, and that due to its higher density and viscosity there is a higher flow resistance resulting in larger power requirements and hence increased operating costs. The Polish Gas Industry[1] studied this problem and came to the following conclusions:

> This (lean) gas could be workable substitute for town gas produced in a small works, only when the gas is being transported over a distance shorter than 150 Km.

> In the case of industrial users like the cement, glass and other building material plants, the use of (lean) gas could pay only when the transport distance does not exceed 200 Km.
>
> For distances over 150 Km. and 200 Km., it would be better to purify the gas before distribution.

Additional advantages stated for the installation of a methane upgrading plant are as follows:

> . . . the combustion of this gas in its crude state would mean only a partial utilisation of a valuable raw material.
>
> If the nitrogen is removed, the purified gas could be transported and distributed by means of the existing system of pipelines for natural gas. This would mean a considerable lowering of the investment cost in utilising this (lean) gas.

The Polish report concludes that since the average distance of the lean gas fields from highly industrialised regions in Poland is between 200-250 Km., the most economic solution is to treat the lean gas close to the well head and to distribute an enriched methane gas.

It is within this context of the favourable economics for upgrading lean gas that, by the choice of a suitable cryogenic process, the small concentration of helium in the lean gas can be recovered and made available for use.

7-4. Lean Gas and Product Characteristics

The characteristics of the natural gas feedstock to the plant are shown in Table 7-1. The interesting features are, first, that the helium composition is 0.4 mol.%. While this is not a high content in comparison with some of the gas fields being utilised in the United States, it is the largest concentration found in any European gas. By comparison the British North Sea gas and Dutch gas contains an average of only 0.05 mol.% helium.

The second feature is that the nitrogen concentration is about 43 mol.%. This nitrogen is not required in the methane product and it will be rejected at atmospheric pressure. The available pressure energy of this nitrogen will be utilised by Joule-Thomson expansion to provide the work of separation and to overcome the process irreversibilities.

Table 7-1. Characteristics of Lean Natural Gas

Composition	*Mol. %*
Helium	0.40
Hydrogen	100 vpm
Nitrogen	42.75
Methane	56.01
Ethane	0.44
Propane	0.02
Butanes	0.01
$Pentanes^+$	600 vpm
Carbon Dioxide	0.30
Sulphur Compounds	6 vpm
Water Dewpoint	0 °C
Pressure	57 Kg/cm^2G
Temperature	5 to 15 °C

The gas is also notable for its low content of hydrocarbons heavier than methane; these range from ethane up to toluene. The other components which are usually present in natural gas, namely carbon dioxide, sulphur compounds and water, have also been detected.

The criteria for the design of the gas processing plant are:

1. Upgrading the lean natural gas into a high pressure methane rich gas;
2. Recovering and liquefying the helium.

The composition of the main process streams are as shown in Table 7-2. These data refer to one production line. Two lean gas processing lines are being installed to upgrade methane, while the helium from only one line will be recovered.

7-5. Process Outline

The process flowscheme is shown in Figure 7-1. It consists of several units, namely:

1. Pretreatment of lean gas;
2. Distillation to recover methane and crude helium;
3. Hydrogen removal from curde helium;
4. Nitrogen removal from crude helium;
5. Helium liquefaction, storage and distribution.

Table 7-2. Compositions of Main Process Streams

Stream	*Lean Feed Gas*	*Methane Product*	*Waste Nitrogen*	*Helium Product*
Flow				
MMSCFD	123	71.5	49.6	0.43
State	Gas	Gas	Gas	Liquid
	Mol. %	Mol. %	Mol. %	Mol. %
Helium	0.40	–	0.09	100.0
Nitrogen	42.75	4.00	98.95	–
Methane	56.02	95.09	0.96	–
Ethane$^+$	0.53	0.91	–	–
CO_2	0.30	–	–	–

Pretreatment of Lean Gas. This consists of three processing steps which are designed to remove from the lean gas those components which would solidify in the colder sections of the plant. The carbon dioxide and hydrogen sulphide are removed by absorption in mono-ethanolamine solution. The water vapour is removed by adsorption. The heavy hydrocarbons of high freezing point and limited solubility are also removed by adsorption prior to the gas entering the low-temperature units.

Distillation to Recover Methane and Crude Helium. This section of the process represents the most significant difference between the current generation of operational helium extraction plants and the Polish plant. The methane-nitrogen separation and helium recovery will be performed by adapting the classical double distillation column process traditionally used in the air separation industry. The low-temperature process operates without any external refrigeration and results in a plant which has a reduced equipment inventory and is inexpensive to operate. The operation of the distillation section will be described in detail.

The lean gas at supercritical pressure from the pretreatment units is cooled in plate-fin heat exchangers against the returning methane and nitrogen streams. It is then expanded, and a two-phase mixture passes into the base of the lower distillation column operating at about 27 atmospheres. An initial separation of the

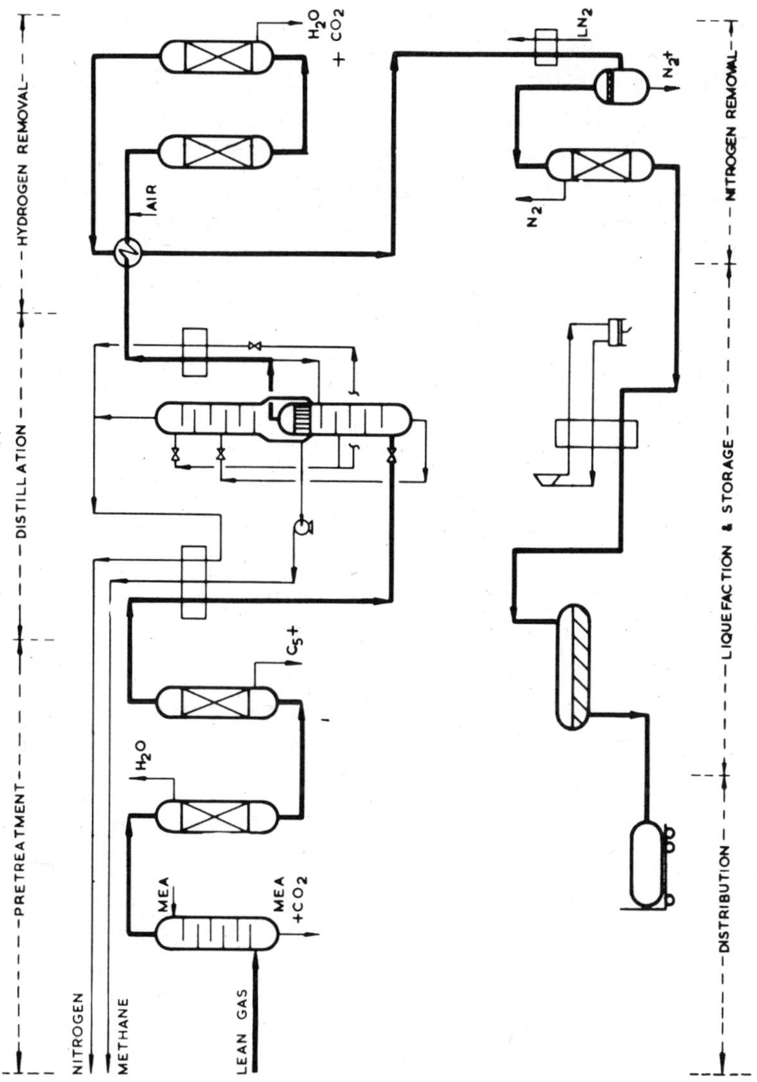

Fig. 7-1. Process Flowscheme

nitrogen and methane components occurs at this point. This is also the first stage in the helium enrichment process. A nitrogen enriched vapour, containing substantially all the helium flows up the column and the contained methane is returned to the base of the column by a nitrogen reflux stream produced in the overhead condenser. This rectification operation results in a vapour at the top of the column consisting mainly of helium and nitrogen and thus represents the second stage of the helium enrichment process. The helium-nitrogen mixture then passes upwards into the overhead condenser where a large portion of the nitrogen is liquefied and returned to the column as reflux. This further enriches the vapour phase which now contains about 10 mol.% helium. The liquid reflux will contain some helium, but as this stream passes back down the column into the warmer regions the helium is stripped out of solution and passes up the column again. The cold requirements for liquefying the nitrogen in the condenser are met by evaporating some of the liquid methane from the base of the upper column.

Two liquid streams are taken from the lower column and become the feed and reflux for the upper distillation column. The feed is an enriched methane stream taken from the base of the lower column and the reflux is a nitrogen rich stream taken off the lower column a few trays below the top. The methane upgrading operation is completed in the upper column. The feed stream is distilled to give a liquid methane product, containing about 4 mol.% nitrogen, which is pumped from the column base and evaporated against the incoming lean gas before passing into the distribution pipeline. The waste nitrogen taken from the top of the upper column is also heated against the incoming lean gas before being rejected from the plant.

The helium purge leaving the top of the lower column, is cooled by flowing upwards through a refluxing exchanger. Nitrogen in the helium stream is condensed in this exchanger by the evaporation at lower pressure of a small portion of the upper column reflux liquid. The condensed nitrogen is returned to the lower column. The vapour leaving the reflux exchanger contains about 88 mol.% helium and this completes the enrichment of crude helium in the Distillation Unit.

Hydrogen and Nitrogen Removal from Crude Helium. The impurities in the helium stream are mainly hydrogen and nitrogen and must be removed before the helium is liquefied. Hydrogen is removed by combining it with oxygen at elevated temperatures in the presence of a platinum catalyst. The oxygen requirement is added in the form of air. While this is a convenient operation it has the disadvantage of introducing into the helium four times as much nitrogen plus the other constituents of air. Prior to recooling the gas for nitrogen removal, the water of reaction and carbon dioxide are removed by molecular sieve adsorption.

In the nitrogen removal operation crude helium is cooled against nitrogen boiling at subatmospheric pressure and its impurity level is reduced to about 1.5 mol.%. The remaining impurities, mainly nitrogen, are removed by adsorption.

Helium Liquefaction, Storage and Distribution. The pure helium stream is now cooled, liquefied and passed into a large storage tank which will hold about seven days production. The liquid helium will be collected and distributed in a similar manner to the present United States practice.

7-6. Process and Equipment Consideration

Basic Data. The first problem confronting the plant designer is that he must have access to the necessary basic data on vapour-liquid phase equilibria, enthalpy predictions and transport properties of all the components in the gas and of their mixtures. The designer needs equilibria and enthalpy data in order to determine the various operating parameters such as temperatures, pressures, compositions, yields, utilities and refrigeration requirements. The transport properties are needed to ensure that the equipment required to achieve the desired process conditions can be designed accurately.

Companies practising in the cryogenic field assemble considerable quantities of basic data either by conducting their own research programmes, by sponsoring research, by monitoring the output of national and university bodies, or by operating plants. This information is then collated to set up various mathematical equations which can be used to design commercial units.

In the design of the Polish helium plant, equations of state which predict thermodynamic functions were used in conjunction with published experimental results and operational data obtained in our development laboratories. Among the more important reports of experimental data are the works of Bloomer and Parent[4] on the phase equilibria of the methane-nitrogen system and of Devaney[5] and co-authors on the phase equilibria of the helium-nitrogen binary system. Having established the various thermodynamic properties to be used in the design, the process engineer can study various schemes and decide which alternative is the most suitable for his particular requirements within the specifications laid down by the eventual plant owner and operator.

Distillation Column. After it had been established that a double distillation column would be used to perform the methane-nitrogen separation, the physical conditions in the upper part of the lower column led to special attention being paid to the distillation tray hydraulic design. This column operates close to the critical pressure of nitrogen and hence the fluid characteristics are not those normally encountered. The principal difficulties arise from very low liquid densities, high vapour densities and low surface tension. After close liaison with the distillation tray vendors and a study of tray dynamics when operating on fluids of similar physical properties, a satisfactory tray design was established.

Crude Helium Purification. The crude helium from the reflux exchanger contains up to 12 mol.% impurities. At this point the plant designer is faced with the problem of deciding what the particular impurities are, their concentrations and by what methods they can be removed. The problem arises from the difficulties in analysing the initial lean gas to determine whether trace quantities of, say, hydrogen or neon are present and to measure accurately their concentrations. The difficulty is compounded by the gas volumes involved.

The lean gas rate to the plant is about 120 MMSCFD. Thus if the hydrogen content were only 100 volume parts per million, the hydrogen flow would be 12,000 SCFD. The hydrogen will pass through the plant along with the helium stream. The crude helium flowrate is 550,000 SCFD, and of this the hydrogen flow is 12,000 SCFD. Therefore the hydrogen concentration has now increased to 2.18 mol.%. The concentration has increased by a factor of almost 220, and what was initially only a trace component has now become a serious removal problem. Since it is difficult to analyse accurately the lean gas for such small concentrations, the helium purification process should be designed to tolerate considerable variations in the concentration of impurities in the crude helium.

Several process schemes were considered for the enrichment and final purification of the helium but were rejected either because of inadequate design information or low helium recovery.

One alternative considered was to remove all the impurities in a single adsorption step. This route was rejected as it was considered that insufficient data were available on the rates of co-adsorption and interference effects when hydrogen and nitrogen are simultaneously adsorbed. It was decided to utilise proven technology and to provide separate facilities for the removal of hydrogen, nitrogen and other impurities.

The hydrogen is removed from the crude helium by converting it to water. This catalytic oxidation process is operated with a slight stoichiometric excess of oxygen, and the hydrogen concentration is reduced to less than 2 vpm (ppm by volume).

The nitrogen removal process is a combination of partial condensation and physical adsorption. The condensation process is straightforward but the adsorption step, in which the final impurities are removed, presents a number of alternatives.

The adsorbent to be used must have certain characteristics, namely:

1. A high adsorption capacity for the impurities so that the equipment size and cost can be minimised;
2. The capability of reducing the concentration of impurities to very low levels so that no solidification can occur in the liquefier;

3. A low adsorption capacity for helium, otherwise large volumes of helium would be lost on adsorber reactivation;
4. The capability of being readily desorbed of the impurities so that the reactivation equipment size and cost can be minimised.

The two adsorbents meeting the above requirements are activated charcoal and molecular sieve. However, by studying the relative adsorption characteristics (Figure 7-2) it becomes apparent that activated charcoal has the larger capacity at high concentrations of nitrogen while molecular sieve has the larger capacity at low concentrations of nitrogen. These characteristics can be utilised best by placing activated charcoal at the inlet end of the adsorber where the impurities concentrations are at the highest, and then using molecular sieve at the outlet end where a very low impurities concentration is required.

To allow continuous operation of the plant, two nitrogen adsorbers are included. One will be on-stream while the second one is being reactivated.

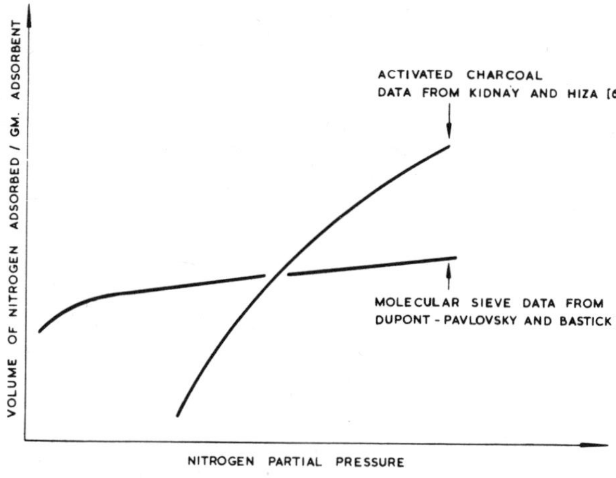

Fig. 7-2. Nitrogen Adsorption Characteristics of Activated Charcoal and Molecular Sieve

Liquefaction Unit. The pure helium passes to this unit as a gas at 86°K and 24 atmospheres. The liquefaction unit is a Claude cycle in which the refrigeration requirements are provided by a closed-circuit, compressor-expansion turbine system. The cooled helium undergoes a double Joule-Thomson expansion, the final expansion being directly into the liquid storage tank, which then acts as a separator/receiver providing adequate vapour-liquid disengagement volume.

The successful operation of a helium liquefier revolves around three main criteria:

1. Adequate basic design data for the helium;
2. Machinery selection;
3. Provision for tolerating occasional maloperations.

The first of these requirements, that of basic data, has been met by the NBS Technical Notes No. 8[8] and 154[9].

The second requirement concerns mainly the refrigeration cycle compressor and the expansion turbines. Since the helium has undergone extensive purification, the equipment design philosophy is to select machinery which eliminates any further contamination. Consequently, it was decided to use a nonlubricated reciprocating compressor and helium-gas-bearing expansion turbines. This decision, while overcoming process problems, introduces certain mechanical limitations. The use of PTFE piston rings for the cycle compressor affects the allowable temperature rise and hence compression ratio on each stage of the machine. The maximum compression ratio is set at about 2:1, and the discharge temperature is limited to 160°C.

The use of gas-bearing expansion turbines on a bulk helium liquefaction plant is a departure from the established United States practice. However, these turbines have a distinct advantage over the oil-lubricated type in that the possibility of oil contamination is completely eliminated. Also the turbine system and installation are simpler and there are no seal gas recovery problems. In 1966 Petrocarbon, in conjunction with Lucas Industrial Equipment Ltd., developed a gas-bearing helium turbine capable of producing refrigeration at about 10°K. The expansion turbines to be installed in the Polish helium plant will incorporate the various control and safety features which were evolved during the development programme. Petrocarbon also pioneered the use of gas-bearing turbines in small air separation plants. About 100 of these units are operating throughout the world. This experience has shown that gas-bearing turbines with a well-developed protection system achieve the same reliability as the oil-lubricated type.

The third requirement for the successful operation of a production plant is that the process and equipment must be capable of accepting, without plant shutdowns or significant reductions in output, the various maloperations which occur from time to time. One of the most critical maloperations is the contamination of the refrigeration cycle helium with air, brought about by the low pressure system operating occasionally at subatmospheric pressure. If this air is allowed to remain in the helium it will freeze out in the liquefier and severely restrict production. Therefore, several guard adsorbers are incorporated into the liquefier to clean up any trace impurities from air ingress or even from a breakthrough on the adsorbers in the nitrogen removal unit.

Storage and Distribution. The on-plant liquid helium storage requirement is a function of production rate, demand pattern, trailer availability and weather conditions. A 30,000 U.S. gallon tank will be installed and the transfer of liquid from the storage tank to the trailer will be by gravity displacement (Figure 7-3). The tank will be elevated 30 feet above the ground and the liquid will flow into the trailer at about 1500 U.S. gph. The vapour displaced from the trailer by the input of liquid will be returned to the storage tank. By adopting this technique the liquid helium loss due to flashing is reduced, and only a small volume of vapour is generated by the heat leak into the transfer lines and trailer.

7-7. Summary

The Polish helium plant, which is due to start up at the end of 1974, will ahve a production capability of 150 MMSCFA. It will be the first bulk helium liquefaction plant to be built in Europe and will play an important role as the major indigenous source of helium in the Eastern Hemisphere.

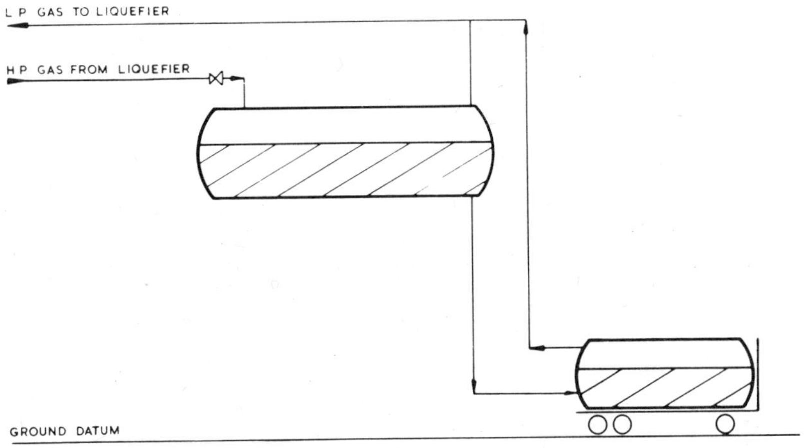

Fig. 7-3. Liquid Helium Storage and Transfer

References

1. Z. Blotny, R. Szwarski, and Cz. Truchan, "Possibilities of Natural Gas Utilisation with High Inert Content," International Gas Union IGY/A10-70, 1970.
2. Gas de France publication No. B **15**, 803, 1968.
3. Gas Council Research Communication GC 150, 1968.
4. O. T. Bloomer and J. D. Parent, Chem. Eng. Progress Symp. Series 49, 6, 1953.
5. W. E. DeVaney, B. J. Dalton, and J. C. Meeks, *J. Chem. Eng. Data* **8.**, 4, 473, 1963.
6. A. J. Kidnay and M. J. Hiza, *A.I. Ch. E. Journal,* **16**, 6, 949, 1970.
7. N. DuPont-Pavlovsky and J. Bastick, *Bull. Societe Chimique de France,* p. 24, 1970.
8. D. B. Mann and R. B. Stewart, NBS Tech. Note 8, 1959.
9. D. B. Mann, NBS Tech. Note 154, 1962.

Acknowledgment

The author's thanks are extended to colleagues who assisted in the preparation of this paper and to the Directors of Petrocarbon Developments Limited for permission to publish.

8

REFRIGERATION AND THERMOMETRY BELOW 1K

A. C. Anderson

^3He - ^4He dilution refrigerators capable of continuous refrigeration to 0.01 K have become an important tool in the research laboratory. Principles of design, operation and application of these refrigerators are discussed. Since the refrigerator operates in a temperature regime where no temperature scale has yet been established, appropriate primary and secondary thermometry is also mentioned.

8-1. Introduction

Temperatures below the ambient may easily be maintained by permitting a cryogenic liquid to evaporate freely into the atmosphere. Normally, however, economics and/or application dictate that the refrigerating gas be contained in a closed-cycle system capable of continuous operation. Closed-cycle refrigerators are commonly built to operate at temperatures ranging from ≈240 K with Freon down to ≈1 K with ^4He or even ≈1/4 K with ^3He. ^4He is the helium isotope obtained from natural gas wells in the Southwest United States and is readily available on the commercial market. ^3He on the other hand is in a sense a man-made isotope of helium and is obtained as a by-product of nuclear reactions. It has been made available for sale to the public by various national atomic energy agencies. ^3He has a higher vapor pressure than ^4He and thus can be used to refrigerate to a slightly lower temperature.

A temperature of ≈1/4 K is the lowest that may be maintained using a standard compression-condensation-evaporation cycle. Lower temperatures have been obtained by other techniques, such as adiabatic demagnetization,[1] but these techniques always involved a single-shot or noncontinuous refrigeration process. Only with the recent development of the ^3He-^4He dilution refrigerator have temperatures to below 0.01 K been produced in a continuous, closed-cycle refrigerator. The principle of operation, design and application of this refrigerator are discussed below.

First it is important to understand the significance or meaning of these very low temperatures. The dilution refrigerator produces temperatures of the order 0.01 K, which is only a fraction of a degree below that produced by a ^3He refrigerator, namely 0.25 K. However it is not the difference in temperature which is important in thermodynamic processes, but rather the ratio of temperatures. Hence the ratio of 25 between 0.25 K and 0.01 K is more significant thermodynamically than the ratio of 15 between room temperature (≈300 K) and liquid hydrogen temperature (20 K).

8-2. The Dilution Refrigerator

The cryogenic fluid used in a dilution refrigerator is a mixture of ^4He and ^3He isotopes. The behavior of these mixtures as a function of ^3He concentration and temperature is depicted in Figure 8-1. Consider as an example a 50/50 mixture of ^3He and ^4He (x_3=0.5 on Figure 8-1). As the temperature is lowered the mixture is first a normal, viscous liquid shown as region I. Near 1.3 K the mixture becomes a superfluid, region II. The properties of the superfluid are similar to those of pure superfluid ^4He.[2] With further reduction in temperature a phase separation occurs near 0.7 K. A ^3He rich phase forms with $x_3 \approx 0.7$ and a ^4He rich phase with $x_3 \approx 0.5$. Since the ^3He is lighter (^3He has only three nucleons in its nucleus vs four for ^4He) the ^3He rich phase floats to the top and the ^4He rich phase settles to the bottom of its container. With further reduction in temperature to 0.1 K, the ^3He rich phase becomes almost 100% pure ^3He while the ^4He rich phase retains about 6% ^3He in solution. Mixtures originally having less than 6% ^3He in ^4He thus do not phase separate.

Hence, if x_3 is greater than 6%, one has at low temperatures a container of essentially pure liquid ^3He floating on top of a mixture of 6% ^3He in superfluid ^4He. Consider now the removal of

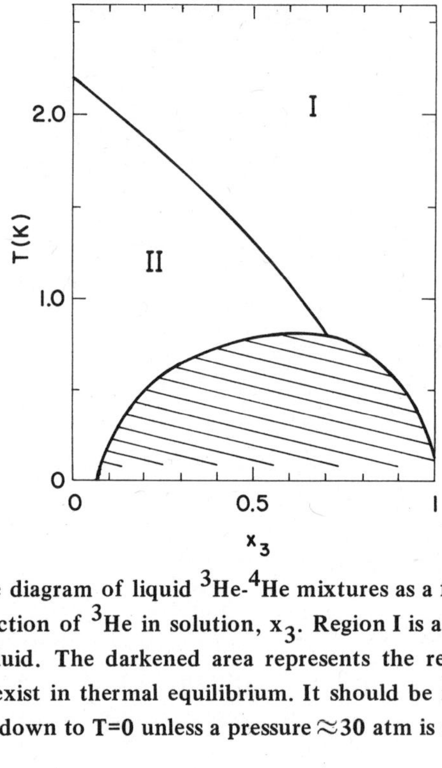

Fig. 8-1. Phase diagram of liquid ^3He-^4He mixtures as a function of temperature T and fraction of ^3He in solution, x_3. Region I is a normal fluid; region II is a superfluid. The darkened area represents the region where a single phase cannot exist in thermal equilibrium. It should be remembered that He remains liquid down to T=0 unless a pressure ≈30 atm is applied.

some ^3He from the top ^3He layer to the bottom ^4He rich layer. Experiment has shown that the pressure and temperature will remain constant during this process only if heat is ***added*** to the container. Thus if we can devise a method of continuously moving ^3He from the top to the bottom layer or phase in the container we have the possibility of continuous absorption of heat, i.e. refrigeration.

A schematic flow diagram of the desired process is shown in Figure 8-2. At M we have the container with the ^4He rich phase at the bottom, the ^3He rich phase floating on top. The ^4He-6% ^3He mixture is removed from the bottom to the separator S. At S the two isotopes are separated. The ^4He is returned to the lower part of the container M; the ^3He, to the top of M. In continuous operation heat $\dot{Q}$ must be added to M to keep the temperature and pressure of M constant. Hence M becomes the cold-finger or heat

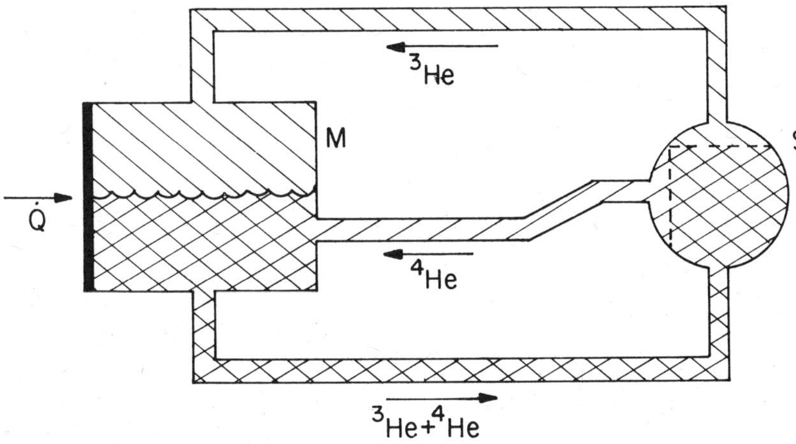

Fig. 8-2. Schematic flow diagram of a ^{3}He-^{4}He dilution process. Refrigeration takes place at M; S is a device used to separate the two isotopes.

sink of the refrigerator. However we have neglected the fact that the ^{4}He is superfluid. Thus the ^{3}He moves freely through the ^{4}He without viscosity, the liquid ^{4}He need not move. The return line of Figure 8-2 labeled ^{4}He may therefore be removed. We are then left with a single loop in which only the ^{3}He atoms circulate. It may be noted that this refrigeration cycle is reminiscent of the ammonia-hydrogen cycle. Alternatively it may be considered to be a standard single-constituent, condensation-evaporation cycle with the ^{3}He "evaporating" downwards into the stationary ^{4}He rather than into an actual vacuum.

Obviously a crucial component in the system of Figure 8-2 is the separator S. In practice the separation of isotopes is accomplished rather easily by vaporizing the He mixture. The ^{3}He has a higher vapor pressure than ^{4}He and thus evaporates preferentially from the mixture. A practical refrigerator incorporating this idea is shown in Figure 8-3. The mixer in Figure 8-3 is the M of Figure 8-2, i.e., the cold finger of the refrigerator. The still is the S of Figure 8-2, where ^{3}He is separated from the stationary ^{4}He by vaporization. The gaseous ^{3}He is then compressed by standard oil diffusion and mechanical vacuum pumps at room temperature to a pressure of 10-100 Torr. The ^{3}He gas is next recondensed at a

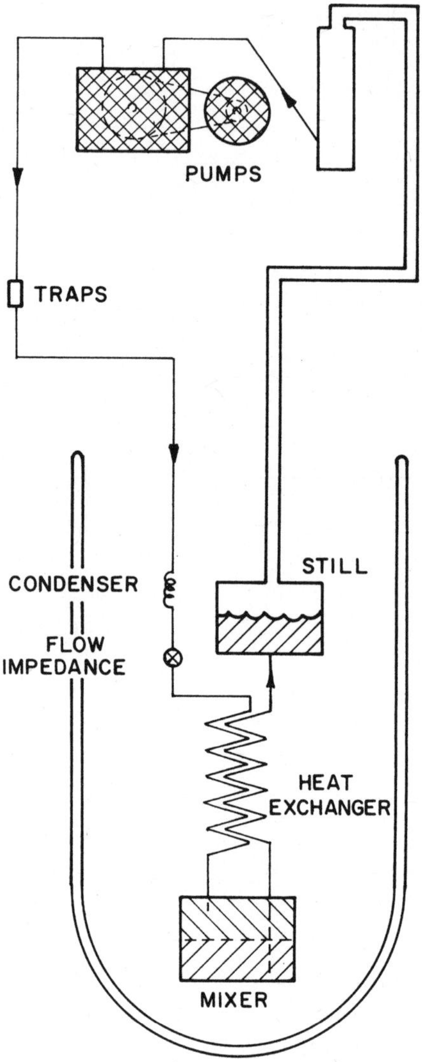

Fig. 8-3. Schematic diagram of a dilution refrigerator. The still, heat exchanger and mixer would be situated in a vacuum, while the condenser normally would be located in a liquid ^{4}He bath. The trap moves pump oil and residual air which otherwise might plug the low-temperature capillary line.

temperature of 1-2 K normally provided by a liquid ^{4}He bath. A heat exchanger provides the usual improvement in efficiency. The flow impedance serves only to limit the liquid ^{3}He flow from the high to low pressure portions of the circuit; it is not an expansion valve. The refrigeration capacity of the device is limited primarily by the flow capacity of the heat exchangers, pumping lines and pumps. Parameters which determine the low temperature limit of the device are more subtle, but include the efficiency and design of the heat exchangers as well as the residual heat leaks common to all ultra low temperature cryostats.

Very simple dilution refrigerators operate reliably to below 0.05 K;[3] more elaborate designs operate to below 0.01 K.[4] Dilution refrigerators of various designs may be purchased from several vendors in the United States and abroad. Several aspects of the operation of dilution refrigerators are still not understood, for example the transfer of heat in heat exchangers at very low temperatures.[5] Hence we may expect future improvements in the physical size, refrigeration capacity and limiting low temperature as we increase our understanding of the physical phenomena involved.

8-3. Application of the Dilution Refrigerator

We have indicated above that one may readily have continuous refrigeration at temperatures down to 0.01 K either through construction or purchase of a ^{3}He-^{4}He dilution refrigerator. It would be quite misleading to leave the subject at this point without some indication of the problems one might encounter in the application of the refrigerator.

The refrigeration capacity of a modest refrigerator may be $\approx 1/2$ microwatt near 0.02 K. To a person having historical roots in the ultra low temperature field, this is a large and welcome (continuous) refrigeration capacity. To the newcomer it is the energy expended by one ordinary housefly doing about ten push-ups per second. Hence the residual heat leak into the cryostat must be kept very small in order for the refrigerator to attain temperatures of the order 0.01 K. To this end the cryostat is designed to be isolated from building vibrations, to have physical supports for the cold finger which have small thermal conductance, and to shield the refrigerator from all radiation, including radio-frequency radiation of nearby F.M., T.V. or other transmitters.

However the magnitude of the residual heat leak is only half the problem. When heat is transferred across the interface between two materials a thermal resistance is observed which varies as I/T^3, T being the average temperature.[5,6] Thus this resistance to heat flow becomes very large as the temperature is lowered. For example a heat flux of 1/10 microwatt per square centimeter from copper to liquid He would leave the copper at a relatively high temperature of ≈0.07 K even if the He had a temperature of absolute zero. Hence considerable care must be exercised in providing good thermal contact between the refrigerator and the items or samples to be cooled.

Since the thermal boundary resistance becomes smaller at high temperatures and, in addition, the refrigeration capacity of the dilution refrigerator increases rapidly with increasing temperature, the problems discussed above become much less important as the temperature is raised above ≈0.1 K.

8-4. Thermometry Below 1 K

Normally an individual operating a refrigerator is interested in the temperature being produced by the instrument. At temperatures above ≈1 K there is no problem since an internationally agreed upon temperature scale exists, accurate to better than 1%, based on the vapor pressures of pure liquid ^{3}He or pure liquid ^{4}He. One may either purchase calibrated thermometers or, alternatively, calibrate his own thermometer against the He vapor pressure scales using a simple manometer-cathetometer arrangement.

Below ≈1 K however there does not exist an international temperature scale and each laboratory must establish its own scale. In order for the scale to be very useful, it should correspond closely to the absolute or Kelvin temperature. One typically picks a thermometric parameter which varies in a predictable manner with temperature, calibrates the thermometer near 1 K, then, trusting the thermometric parameter, extrapolates the international scale below 1 K. The parameter and material most frequently used has been the magnetism of cerous magnesium nitrate, a hydrated salt which is about as magnetic as window glass. Nevertheless its weak magnetism M varies as $M=(C/T)$ down to ≈ 0.02 K. The constant C of the thermometer is determined by comparison to the He vapor pressure scale or to a calibrated thermometer near 1 K.[7] The even weaker nuclear magnetism of copper or other metals may be used to even lower temperatures, although specialized techniques are required to read the extremely weak magnetic signal.[8]

Any extrapolation procedure like that discussed above is subject to uncertainties unless considerable care and cross checking is exercised. As a result attempts are being made to find thermometric parameters which are related directly to the absolute temperature, thus eliminating the need for calibration, extrapolation, or any reliance on the temperature scales above 1 K. Various techniques are being developed, including the use of gamma ray anisotropy, the Mössbauer effect, and nuclear magnetic resonance.[8] Perhaps the most useful technique in that it is not limited to a narrow temperature range is the measurement of the so-called thermal or Johnson noise voltage E produced by any electrical resistance R.[8] Since $E = B\sqrt{TR}$ and the constant B is known, a measure of E gives immediately the absolute temperature. Unfortunately the voltage E is very small and sophisticated techniques must be used to measure it. Hence it is likely that one or more of the above techniques will serve only as primary thermometers with which to calibrate secondary thermometers, for example resistance thermometers.[8] Such secondary thermometers could eventually be provided on a commercial basis as is now done for the temperature range above ≈ 1 K.

8-5. Summary

Temperatures down to 0.01 K are readily available to anyone simply through the use of ^3He-^4He dilution refrigeration. The complications enter in the application of the instrument, that is in the transfer of heat to the refrigerator from the material to be cooled, and in the determination of the temperature at which the refrigerator operates or to which some material or device has been cooled by the refrigerator. In spite of these problems the application of the dilution refrigerator has been rapid and widespread, including the polarization of targets for nuclear and high-energy particle research, improvement of signal to noise in gravity-wave detectors, and for an extensive range of research on the properties of materials at ultra low temperatures.

References

1. F. E. Hoare, L. C. Jackson, and N. Kurti, editors, *Experimental Cryophysics,* Butterworth and Co., London, 1961.
2. J. Wilks, *The Properties of Liquid and Solid Helium,* Clarendon Press, Oxford, 1967.
3. A. C. Anderson, *Rev. Sci. Instrum.* **41**, 1446, 1970.
4. J. C. Wheatley, R. E. Rapp, and R. T. Johnson, *J. Low Temp. Phys.* **4**, 1, 1971.
5. A. C. Anderson and W. L. Johnson, *J. Low Temp. Phys.* **7**, 1, 1972.
6. A. C. Anderson and R. E. Peterson, *Cryogenics* **10**, 430, 1970.
7. A. C. Anderson, R. E. Peterson, and J. E. Robichaux, *Rev. Sci. Instrum.* **41**, 528, 1970.
8. See: *Temperature, Its Measurement and Control in Science and Industry,* **4**, Instrument of Society of America, Pittsburgh, 1972.

Acknowledgment

This work was supported in part by the National Science Foundation Grant GH-33634.

9

APPLICATIONS OF CRYOGENICS IN ELECTRON MICROSCOPY

H. Fernández-Morán

Results of earlier work with superconducting lenses (objective lens excited with 27,220 ampere-turns) indicated some unique advantages inherent in systematic application of this approach in high resolution electron microscopy. Operating in the persistent current mode made possible resolutions of 10Å to 20Å during longer exposure times at lower beam intensities, thus reducing specimen radiation damage.

With the first prototype of a cryoelectron microscope system (i.e. large-scale, 20 watt capacity, Collins closed-cycle superfluid helium refrigerator with 36-foot transfer lines and novel heat exchanger integrated with a modified 200 kV electron microscope), significantly reduced radiation damage, contamination and thermal noise are available in prolonged vibration-free examination of specimens at 1.8° to 4.2°K. Simultaneously, the system provides high penetration power, ultra-high vacuums, decreased spherical and chromatic lens aberrations, and enhanced image contrast. Specially-adapted vacuum-tight microchambers have been used in studies of ice-crystal structure, wet membranes and new types of organometallic superconducting compounds. Consistent resolutions of 8Å to 16Å are achieved, compared with corresponding low temperature resolutions of about 50Å one year ago.

Descriptions of these instrumental developments are given together with results of their applications and implications in specific research areas, particularly membrane ultrastructure, cryobiology, microelectronics and general superconducting work.

9-1. Introduction

The inherently high resolving power of the electron microscope has put within investigators' grasp the potential to visualize directly structures of molecular and pauciatomic dimensions and atomic spacings in crystalline lattices. Consistent resolutions on the order of 2Å to 3Å should be possible. However, to achieve this ultimate theoretical resolution, several major problems have to be overcome, including: (1) correction of spherical and chromatic lens aberrations, (2) stabilization of lens excitation current and accelerating voltage, and (3) reduction of specimen radiation damage and thermal noise.

In an effort to deal with these problems, during the past decade we have developed, tested and routinely implemented improved point cathode sources, superconducting lenses and the Collins closed-cycle superfluid helium refrigeration unit in our specially-designed clean room facilities at The University of Chicago. The result is the world's first functioning prototype for a large-scale high voltage cryoelectron microscope system capable of consistent operation at 1.8° to 4.2°K. (Figure 9-1.)

9-2. Chronological Overview

Early Low Temperature Research. When I began my work in biological ultrastructure a quarter century ago, most methods for ultra-thin sectioning of tissue involved dehydration and embedding. These processes not only deprived biological specimens of their inherent hydrated components, but they also effectively limited contrast and made it impossible to carry out chemical and other correlated studies which required untreated specimens.

While the advantages of solidifying tissue by freezing were generally recognized as important means for approaching the natural condition in tissue specimens, the problem of obtaining sections of required thinness for use with the electron microscope remained.

In 1951, it was shown that well preserved ultra-thin frozen sections of unembedded fixed or fresh material could be cut from small tissue blocks in a modified microtome. Further improvements in preparation and mounting of serial frozen sections

Fig. 9-1. Collins heat exchanger with control panel and special dewar attached to modified HU-200 electron microscope for closed-cycle refrigeration at 1.85°K.

combined with freeze-drying procedures carried out directly on the ultra-thin (about 0.5 to 0.1 micron) sections also made it possible to examine fresh biological tissues with the electron microscope in a manner which preserved their three-dimensional structure. Beyond this, the results of such studies could be correlated with polarized light and phase contrast investigations.[1] Shortly thereafter, my conception and invention of the diamond knife and associated microtome made possible even thinner sections (about .001 to .004 microns), thus broadening the practicality of freeze-sectioning for electron microscopy.[2]

In the late 1950s, working with Dr. William Sweet and Professor Samuel C. Collins at the Massachusetts General Hospital and M.I.T., I began preliminary experiments involving cryofixation, ultra-rapid freezing with liquid helium, and related low temperature preparation techniques in studies of serially arrested states of activity in biological systems and of ice-crystal formation.[3] The results indicated that significant advances in the study of life processes under conditions of minimum perturbation might well be in large measure dependent on further development of the unique potentialities inherent in the low-temperature domain.

The suggestion was then made that systematic research in cryobiology could reveal new phenomena derived from the cumulative effect frequently associated with the high degree of molecular order characteristic of the solid state induced in biological specimens at temperatures close to absolute zero.

Pointing out the complementary interrelationship between the development of adequate low temperature preparation techniques (necessary for successful application of electron microscopy to problems of ultrastructural research) and the systematic study of the fundamental mechanisms of biological freezing, radiation damage and tissue preservation at the molecular level, it was also possible to make some comments on the relationship of low temperatures and radiation damage.

Many unstable chemical species such as free radicals are still considerably reactive at temperatures as low as 70°K, and they play a key role in enzymatic reactions and metabolic electron transfer. In addition, the generation of abundant free radicals is inextricably linked with all biological effects of ionizing radiation.

Broida had shown that at liquid helium temperatures, free radicals could be effectively trapped, thereby reducing one of the most deleterious forms of specimen radiation damage.[4] Later, Box and his associates reported that by maintaining organic compounds at 1.8° to 4.2°K many or perhaps most irradiated molecules could be preserved as ions.[5] Single crystal electron spin resonance spectroscopy study of various organic crystals (i.e. glycine, ice, disulfides, etc.) demonstrated that specific positive and negative ions are stabilized at low temperatures.

Thus, this work made clear that systematic research at liquid helium temperatures would not only enable us to trap free radicals and stabilize ions, thereby reducing specimen radiation damage, but also to begin to study the initial stages of radiation damage at the molecular level in a controlled manner.

Development and Application of Superconducting Lenses. Developing the electron microscope from its infant stage to the entirely new generation of instruments in use today has been a 30-year effort of physicists, engineers and biologists from many countries. A recent review of superconducting technology[6] referred to the announcement of the first applications of superconducting solenoids in 1961[7] as the "dawn of a new era of electrical technology" which has had important implications for both the electron microscopist and the low temperature researcher.

In preliminary experiments[9,10] large, accurate (i.e. better than one part in 10^8), uniform magnetic fields in excess of 60 kilogauss were obtained using high-field superconducting solenoids of NbSn and NbZr. Further work showed that these magnetic fields were extremely stable and noise-free when operated under controlled conditions in the persistent current mode at liquid helium temperatures.

A series of studies were made on a simple electron microscope without pole pieces, but with high-field superconducting NbZr solenoid lenses in an open-air core, liquid helium Dewar (Figure 9-2). These showed that electron microscopic images of test specimens could be recorded while operating at 32.2 kilogauss in the persistent current mode, with accelerating potentials of 4 to 8 kV. The electron micrographs demonstrated both the exceptional stability of images, both long-term and short-term over a period of four to eight hours, and the relatively high quality of the images at magnifications of 50X to 100X.[10,11]

This initial work brought to light several problems which remained to be solved before the specific advantages of this approach could be established for high resolution electron microscopy. These problems included: (1) incorporation of superconducting solenoid lenses and associated cryogenic components into high

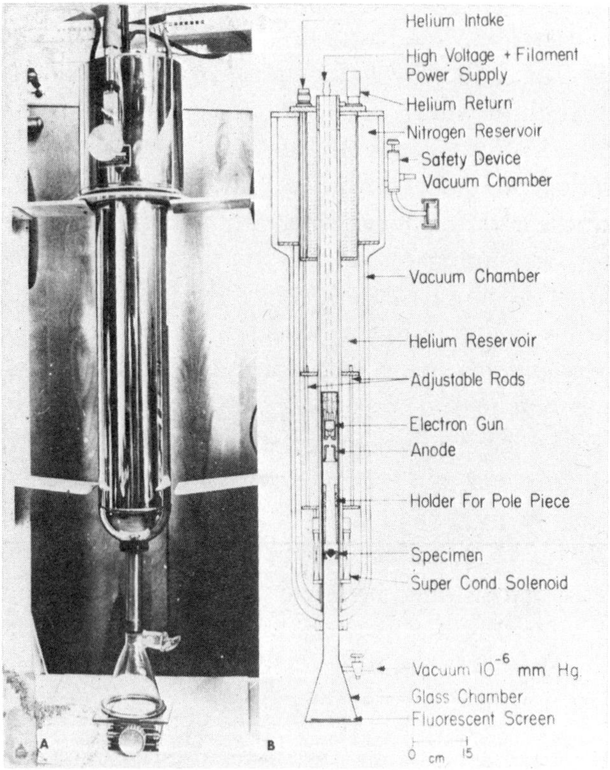

Fig. 9-2. Photograph and diagram of cryoelectron optical bench system I comprising miniaturized electron microscope with superconducting solenoid in liquid helium dewar.

performance electron microscope systems; (2) precise control and reproducible current settability for focusing superconducting solenoid lenses; (3) satisfactory specimen mounting to prevent temperature drift and achieve high degrees of stability during irradiation; (4) stabilization of lens excitation currents and accelerating voltage, and improved electron source characteristics; (5) reduction of magnetic, electrical and mechanical perturbations under carefully controlled conditions in the necessary cryogenic environment; and (6) adequate continuous recording of images without breaking high vacuum under cryogenic conditions.

To meet these specific needs, a 25 amp regulated power supply was improved and used in conjunction with additional current changes of 10^{-9}, resulting in the capacity for reproducible superfine focusing several orders of magnitude better than that achieved with conventional systems. Highly stabilized 50 kV power supplies connected to a central regulated motor generator set with solid-state regulators provided better than 0.1% voltage stability. A special superconducting objective lens in a liquid helium cryostat (Figure 9-3), consisting of a main NbZr coil (27,220 ampere-turns) with vernier coil, superconducting stigmators, persistent current switches and improved current control devices was also designed and incorporated into the microscope system.[11-13] Conventional objective pole pieces with focal lengths of 1.6 to 2.6 mm were also used.

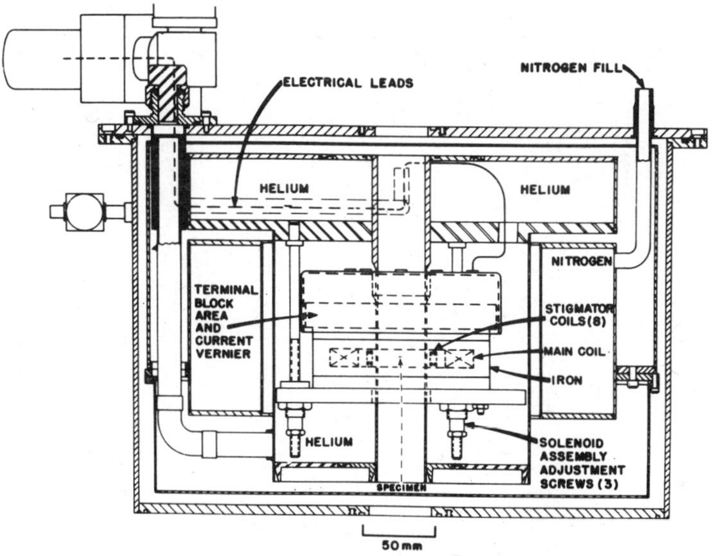

Fig. 9-3. Sketch of superconducting objective lens with stigmators and vernier coils of special design in liquid helium cold stage assembly for cryoelectron microscope.

With this prototype superconducting electron microscope exceptionally long-term stability and high quality electron microscopic images could be recorded directly at 200X to 20,000X while operating in the persistent current mode at 4 to 32 kilogauss with 4 to 50 kV accelerating potential. Reproducible resolutions of 10Å to 20Å were obtained in various biological specimens recorded at 4.2°K,[12-15] see Figure 9-4. Figure 9-5 illustrates the types of results which can be achieved with superconducting lenses at accelerating voltages of 100 kV to 1000 kV.

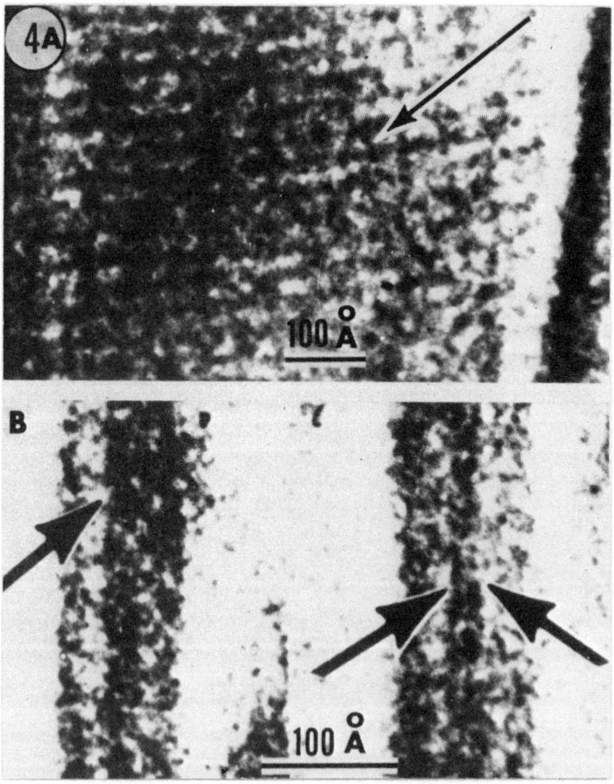

Fig. 9-4. Electron micrographs of catalase and asbestos specimens recorded at 4.2°K using superconducting objective lens in persistent current mode and electron microbeam illumination. 5A Mag.=1,100,000X; 5B Mag.=1,850,000X.

Voltage kV	C_s–limit	70 Contrast below background				Half-width of image fig.	Interatomic spacing dmin for 5% contrast
		C	NC	MO	SU		
100	.40 mm	7	18	24	35	1.8Å	2.44Å
200	.54	6	18.5	24	35	1.5	2.00
500	.88	8.5	22.5	30.3	45	1.1	1.44
1000	1.30	5	27	37	56	0.9	1.20
	C_c2. mm						

Reasons for Superconducting Objective Lens for High Resolution Electron Microscopy

A lower spherical aberration coefficient is obtainable with the higher fields given by superconducting material. The use of liquid helium temperatures in the objective system has additional essential advantages: the specimen is kept at a temperature where the zero point vibrational amplitudes are giving the motion of the atoms, and the specimen section is cryo pumped.

Fig. 9-5. Contrast and Resolution of Single Atoms with Superconducting Lenses

Special pole pieces of dysprosium and holmium have also been used for high-field superconducting lenses in the cryoelectron microscope. These rare earth metals become ferromagnetic at liquid helium temperatures, and because their saturation magnetization is reached only in fields of 50 to 80 kilogauss they act as exceptional magnetic flux concentrators, producing high fields in small volumes. The resulting high peak field of these superconducting objective lenses leads to optimized performance for both low- and high-voltage electron microscopes.

The use of liquid helium temperatures in the objective system has permitted us to extend our earlier superconducting research. Characteristic electron optical phenomena associated with trapped fluxes in thin superconducting films have been observed.[12] In addition, a temperature-dependent anomalous electron transparency effect has been demonstrated in 200 kV studies of thick (i.e. 1000Å) lead films, Figures 9-6 and 9-7. This may well represent

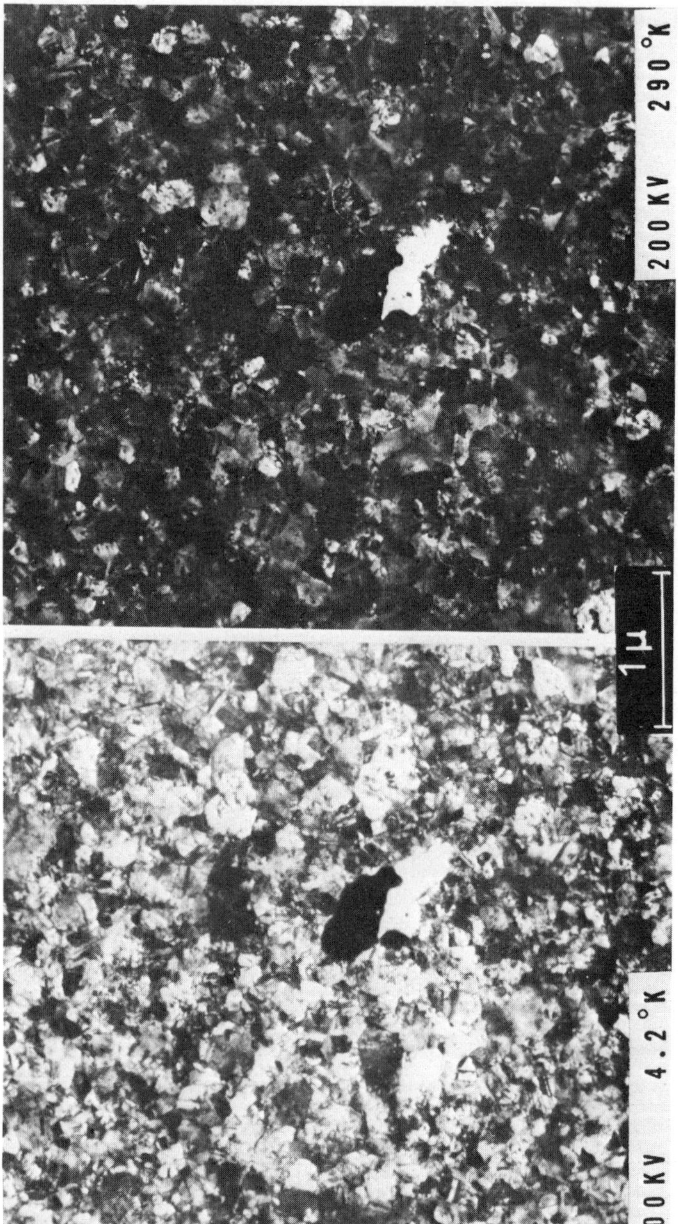

Fig. 9-6. High resolution electron micrographs illustrating marked temperature dependence of the transparency of 1000Å thick lead film for 200 kV electrons as shown by recording the same specimen area under identical conditions at 4.2°K (left) and 290°K (right). Mag. 20,000X

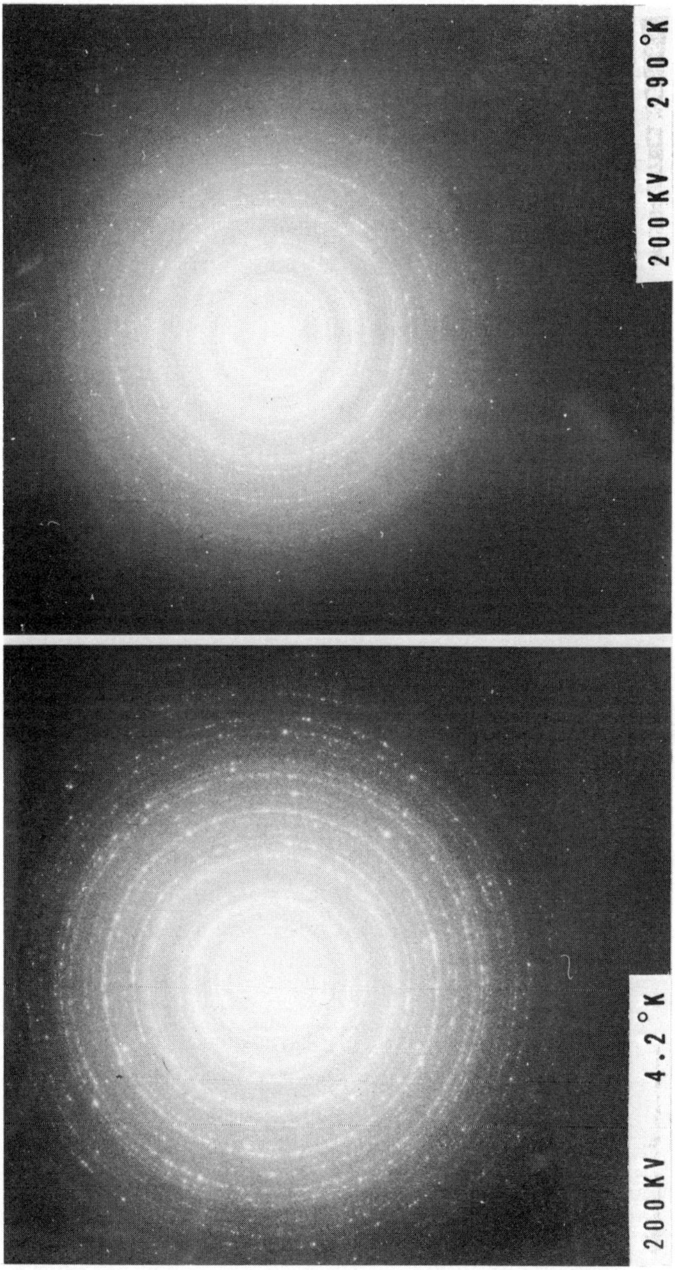

Fig. 9-7. High resolution electron diffraction patterns illustrating marked temperature dependence of the transparency of 1000A thick lead film for 200 kV electrons as shown by recording the same specimen area under identical conditions at 4.2°K (left) and 290°K (right).

a whole range of new phenomena which can be visualized directly only at temperatures in the liquid helium range.[11, 16-18]

The Collins Closed-Cycle Superfluid Helium Refrigeration System. The work of Professor Samuel C. Collins, who conceived and built a helium cryostat simpler in design and more efficient than any previous liquefier, resulted in a reliable piece of standard equipment now in use throughout the world. Collins and his associates made large quantities of superfluid helium available for the first time by reproducibly achieving temperatures of 1.8° to 4.2°K with relative efficiencies close to 8% to 10% of the Carnot cycle. This is comparable with the efficiency ratings of machines operating at much higher temperatures.[19]

The unique nonviscous flow properties of the superfluid helium and its singularly high thermal conductivity at such low temperatures demand that the entire system, including over 36 feet of transfer lines and a novel heat exchanger installed in parts of a 5-story building, be super-leak-proof (Figures 9-8, 9-9, 9-10). The main heat exchanger consists of two stainless steel shells, each of which is the frustrum of a cone. Finned tubing is wound around the inner shell. The outside of the outer shell is helically wound with wire to form a channel through which the helium can pass during liquefication and refrigeration. In addition, the system encompasses two compressor units, each with appropriate piping and isolating valves.[20,21]

With the assistance of Professor Collins, M. Streeter and R. Osburn of Cryogenic Technology, Inc., the large-scale Collins closed-cycle superfluid helium refrigerator has been fully integrated with the modified high voltage (200 kV) electron microscope in our laboratory (Figures 9-1 and 9-11). This cryoelectron microscope system has been used routinely in over 165 successful experiments to significantly reduce specimen radiation damage, contamination and thermal noise during prolonged vibration-free examination of specimens at 1.8° to 4.2°K. At the same time, it makes possible to combine these advantages with high penetration power, ultra-high vacuums, decreased spherical and chromatic lens aberrations and enhanced image contrast. At 1.9°K, this closed-cycle superfluid helium unit, with a 20-watt refrigeration capacity, can supply liquid and superfluid helium to the specially-designed microscope cryostat at rates of 5 to 8 liters per hour.[16] With pre-cooling, the rate of liquification approaches 216 liters per day.[21]

Preliminary data indicate that the highest recorded resolutions at liquid helium temperatures have been attained in the course of this developmental work. At 1.8° to 4.2°K, consistent resolutions of 8Å to 16Å have been achieved under ideal conditions.[17] This compares with resolutions of about 50Å at low temperatures only a little over a year ago.[18] In addition, this low temperature approach may make controlled manipulation of molecular structure under conditions of reduced entropy a real possibility.

A prolonged maintenance of liquid helium temperatures has been observed in the cold stage of the microscope connected with the superfluid closed-cycle system for periods of up to 29 minutes after all of the refrigeration equipment has been completely shut off! This unexpected result means that, after the initial start-up,

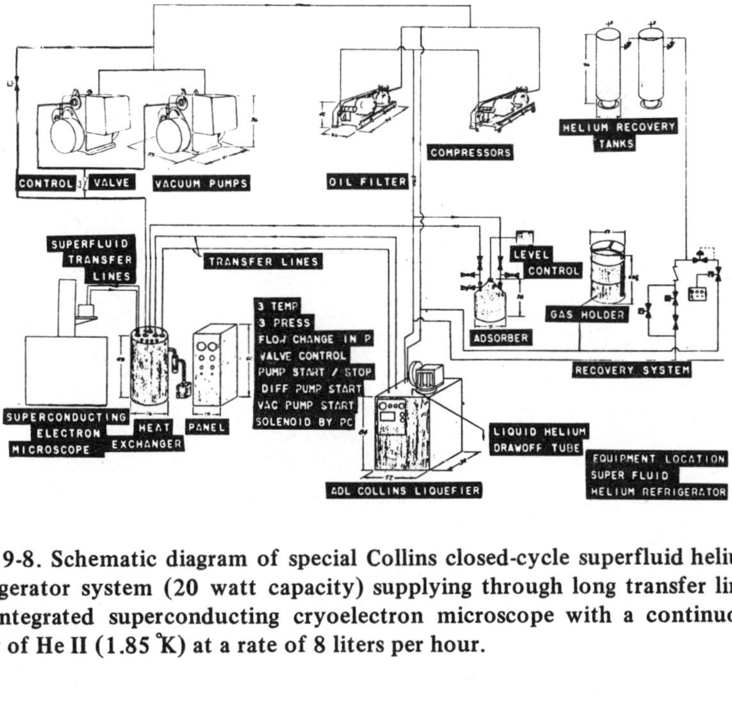

Fig. 9-8. Schematic diagram of special Collins closed-cycle superfluid helium refrigerator system (20 watt capacity) supplying through long transfer lines an integrated superconducting cryoelectron microscope with a continuous flow of He II (1.85°K) at a rate of 8 liters per hour.

Fig. 9-9. Special ADL/Collins helium liquifier at The University of Chicago.

liquid and superfluid helium can be recycled continuously for an indefinite period to provide the requisite specimen cooling. It also has important implications for the loss-less transmission of large blocks of electrical energy over long distances using superconducting cables.[17,31]

It is in large measure based on this demonstration of the practicality of the low-temperature techniques for achieving the highest theoretical and instrumental resolution (i.e. 3Å to 7Å) under near ideal conditions that Dr. I. Dietrich in Germany and researchers in France, Holland and other countries are now constructing similar facilities following the design features of the Chicago laboratories.

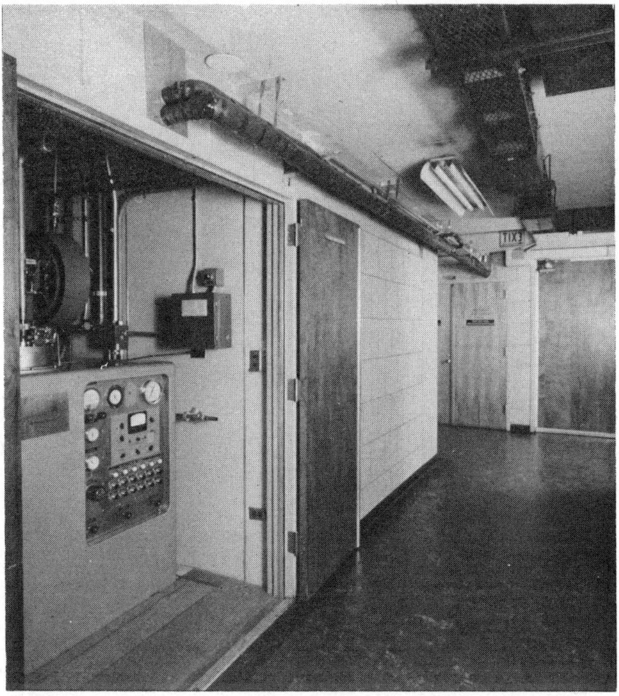

Fig. 9-10. ADL/Collins liquifier on vibration mount with special 36-foot transfer lines connected to high voltage (200 kV) electron microscope system.

Related Work in Other Countries. In a study of amino acid specimens using the 3 MeV electron microscope in the laboratories of Doctors G. DuPouy and F. Perrier in Toulouse, France, Doctors Gareth Thomas and Robert Glaeser reportedly have observed important new phenomena relating to the problems of specimen radiation damage.

At a given electron beam current, Thomas could examine specimens at 2.5 MeV for ten times longer than he could with the same amount of damage at 1 MeV. The experiments indicated that a decrease in the damage from 1 MeV to 2.5 MeV should continue through 5 MeV.[22]

Another recent report on the construction and performance of a new 3 MeV electron microscope in Japan by Doctors S. Ozasa, E. Sugata et al. states that: "The drift and ripple of high voltage are less than 1 x 10^{-5}/min. and 1 x 10^{-6} respectively. The calculated resolving power of this electron microscope is 2Å. The objective lens is excited with 45,000 ampere-turns at 3 MeV."[23]

In 1966 we demonstrated the performance of special superconducting objective lenses with a main NbZr coil of 27,220 ampere-turns.[9,11,13] Thus, six years ago, we were already approaching the figure quoted in the 1972 Japanese report.

Since that time we have built, but not yet tested, a superconducting objective lens of 100,000 ampere-turns for work in a superfluid helium cryostat. During the Grenoble Congress on Electron Microscopy in 1970 we reported several advantages inherent in this type of combined high-voltage, superconducting cryoelectron microscope approach:

> The large capacity, unique heat transfer and quantized rotational properties of the superfluid helium system enable it to supply both the lens cryostats and a compact superconducting linear accelerator (based on one built by Fairbanks at Stanford with 6 MeV electron beam stabilized to better than one part per million) used to provide high voltage electron sources. **Such integrated cryo-electron microscopes would exhibit unmatched stability of lens excitation current and high voltage (5 kV to 10 MeV),** correction of spherical and chromatic lens aberrations including trapped flux, phase zone apertures, and coherent electron point cathodes operating in cryogenic ultra-high vacuums to decisively reduce radiation damage, contamination and thermal noise while enhancing image quality with an optimized image intensifier operating at liquid helium temperatures.[16]

9-3. Applications of High Voltage Cryoelectron Microscopy

Biological Research. Systematically using the modified high voltage cryoelectron microscope, together with improved low temperature preparation techniques such as diamond knife cryoultramicrotomy, we have been able to approach observation of the native hydrated state in investigations of various biological specimens. It is now possible to directly observe increasing amounts of meaningful information and structural detail in "wet"

Fig. 9-11. High voltage 200 kV electron microscope with specially designed helium cryostat specimen stage attached to closed-cycle superfluid liquid helium refrigerator with Collins heat exchanger and accessories for superconducting high voltage microscopy.

biological specimens and various forms of ice and organized water, and to further our understanding of the crucial structural and functional relationships in the molecular and pauciatomic domain of "living" systems.

Based on earlier work[24,25] and on recent technical improvements to overcome the limitations of vacuum sublimination, electron bombardment and contamination, it has been possible to directly study ice-crystal formation and growth. With special shielding devices to protect the specimens from contamination, we have been able to record electron micrographs and electron diffraction patterns of minute ice crystals which have been stabilized by cooling.

Critical investigations have recently been carried out on ice crystals formed by deposition of water vapor on cold, thin carbon and single-crystal gold films in order to determine the existence of cubic and other ice modifications at temperatures of 1.8° to 4.2°K (Figures 9-12 and 9-13). The results of these studies will have great implications for the preservation of biological systems at low temperatures and in understanding the state of water in frozen materials. They may also have a direct bearing on the concept of selective permeability of nerve membranes envisaged in terms of molecular "pores" lined with highly ordered water.[17,18]

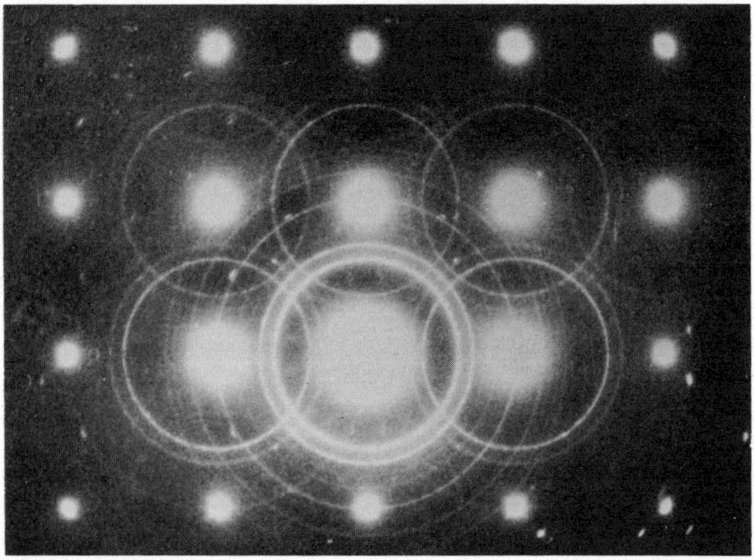

Fig. 9-12. High resolution electron diffraction patterns recorded from single crystal gold film displaying characteristic features of ring patterns observed at 1.8°K and 200 kV.

Extending earlier investigations,[26] specimens of DNA, nerve myelin, cock retinal rod outer segments and ice have been sandwiched between impermeable ultra-thin films in specially-adapted vacuum-tight microchambers[8,24] to trap their hydrated components at liquid helium temperatures. Preliminary data indicate that new details of wet membrane structure and dimensions can be

hkl	*Estimated Intensity*	*d* Å	*Values of Cubic Normal Ice (König et al.)*
111	v. strong	3.70	3.67
220	strong	2.26	2.25
311	medium	1.93	1.92
331	medium	1.47	1.46
422	medium	1.31	1.30
511	weak	1.23	1.22
531	medium	1.07	1.12

Fig. 9-13. Measurement and preliminary indexing of electron diffraction patterns recorded on gold single crystal film at 1.8°K and 4.2°K. (Experiment No. 114)

visualized with repeatable resolutions of 8Å to 16Å (Figure 9-14). Supplementary dark field studies of unstained shadow cast myelin and cock retinal rod outer segments have been conducted with consistent resolutions on the order of 8Å to 10Å at 1.8° to 4.2°K. This represents the highest resolution attained to date with any high-voltage instrument at low temperatures.

Low Temperature Studies of New Classes of Organometallic Superconducting Compounds. Representative samples of new classes of layered, transition metal dichalcogenide intercalation complexes with unique superconducting properties have been examined in correlated high resolution, high-voltage cryoelectron microscopy and diffraction studies. These exceptionally anisotropic, crystalline, metallic compounds, formed by inserting organic molecules of pyridine, aniline, and N_1N-Dimethylaniline between metallic layers of tantalum disulfide, niobium disulfide and niobium diselenide by F. R. Gamble and his colleagues,[27] are bulk superconductors which may have useful high magnetic field properties.*

* See also chapter 12, section 12-5, in this book.

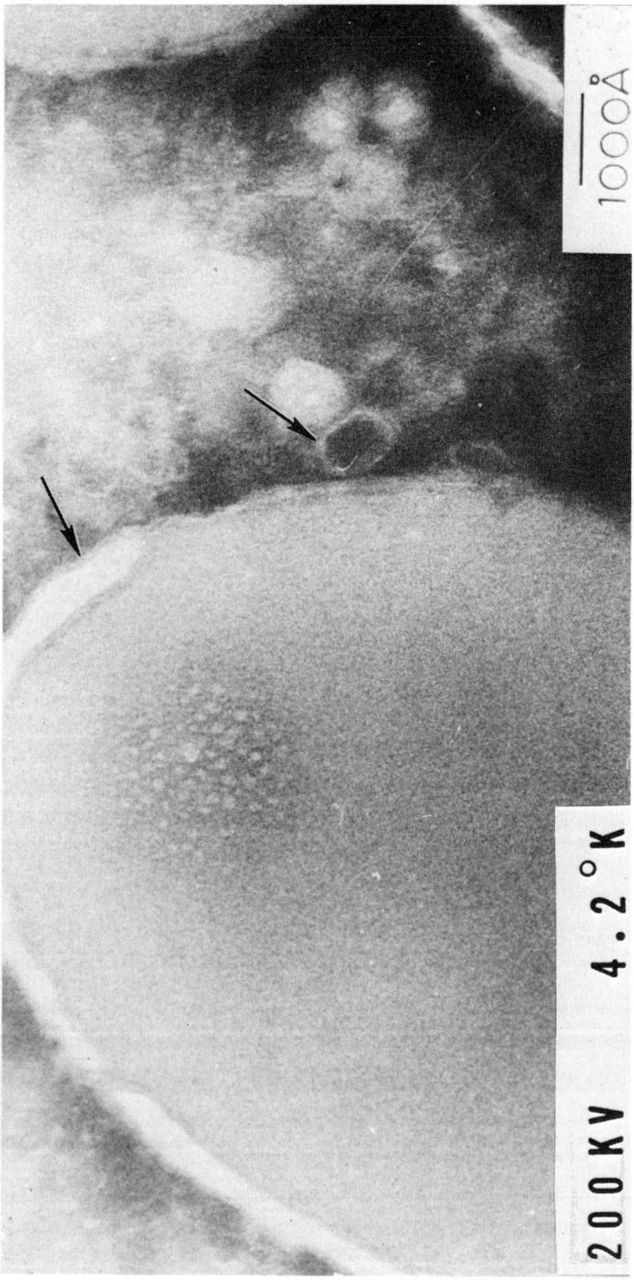

Fig. 9-14. High resolution electron micrograph of hydrated bacteria and bacteriophage examined with 200 kV electron microscope with specimen cooling at 4.2°K in a specially-designed vacuum-tight microchamber. New details of hydrated membrane structure and bacteriophage (arrows) can be observed with resolutions on the order of 10Å to 20Å. Mag. 32,000X.

Extending our previous studies at room temperatures,[28] improved low-temperature instrumentation and preparation techniques have been used to resolve directly significant details of the atomic lattice in the microcrystalline specimens for the first time at 1.8° to 4.2°K (Figure 9-15). At these temperatures characteristic changes have been observed repeatedly in the fine structure of the complexes, and the relationship of these results to the singularly high anisotropic nature of these organometallic superconductors is now being investigated.

Recent work has centered on 2H-NbS_2 and related compounds examined by high resolution electron microscopy and diffraction with specimen cooling at 4.2° to 1.8°K.[29] When the thin layers (200Å to 600Å) of either 2H-NbS_2 or 2H-$NbSe_2$ are imaged with the incident 200 kV electron beam normal to the layers, characteristic doughnut-shaped structures about 100Å to 200Å in diameter are regularly observed (Figure 9-16). These doughnut-shaped structures have a typical fine structure, exhibiting an electron-dense ring 20A to 40A in diameter and a light core (Figure 9-17). In addition, these characteristic ring structures appear only when the specimen is observed at liquid helium temperatures; they disappear as soon as the specimen warms up above 4.2°K and reappear again when the same specimen is cooled down to 4.2°K.

In many areas, the doughnut-shaped structures seem to be arranged in regular patterns, but this is only a preliminary observation. However, in view of the reproducibility of these images and results, recorded with specimen cooling at 4.2°K and in a magnetic field of about 6000 to 8000 gauss, we believe we may be dealing with electron microscope images of fluxoids in the form of quantized vortexes.

Model experiments are now being conducted with evaporated thin films of lead-bismuth, as suggested by Doctors T. Geballe and C. M. Varma. The results of this work may permit us to characterize further the highly anisotropic nature and other properties of these new superconductors.

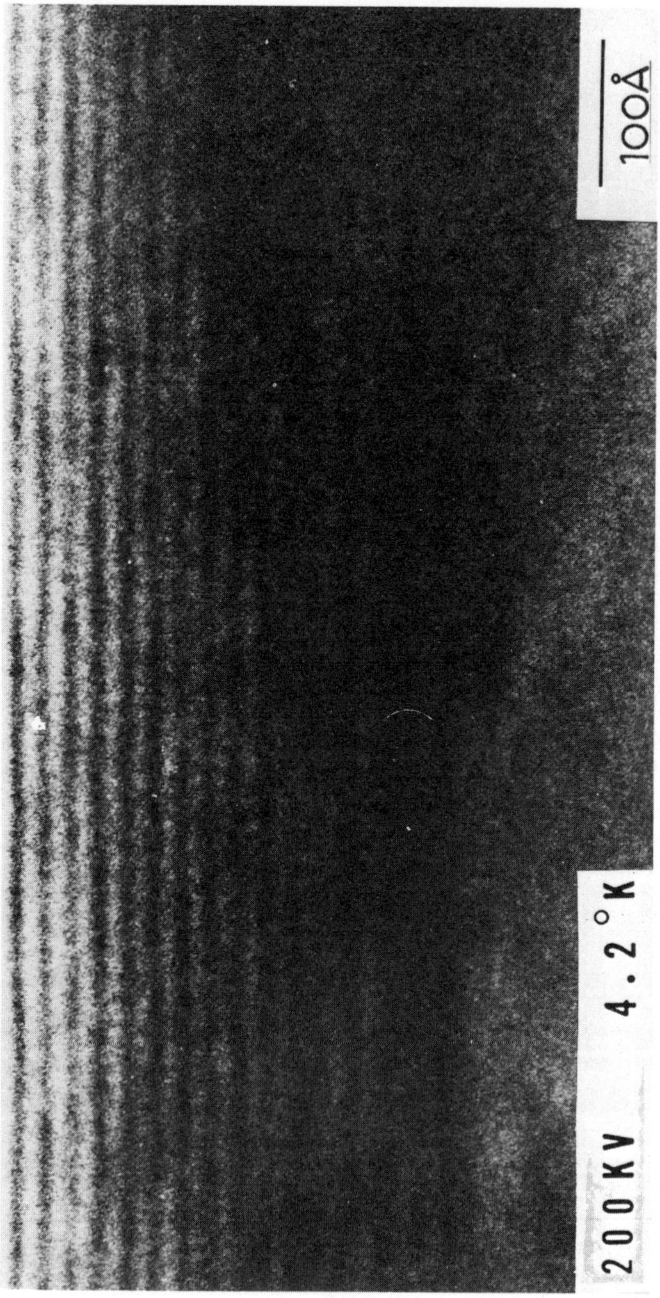

Fig. 9-15. High resolution electron micrograph of $TaS_2(C_5H_5N)_{1/4}$ recorded at 4.2 °K with the 200 kV cryoelectron microscope showing highly ordered, atomically thin, layered structures in suitably oriented crystals. Mag. 2,350,000X.

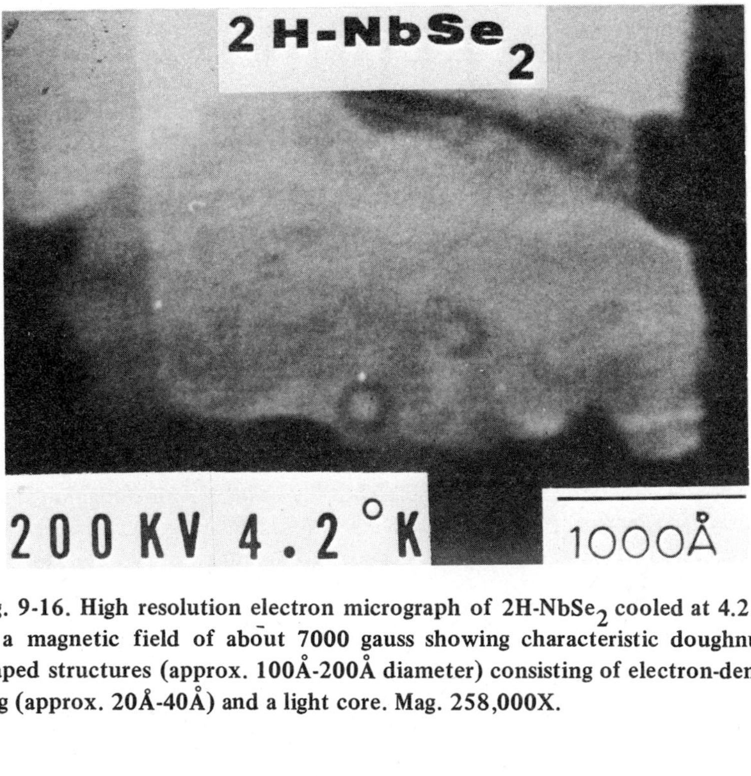

Fig. 9-16. High resolution electron micrograph of 2H-$NbSe_2$ cooled at 4.2°K in a magnetic field of about 7000 gauss showing characteristic doughnut-shaped structures (approx. 100Å-200Å diameter) consisting of electron-dense ring (approx. 20Å-40Å) and a light core. Mag. 258,000X.

9-4. A Look to the Future

Results of our recent work indicate that further development of the superconducting cryoelectron microscope operating at liquid helium temperatures, preferably at high voltages, represents a significant approach to the problems now confronting the biological, biomedical and physical scientists. It provides, in a single system, optimized electron optical conditions with minimized specimen damage and enhanced image quality.

Improved liquid helium cold stage assemblies would enable us to extend the temperature range from about 1° to 0.06°K using

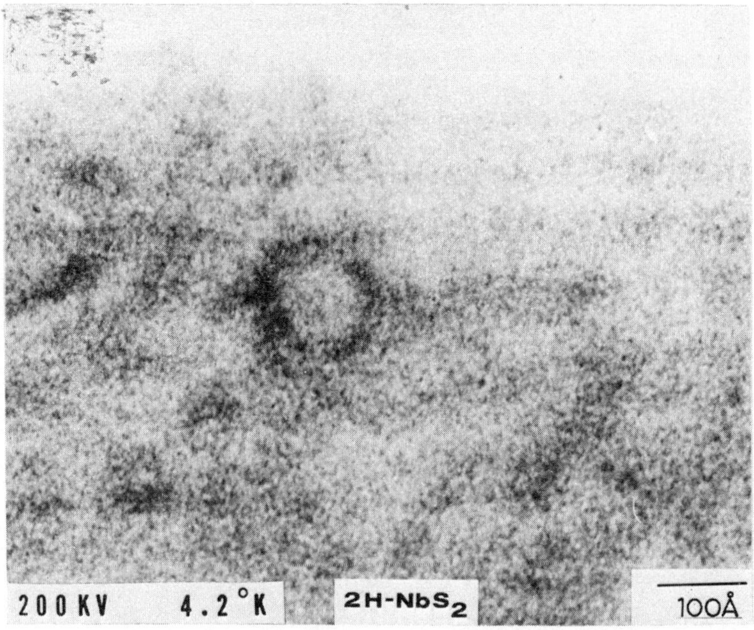

Fig. 9-17. High resolution electron micrograph of 2H-NbS_2 cooled at 4.2°K in a magnetic field of about 7000 gauss showing characteristic doughnut-shaped structures (approx. 100Å-200Å diameter) consisting of an electron-dense ring (approx. 20Å-40Å) and a light core. Mag. 1,200,000X.

$^3He/^4He$ dilution refrigerators. Lower temperatures in the microdegree range could be attained by adiabatic demagnetization of nuclear spins, using high field superconducting magnets of objective lenses as dual imaging components and high magnetic field sources coupled with cryostats containing paramagnetic specimen holders. Such developments could conceivably reduce specimen radiation damage and thermal noise by an order of magnitude.

This logical extension of the demonstrated properties of such cryoelectron microscope may allow researchers to apply techniques of direct electron-optical observation to the visualization of unexplored domains of basic phenomena related to the properties of matter subjected to the ultimate "entropy squeezer." In turn, cryoelectron optics could well become the most refined and sophisticated direct probe into the realm of vanishing entropy—for the physicist and the biophysicist who has yet to fathom living matter confined to the boundaries of the quantum state.

The development of a new cryo-ultramicrotome for operation at liquid helium temperatures and of complementary low temperature preparation techniques have now made it practical to fully explore the application of the inherently high resolving power of the cryoelectron microscope to the study of hydrated biological specimens. With this new instrument, it is also possible to carry out precision machining of metals at 1.8° to 4.2°K, thus eliminating problems of melting introduced at room temperatures.

The establishment of the ability of organisms to survive almost indefinitely at sufficiently low temperatures is of major importance to cryobiologists. If the independence of radiation sensitivity at temperatures of 1.8° to 4.2°K can be demonstrated for various types of cells, the conservation of biological systems at low temperatures would become of key operational significance. It would be feasible to set up storage banks of important cells, maintained at liquid helium temperatures and adequately shielded. Bone marrow, whole blood, spermatozoa, ova, etc. could then be available for effecting critical transplantations in future generations.[3]

Ample experimentation has been conducted with miniaturized electronic circuitry at liquid helium temperatures. The results indicate that this approach extends the range of basic limits of microelectronics at room temperatures, particularly in regard to:

1. Resistive direct current attenuation when dealing with extremely small circuit elements.
2. The thermal activation failures and most importantly the electromigration defects which have now been found to set in, especially in metal lines less than 1 micron wide where typical current densities of about 10 million amperes per square centimeter lead to breakdown of the circuit.

Beyond this, research at liquid helium temperatures is bound to uncover new physical and physical-chemical phenomena which will yield new approaches in fabrication technologies and related systems applications.[30]

This extension of basic cryogenic research into the field of microelectronics is but one aspect of the synergistic interrelationship between systematic low temperature ultrastructural studies and other important fields of investigation. In a recent paper presented to Congress, the author described how our work with the Collins closed-cycle superfluid helium refrigeration system and precision diamond knife cryo-ultramicrotomy may hold important implications for solving the energy crisis which confronts us today.[31]

What is needed now to realize the full potential and benefits to be derived from basic cryogenic research is a closer working relationship between physicists, engineers, biologists and industrial fabricators. In this way, we may hope to make important contributions to all areas of science from the tracking of the cancer virus and its characterization to the construction of sophisticated ultraminiaturized computer systems which begin to rival the human brain in speed, density of information storage and performance.

References

1. H Fernández-Morán, "Applications of Ultrathin Freezing-Selectioning Technique to the Study of Cell Structures with the Electron Microscope," *Arkiv for Fysik,* IV, 3, 471-483, 1952.
2. H. Fernández-Morán, "A Diamond Knife for Ultrathin Sectioning," *Exptl. Cell Res.,* V, 255-256, 1953.
3. H. Fernández-Morán, "Low Temperature Preparation Techniques for Electron Microscopy of Biological Specimens Based on Rapid Freezing with Liquid Helium II," *Annals of N.Y. Acad. Sci.,* LXXXV, 689-713, 1960.
4. H. P. Broida, "Stabilization of Free Radicals at Low Temperatures," *Annals of N.Y. Acad. Sci.,* LXVII, 530-545, 1957.
5. H. C. Box, H. G. Freund, K. T. Lilgia, and E. E. Budzinski, "Magnetic Resonance Studies of the Oxidation and Reduction of Organic Molecules by Ionizing Radiation," *J. Phys. Chem.,* LXXIV, 40-52, 1970.
6. John K. Hulm, Don J. Kasun, and Edward Mullan, "Superconducting Magnets," *Physics Today,* XXIV, 48-57, Aug. 1971.

7. H. Fernández-Morán, "Biological Applications of Magnetic Fields: Cryo-Electron Microscope Using Superconducting Electromagnetic Lenses at Liquid Helium Temperatures," Paper presented at the International Conference on High Magnetic Fields, Massachusetts Institute of Technology, Cambridge, Mass., 1961.
8. H. Fernández-Morán, "General Design Concepts of Cryo-Electron Microscope Using Superconducting Electromagnetic Lenses, Ultra-High Vacuum, Pointed Filament Source, and Improved Image Viewing Devices," (Typewritten), Sept. 18, 1961.
9. H. Fernández-Morán, "Electron Microscopy with High-Field Superconducting Solenoid Lenses," *Proc. Natl. Acad. Sci. U.S.*, LIII, 445-451, 1965.
10. H. Fernández-Morán, "Application of High Field Superconducting Solenoid Lenses in Electron Microscopy," *Science*, CXLVII, 665, 1965.
11. H. Fernández-Morán, "High Resolution Electron Microscopy with Superconducting Lenses at Liquid Helium Temperatures," *Proc. Natl. Acad. Sci. U.S.*, LVI, 801-808, 1966.
12. H. Fernández-Morán, "Electron Microscopy with Superconducting Lenses, *Proc. Argonne National Laboratory High Voltage Electron Microscope Work Shop*, Argonne, Ill., n.p., 1966.
13. H. Fernández-Morán, "Low Temperature Electron Microscopy with High Field Superconducting Lenses," *Proc. 6th Internatl. Cong. EM*, Vol. I, Tokyo, Japan: Maruzen, Ltd., 1966.
14. H. Fernández-Morán, "Electron Microscopy with Superconducting Lenses," *McGraw Hill Yearbook of Science and Technology*, McGraw Hill Book Company, Inc., N.Y., 1966.
15. H. Fernández-Morán, "Electron Microscopy in the Future," *Proc. EMSA 25th Anniversary Meeting*, Claude Arceneaux, ed., Claitors, Baton Rouge, La., 1967.
16. H. Fernández-Morán, "High Voltage Electron Microscopy at Liquid Helium Temperatures," *Microscopie Electronique: Proc. 7th Internatl. Cong. EM*, Vol. II, Grenoble, France, n.p., 1970.
17. H. Fernández-Morán, "Electron Microscopy: Glimpse Into the Future," *Annals N.Y. Acad. Sci.*, CXCV, 376-389, June 1972.
18. H. Fernández-Morán, "Cell Fine Structure and Function: Past and Present," *Exptl. Cell Res.*, LXII, 90-101, 1970.
19. John G. Daunt, "Cryogenic Refrigerators," *Proc. Workshop on Naval Applications of Superconductivity*, NRL Report 7302, 40-61, Naval Research Laboratory, Washington, 1970.
20. Samuel C. Collins, Series of Personal Communications, 1960-1972.
21. A. D. Little, Inc. *Operation and Maintenance Manual: ADL-Collins Helium Liquifier and Related Systems*, ADL, Inc., Cambridge, Mass., 1966.
22. "Science and the Citizen," *Scientific American*, CCXXVII, 46-47, Oct. 1972.

23. S. Ozasa, Y. Kato, H. Todokoro, S. Kasai, S. Katagiri, H. Kimura, E. Sugata, K. Fukai, H. Fuhita, and K. Ura, "3 Million Volt Electron Microscope," *Japanese Journ. E.M.*, XXI, 109-118, 1972.
24. H. Fernández-Morán, "Direct Study of Ice Crystals and of Hydrated Systems by Low Temperature Electron Microscopy," *J. Appl. Phys.*, XXXI, 1840, 1960.
25. H. Fernández-Morán, "Forms of Water in Biologic Systems and the Organization of Membranes," *Annals of N.Y. Acad. Sci.*, CXXV, 739-752, 1965.
26. H. Fernández-Morán, "High Resolution Electron Microscopy of Hydrated Biological Systems," Paper presented at the International Biophysics Congress, Stockholm, Sweden, July 31-Aug. 4, 1961.
27. F. R. Gamble, J. H. Osiecki, M. Cais, R. Pisharody, F. J. DiSalvo, and T. H. Geballe, "Intercalation Complexes of Lewis Basis and Layered Sulfides: A Large Class of New Superconductors," *Science*, CLXXIV, 493-498, 1971.
28. H. Fernández-Morán, M. Ohtsuki, A. Hibino, and C. Hough, "Electron Microscopy and Diffraction of Layered Superconducting Intercalation Complexes," *Science*, CLXXIV, 498-500, 1971.
29. H. Fernández-Morán, "Electron Microscopy and Diffraction of Layered Superconducting Intercalation Complexes at Liquid Helium Temperatures," Paper presented at the Office of Naval Research Conference on the Physics and Chemistry of Layered Compounds, Monterey, Calif., Aug. 17-18, 1972.
30. H. Fernández-Morán, "Microelectronics of Liquid Helium Temperatures," Paper presented at the 18th Annual International Electron Devices Meeting, Washington, D.C., Dec. 4, 1972.
31. H. Fernández-Morán and S. J. Rowe, "New Approaches in Energy Research and Development," *Hearings on Energy Research and Development, Subcommittee on Science, Research, and Development, U.S. House of Representatives,* Ninety-Second Congress, Second Session, No. 24, 461-469, U. S. Government Printing Office, Washington, D.C., May 1972.

Acknowledgments

The research reported here represents a continuous fifteen year effort in which I have been privileged and grateful to work with Professor Samuel C. Collins of M.I.T. and the Naval Research Laboratory, Dr. F.O. Schmitt of M.I.T., Dr. W. Sweet of the Massachusetts General Hospital, Doctors T. Halpern, H. Stanley Bennett, W. Bloom, D. Yada, R. Zirkle and L. Meyer, all associated with the University of Chicago during the course of this work, Mr. R. Szara and his associates of the Low Temperature Department, Dr. D. Cohen of the National Magnet Laboratory, Dr. C. Laverick of Argonne National Laboratory, Messrs. M. Streeter and R. Osburn of Cryogenic Technology, Inc., F. Lins, A. Peuron, D. Kasun and Dr. S. Autler of Westinghouse Cryogenics Division. These men provided helpful suggestions and assistance in the development of improved instrumentation, particularly superconducting lenses, the Collins closed-cycle superfluid helium refrigerator and the integrated cryoelectron microscope.

I am also indebted to Doctors F. R. Gamble of Esso Corporate Research Laboratories, T. Geballe of Stanford University, C. M. Varma of Bell Telephone Laboratories, and I. Dietrich of Siemens Corp. in Munich, Germany for suggestions and discussions relating to organometallic superconducting compounds and associated low temperature electron optical phenomena.

The major portion of the development and research has been carried out in a systematic team effort by the members of our interdisciplinary research group, particularly M. Ohtsuki, C. Hough, H. Krebs, R. Moses, R. Vicario, G. Arcuri, J. Richardson and G. Bowie, together with the skilled men of the Physical Sciences Developmental Workshop. I wish to thank S. Rowe for the compilation and technical editing of this manuscript, S. Erikson for administrative assistance, and the many outstanding men and women who have been associated with our laboratory during the course of this research.

The work described in this paper was carried out under NIH Grants USPHS GM 13243 and USPHS GM 18236, NASA Grants NGL 14-001-012 and NGR 14-001-166, The Spastic Paralysis Research Foundation, and the Pritzker Fund, the Otho Sprague Fund and the L. Block Fund of The University of Chicago.

10

THE STATE OF THE ART IN CRYOBIOLOGY

E. F. Graham

The roots of cryobiology can be traced to the 18th century, but it is only in the last two decades that this interdisciplinary subject has blossomed and begun to earn its inclusion among the sciences. Diverse, and largely empirical, applications of cryobiology range from the mundane (such as frozen foods) to the futuristic (such as suspended animation and "spare parts banks").

This paper urges that as understanding and methods of preservation are improved, a biological-genetic reservoir should be established for all the earth's life. In this way only, can future generations of scientists observe in detail the history of life from our time forth.

10-1. History of Cryobiology

The application of techniques of cryogenics to the field of biology has resulted in a new unique field of study—cryobiology.

Throughout the millenia of time, animal and plant species have had to adapt to "cold." Only in the last few centuries has man suspected that cold could be used to his advantage, and only in the last few decades has man utilized this advantage in biology.

What characterizes these last few decades is the intensification and organization of the research and the accumulation of information on many multiple aspects of biology at low temperature. Although some scattered, independent work had been going on for several centuries in low-temperature biology, it did not evolve as a visible entity until its inception as a dedicated society in 1963. The birth and evolution of this new science has an interesting genealogy. Although cryobiology has well-defined parents of biology and cryogenics, parental guidance has been minimal.

Before a new science can be considered as developed, it should possess certain credentials. These should include a respectable history of serious scientific interest, a well-developed theoretical basis founded in the physical sciences, a substantial system of measurement which can be repeated and confirmed, a relatively standardized methodology, and a hard core of well-trained investigators dedicated to continued development of both the theory and application.

Cryobiology could hardly lay claim to any of these criteria before 1940, with the possible exception of the effect of frost and ice on plants. In 1936 Harvey, in his bibliography on the "Low Temperature Relations in Plants," lists some 4000 titles. Despite historic concern with temperature, early scientific efforts were directed primarily at empirical applications of cold rather than toward a thorough scientific investigation of its biological effects. In the bibliography of the classic volume *Life and Death at Low Temperatures* by Luyet and Gehenio, only twelve references to the preservation of life or the mechanism of injury by freezing are cited prior to 1850, and only ninety-seven prior to 1900. Virtually all of those investigators were concerned simply with qualitative observations of death or survival following freezing. Even well into the twentieth century the many theories of the nature of freezing injury were largely unsubstantiated by experiment and vague even to the point of resorting to the mysterious "direct action of cold." No standardized methodology had been developed for controlling or for measuring freezing and thawing rates, or for quantifying the physical and chemical effects of freezing. The early emphasis was of practical nature and principally concerned with agricultural applications. Other than the pioneering studies by members of the Low Temperature Institute of Cambridge University under the direction of Sir William Hardy, and those of Luyet and his associates in the United States and Nord in Germany, it would have been difficult in 1940 to find investigations directed primarily toward the fundamental questions of biological freezing.

10-2. The Last Two Decades

It would be comforting to report that the increased interest in cryobiology of the last two decades has rectified all these defects, but unfortunately this is not wholly true. Despite the fact that over the last twenty years there have been many symposia, several books, the formation of a Cryobiology Society and publication of a number of related scientific papers, the theoretical basis of freezing injury is only barely beginning to develop. Many theories have been advocated, but those theories still remain diverse, crude and largely unsupported by experimental evidence. Even as basic a parameter as freezing rate still remains undefined and often even unmeasured. The number of investigators primarily concerned with the fundamental nature of freezing injury are still but few.

The possibilities of cryobiology are immense, and empirically devised procedures have proven fruitful in solving specific problems, but little time has been spent on establishing a sound theoretical base that would enhance the field.

The recent surge of interest in cryobiology can be attributed to the discovery by Polge, Smith and Parkes in 1949 of the protective effect of glycerol. From this discovery have come the empirical procedures which have enabled the preservation of a wide variety of cells and tissues.

The theories set forth for freeze damage are vague and in some instances poorly substantiated. The most accepted theories include: (1) mechanical rupture of structural elements through ice crystal growth and recrystallization; (2) physiological differences resulting in denaturation from electrolyte concentration; (3) failure to survive due to osmotic stress; (4) pH changes; (5) dehydration sufficient to precipitate proteins from solution; and (6) the direct effect of the removal of structurally important water. Since the reason for its success is not understood, it was only by chance and error that the cryoprotective qualities of glycerol were rediscovered.

Cryobiology is at an early state in its development, propelled headlong by its applications far beyond the limits of current basic understanding. What cryobiology needs now is the participation of more than the present limited number of investigators working independently throughout different laboratories and departments. In particular, it needs the support of all physical and biological scientists, and specifically of physicists, biophysicists, biochemists, botanists, physiologists, zoologists, molecular biologists and engineers trained and experienced in research at the level of cell function where the effects of freezing become lethal. Secondly, cryobiology needs a long-range goal that will inspire these scientists of diverse backgrounds to draw together and pool their talents and resources.

10-3. Cryogenic Preservation

The keynote in defining cryobiology is "The Preservation of Living Tissue," be it animal or plant, either through freezing, drying or freeze-drying. The basic object in storing living cells is to slow the processes of aging and degeneration. When living cells are cooled, there is a slowing down of the biochemical processes involved in respiration, metabolism and all the other interactions between cytoplasm of the cells and their environment. If they are cooled and stored below -60°C, all chemical changes are either slowed to a minute fraction of the normal rate or are completely arrested. Aging should not occur, and it should be possible to preserve biological materials for indefinitely long periods of time. After rewarming the cells should resume full activity and complete their normal life span—provided they have not been damaged either during cooling, drying or freeze-drying after resuscitation.

The whole concept of suspended animation, the interruption of time by use of low temperature is an exciting one and holds for many of us the same emotional charge that is found in such other examples of science fiction come true as space travel and energy from the atom. But herein lies one of the hazards of cryobiology. The belief that because we can freeze certain single cells carries the implicit implication that complex systems such as tissue, organs or whole animals can be placed in suspended animation. Today, there are organized societies that offer the service of freezing the whole human body after death. The purpose is to be rejuvenated after the cure for the cause of death has been found, for a second chance at life. Membership fees are high. The saving grace of these societies is that one must be dead to become a solid member.

10-4. Diversity of Modern Cryobiology

Cryobiology includes a diversity of study and it is this very diversity of its applications that helps perpetuate its state of undevelopment. Cryobiology means different things to different researchers. To the physicist it means the physics of ice, superstate physics, the study of properties at low temperatures, the dynamics of low temperatures, biophysics, and cryogenics. To the engineer it means hardware, measurement and thermal analysis. To the physical chemist it means phase transitions, gases, pressure, and electrical potentials. To the biochemist enzymatic reaction and protein denaturation. To the crystallographer it means the study of the structure of ice, its formation and re-formation. To the meteorologist it means snow, clouds and the seeding of clouds. To the metallurgist it means low-temperature treatments of metals for fitness and heat transfer. To the geologist it means the study of ice ages. To geography and exploratory science it means polar and high altitudes. To plant and animal physiologists it means acclimatization and preservation of bacteria, protophytes, higher plants, protozoa, metazoa, spermatozoa, ova and zygotes. To the ecologist it means arctic biology. To the cytologist it means tissue culture and electronmicroscopy. To the geneticist it means selection for frost resistance. To the paleontologist it means the study of hibernation. To medicine cryobiology means frostbite, surgery by freezing, hypothermia, the preservation of blood, tissues and organs. To the public in general it means frozen foods.

In spite of this diversity and separatism, many types of organisms have responded to the application of empirical techniques of the freeze-thaw process. These include: (1) at least 144 species of microorganisms (moulds, bacilli, cocci, yeasts and spores); (2) blood cells (erythrocytes, lymphocytes, leucocytes, bone marrow cells); (3) at least 33 species of protozoans, pathogens and parasites;

(4) at least 64 types of eggs (mammalian and insect); (5) embryonic tissue (chick hearts, endocrine tissue, skin, cornea, epithelial cells; (6) whole tissue (cornea, ovarian, testicular, adrenal cortex, thyroid, spleen, aorta); (7) many whole insects; and (8) several whole animals (fish, chicken, hamster) to 0°C or below.

10-5. Preserving the Earth's Genetic Heritage

Man has stored large quantities of "spare parts" for conveniences to recall for his mechanical devices, food stores for his use, materials for his warmth and comfort, and medicine for his ailments. Man has not stored "spare parts" of himself or preserved the vast genetic pool that has evolved and become extinct, nor has there been much effort in the preservation of the genetic pool still available.

The usefulness of the preservation of tissue and organs of man himself are evident and far-reaching. The applications are direct and understandable, therefore require little explanation. Success has been reported on the freezing and storage of blood, lymphocytes, the aorta and heart valves. Research has commenced in the freezing and storage of the heart, spleen and pancreas. Recent work in our laboratory reports success in freezing the canine kidney to -20°C with subsequent transplantation which sustains the life of the dog.

The capturing, harnessing, preserving and recalling, at will, the vast genetic pool of material man needs for his use is less direct. When accomplished, it will allow researchers in all areas of endeavor to select, test and improve the natural resources available at present and in the future. The preservation of germ plasm would allow the ultimate in "genetic engineering" for the study of man's most necessary ingredient, the nutrient.

The study of the nutrient cannot be adequately evaluated without the opportunity of drawing upon a vast genetic pool of different species, varieties or strains of material. The survival of life, therefore, is an intriguing mass of interrelationships that must be studied simultaneously or at least be given the opportunity of being preserved for future use.

The vast majority of all the living things that have evolved on earth throughout its history are now extinct. The genetic instructions that endowed them with a multitude of traits which suited them well for survival in their particular niche in time and space are now lost to the ages.

We are confronted, therefore, with the unavoidable reality that a high degree of resolution in reconstructing the evolution of life on earth is forever out of man's reach. We are now witnessing what is probably the premature extinction, brought about by man and his civilization, of a multitude of species of organisms. Species that appear eminently well-suited for survival, save for the presence of man, are being erased in what normally would have required millennia instead of decades.

Our only recourse is to recognize the scope of our impact, take stock of the value we place on the beautiful diversity of life that evolution has provided, and if possible use our manipulative talents to rectify what we are doing.

If 150 million years ago man could have preserved the germ plasm of our now extinct species, we could recall the Brontosaurus, a plant eater that weighed 70 tons. From 30 million years ago we might give life to the wooly Mastidon that weighed eight tons, of which present day man has not even tasted the meat. In years gone by many species of plants and animals that might have been important to mankind either in production or pleasure are gone.

Technology in cryobiology has now provided us with the tools that will enable us to retain and preserve the genetic instructions for the volumes of life that rise and fall to the music of natural selection or are trampled under the foot of man's civilization.

The idea of preservation of germ plasm is new but old. Man preserved agronomic seeds in Biblical days. The earliest suggestion for practical use of the storage of spermatozoa was made by Montegazza in 1866 on his observation that human spermatozoa survived after cooling and storage at -17°C. He speculated that in the future frozen semen might be used in animal husbandry. Earlier observations, by Spallinzani in 1787, on the effect of cold on spermatozoa may have contained the underlying thought of preservation.

Present day methods employed on low-temperature studies for the preservation of germ plasm include:

1. *Background studies*—(a) detailed study of the physical, chemical and physicochemical properties of the cell and the natural carrying medium; (b) detailed study of the morphology of the cell.
2. *Protective mechanisms*—(a) colligative additives such as the classical alcohols; (b) polymeric additives such as the higher molecular weight compounds PVP, HES, dextrans; (c) Glyco-Lipo proteins; (d) pressure for depressing the freezing point; (e) buffer systems (primarily zwitter ionic) that function at the proper pK; (f) chelating and surface active agents for controlling, stabilizing and destabilizing ions; (g) direct action of ionic strength; (h) freeze rates; (i) thaw rates; and (j) the interrelationship of each on the mass of the sample.
3. *Assessment of cellular damage*—(a) viability (motility, fertility, growth); (b) glycolysis; (c) respiration; (d) chemical leakage (ions, enzymes); (e) enzyme stability; (f) osmotic changes; (g) morphological changes.

Using the above criteria some progress is being made in our laboratory on germ-plasm preservation (Table 10-1) with the goal of establishing a "Library of Life."

With the use of frozen, stored spermatozoa artificial insemination is now a reality in several species. For the preservation of species, we should not be satisfied with the preservation of only the male gamete, but must pursue research on the preservation of the female gamete or the zygote as well.

Efforts on the freezing and thawing of viable eggs has met with less success than those on sperm. This is probably due largely to the much greater volume of the egg and to its content of intracellular water, which forms damaging crystals upon freezing. However, eggs of a number of animals such as chickens, rats, mice and rabbits have been frozen and thawed, with their viability retained.

Table 10-1. Summary of Results After Freezing Spermatozoa to -196°C (University of Minnesota)

Animal Species	*Mean % Motility*	*Comparative Normal Acrosomes*	*Comparative Enzyme Release*	*Comparative Fertilizing*	*Work Commenced*
Bull	55	High	Medium	Normal (72%)	1952
Boar	30	Medium	High	Normal (70%)	1960
Stallion	40	Medium	Medium	Normal (60%)	1968
Ram	60	High	Medium	Medium (68%)	1969
Man	60	High	Low	? (33%)	1958
Goat	55	High	Medium	Medium (68%)	1962
Turkey	35	–	Medium	Low (38%)	1967
Camel	50	High	Medium	NOT TESTED	1964
Moose	55	High	Medium	”	1971
Deer	45	–	–	”	1970
Bison	50	Medium	Medium	”	1965
Llama	55	–	–	”	1969
Yak	45	–	–	”	1972
Monkey	55	High	–	”	1958
Black Bear	30	–	–	”	1966
Chinchilla	65	High	Medium	”	1970
Dog	60	High	Medium	”	1972

In some respects a zygote or fertilized egg is more biologically stable than an unfertilized egg and may be easier to freeze and thaw with retention of viability. The preservation of zygotes would also eliminate the need to provide the special conditions necessary for in vitro fertilization.

At the present time the viable gametes of larger animals require a parent organism in which to develop normally after fertilization has occurred. However, in time this problem will also be solved so that the fertilized egg will be grown in vitro and develop externally. Certainly with the correct amount of emphasis and effort we will be able to simulate the in vivo conditions in vitro (outside of the parent animal) within the next few decades. This technological breakthrough will come too late for many species now near extinction, but their extinction can be prevented if their germ plasm is saved now.

10-6. Conclusions

The benefits of a germ-plasm reservoir would be far-reaching and only curtailed by the lack of imagination and ingenuity of man. Through applications of cryobiology we can give our future generations of scientists the ability to observe first-hand the history of life from our time forth, in every intimate detail.

Cryobiology is a unique new field of science and one which is certain to expand in biology and the physical disciplines in the future. It is a worthwhile area of specialization; the problems and payoffs are both relevant and researchable. Cryobiology is an interdisciplinary field and will not function in any conventional department. Today a rare opportunity exists to provide national leadership in its development, if cooperation, interplay and communications can be obtained by all disciplines involved.

References

1. R. B. Harvey, *An Annotated Bibliography of the Low Temperature Relations of Plants*, Burgess, Minneapolis, 1935.
2. B. J. Luyet and P. M. Gehenio, *Life and Death at Low Temperatures*, Biodynamica, Normandy, Mo., 1940.

3. H. T. Meryman, *Cryobiology,* Academic Press, New York, 1966.
4. C. Polge, A. U. Smith and A. S. Parks, "Revival of spermatozoa after vitrification and dehydration at low temperatures," *Nature* (London) 164:666, 1949.
5. A. U. Smith, *Biological Effects of Freezing and Supercooling,* Arnold, London, 1961.
6. A. U. Smith, *Current Trends in Cryobiology,* Plenum Press, New York-London, 1970.

11

COMMERCIAL USES OF SUPERCONDUCTIVITY: PRESENT AND FUTURE

W. D. Gregory, W. N. Mathews, Jr.,
R. D. Ladd, and D. A. Krieger

Most uses of superconductivity fall into two categories: the very large (e.g. power transmission; magnets for MHD generators, fusion reactors, levitated trains) and the very small (e.g. quantized flux measurement of weak magnetic fields). A modest precommercial market now exists in both areas, but development of a true commercial market for superconductivity will depend on social, economic, and political factors as well as on technical feasibility.

11-1. Background

The information contained in this article has been collected from a large number of individuals and sources too numerous to list in detail. However, there were two ***main*** sources of the material that can be referenced easily. One of these was a two-week intensive course entitled "The Science and Technology of Superconductivity," held August 13-26, 1971 at Georgetown University, Washington, D.C. The twenty-five or so lecturers who contributed to this course wrote up their lectures for a two-volume publication of the same name.[1] Many of the observations made below are the result of both the formal lectures and the informal discussion sessions that were part of this course.

The second source of information was a short study of business opportunities in superconductivity, sponsored by Carpenter Technology Corporation and conducted by the authors through Georgetown Instruments, Inc. During this study, numerous people active in the application of superconductivity to technology were contacted, and the various projects they were involved in were scrutinized. The authors wish to thank all of those who contributed their time and thoughts through both of these sources. Any errors promulgated herein are the fault of the authors and should not be attributed to the people who helped by participating in either the course or the CarTech study.

11-2. Scope, Limits and Boundary Conditions of This Study

At the time of its oral presentation at CRYO '72 this paper carried the title "Commercial Uses of Superconductivity in the Next Decade." However, in the retrospective view offered by publication the revised title appeared more appropriate. As will be seen, most of the paper deals with present noncommercial uses for superconductivity. Precious few commercial uses are predicted within the coming decade.

The bulk of the study on which this paper is based was made during a few months of late 1971. Six workers were involved: half of them with a technical background, and half with business and marketing training. The authors served primarily in the capacity of editors, as this report is more a survey and collection of other people's data than an original investigation.

Because of limitations of time and funds, actual "site visits," i.e. visits to locations, were limited to the United States. A fairly extensive literature search and interviews with persons who had inspected outside the United States had to suffice for data collection regarding foreign projects.

It also should be pointed out that the commercial survey for which most of the formal work on this report was conducted was performed for a specific type of company. This fact both consciously and unconsciously caused us to emphasize the investigation of projects compatible with that corporation's goals and capabilities.

These, then, are some of the limitations. With these limitations in mind, we set certain goals and boundary conditions for the conduct of the study. The following operating principles were adopted:

1. A literature search would be performed for the purpose of collecting as much relevant literature dealing with ***applied*** superconductivity as practical. The search was limited primarily to the past ten years.
2. From the literature collected (and from subsequent interviews) a list of things one could make or do with superconductivity—both commercial and noncommercial—would be compiled.
3. Particularly for those uses with commercial possibilities, further attempts would be made to estimate the present technical and funding status. Discussions and interviews with people involved in relevant development programs would be the primary source of data.
4. In addition, information would be sought during the interviews pertaining to unsolved technical, funding, and sociopolitical problems that might inhibit a given use of superconductivity.
5. The ***need*** for a given technology would be investigated and the solution offered by superconductivity would be compared to nonsuperconducting solutions. (However, because of limitations in time and funds, the alternative technologies would ***not*** be investigated in detail.)
6. Finally, estimates of markets and future R&D funding would be sought from the interviewees, independent of our own analysis.

For the purposes of a public report, an obvious extra condition was necessary. We felt that since both economic and technical information was obtained from the interviewees, it would be desirable not to reference specific details of any given research program. This, we hoped, would encourage more candid and detailed reporting during interviews.

11-3. Results of Literature Search—Some Uses for Superconductivity

With even a preliminary search of the literature, one is struck by the fact that the possible uses of superconductivity fall into two categories: ***large-scale*** systems, such as large magnets, motors and power transmission lines; and ***small-scale*** systems, capable of measuring very small physical quantities, such as Josephson junctions and flux quantum devices. These two categories have also been called "macro-superconductivity" and "micro-superconductivity," respectively, by Professor Donald Langenberg of the University of Pennsylvania. The large-scale uses stem primarily from the ability of superconductors to carry large amounts of current without resistance (zero resistance phenomenon), while the small-scale uses are the result of practical paplication of the Josephson effects and of flux quantization, both effects derived from the wave property of the superconducting state.

Table 11-1 lists only some of the uses of superconductivity, not each and every gadget one can make with a superconductor. Some of the commercially more interesting gadgets will be discussed in detail in the next two sections. In addition, many other uses of superconductivity were identified that did not appear to have a commercial market in the near future. These are listed in Table 11-2.

11-4. Large-Scale Uses of Superconductivity

Magnets. Magnets for laboratory research, some of giant size such as those used in high-energy physics, make up one of only two actually ***existing*** (not projected) markets for superconductivity. (The other market already existing is for sensitive laboratory instruments and is discussed in Section 11-5.) Furthermore, most other large-scale uses for superconductors require a magnet as a prime component, and so it is most natural to discuss magnets first.

Leaving the problems of magnets with special nonsolenoidal configurations for discussion under other headings where they are used, we will look first at the "state of the art" capabilities for

Table 11-1. Some Uses of Superconductivity

Large Scale (*zero resistance*)	*Small Scale* (*Josephson junction and flux quantization*)
Magnets for:	radiation detection and generation
MHD (magneto-hydrodynamics)*	
fusion reactors*	fundamental or very sensitive measurements
motors	of current, voltage, magnetic field,
generators	and magnetic field gradients
levitated trains*	computer elements
Power transmission	

Table 11-2. Some Uses of Superconductivity Not Discussed

radio-frequency circuits ($Q \approx 10^{11}$)
computer elements
nuclear detectors*
temperature standards
radiometers
accelerometers
low-frequency receivers
magnetic ore separation
energy storage
electron microscopy*
epr and nmr magnets
far infrared spectroscopy

*These are topics of separate papers either elsewhere in this book or in volume 4 of this series.

producing simple, high-field solenoids. At the present time, the art is fairly advanced. Many early problems related to magnetic and thermal stability of these magnets have been solved by using superconducting magnet wire containing large amounts of normal metal for stability and taking care in the mechanical design of the support structure for the magnet. Typically, fields of up to 150,000 gauss can be obtained in 1.5-inch diameter bores with Nb_3Sn ribbon. In addition, a magnet has been built using Nb-Ti wire for the National Accelerator Laboratory of 3.6m diameter with a field strength of about 18,000 gauss.

Except for the larger scale uses associated with particle accelerators, it is hard to place dollar values on the research magnet market, since this involves the estimate of a large number of smaller purchases. However, some feeling for the market may be obtained from the fact that ten to fifteen companies worldwide now produce superconducting magnets on a custom basis, with the cost now only 15% to 20% of the "dollar-a-gauss" figure usually quoted when the market first developed a few years ago. With the exception of the larger high energy uses, the yearly sales for these smaller systems is probably of the order of $\$10^6$, almost entirely associated with R&D and with no spectacular growth indicated.

As far as ***d.c.*** magnets are concerned, no really difficult problems remain with existing wire materials, now that stability has been taken care of. As better materials are developed with higher critical fields and currents, a further refinement of the techniques developed so far will undoubtedly be necessary. A more serious problem occurs with ***a.c.*** systems since the motion of flux lines in an a.c. system produces losses and instability problems not encountered in d.c. operation. The a.c. loss problem is not completely understood, much less solved, and will limit uses of magnets until solutions are developed.

MHD and Fusion. A generally accepted estimate made some time ago by the Boston Edison Company places the extra power required in the United States by 1990 (at our present rate of expansion) at 10^9 KVA! Since this would require a complete duplication of our present energy sources and distribution system, there has been obvious interest of late in alternate sources of energy. Two of the more promising new methods involve superconductivity-controlled nuclear fusion and MHD (magneto-hydrodynamic) generators.

The case for fusion generators can be opened and closed rather easily. The principle, of course, is to contain a plasma of hydrogen (deuterium) gas in a "magnetic bottle," with pressure, temperature and density sufficiently high that the nuclei fuse, releasing enormous amounts of energy. The magnetic fields required are such that only superconducting magnets will suffice. In fact, the problems of superconductivity and fusion are so interrelated that the U.S.S.R. has attacked the problem of the use of fusion with a joint program of fusion and superconductivity research. Recently, the conditions necessary for controlled fusion were achieved over reasonable time spans in both Russia and the United States. Nevertheless, realistic estimates place the realization of true long-term controlled fusion sometime after the year 2000. Thus this field cannot be considered a valid market during the next decade, except for possible R&D funding.

MHD generators, on the other hand, are being tested in the U.S.A., Germany, and Russia, and in the next decade we should expect to see some working systems in prototype form at least. Some programs in the United States are expected to be funded in the $\$10^7$ plus range in the near future. Most of the problems associated with MHD are due to the tortured magnet configurations and the basic problems associated with MHD alone. For the ***early*** low-powered ground-based systems, one study performed at M.I.T., sponsored by the Iron and Steel Institute, shows that a need for superconductivity is not indicated, since conventional copper magnets will suffice. However, for later systems and for early airborne systems the favorable power/weight ratio afforded by superconducting magnet systems would win out. The importance of this latter fact for superconductivity is that much of the present development money in MHD is for airborne systems.

Motors and Generators. Even if conventional power sources continue to be used (which they will), the increasing power requirement will place demands on existing generator systems that simply cannot be met with conventional magnets. The reason is that the magnetic fields now generated are as large as conventional copper coils will go, while rotor speed cannot be increased for fear of breakup of the rotor due to stress from centrifugal forces. The larger fields available with superconducting magnets provide an alternative. In the reverse case, i.e. that of motors, the superconducting technology offers a substantial improvement in power/weight ratio for large systems. With it comes considerable flexibility, torque conversion, and efficiency, which are useful in such applications as ship propulsion and power generation. (A group at Westinghouse likes to point out that if one used the large system estimates, a superconducting motor scaled to fit in a station wagon would propel the wagon several hundred miles per hour!)

There is sufficient interest in shipboard propulsion with superconducting motors that funds of the order of $\$10^7$ are being sought for a three-six year development program. At present one ship-based motor system has been operated in England and a number of land-based systems (motors ***and*** generators) have been operated by I.R.D.C. at Frawley in England and at Dr. Joseph Smith's laboratories at M.I.T. in the United States. The I.R.D.C. system was a 2.5 MW d.c. motor. Other systems are under development commercially in the United States and the U.S.S.R.

The major problems associated with motors and generators involve: the use of rotating seals, tight to liquid helium; the need for high current density brushes; the ever present a.c. losses; fatigue and material strength under stress at low temperatures; and the attainment of appropriate speed, voltage, and current conditions on commutators. The next decade appears to be a reasonable time frame for the solution of these problems.

Power Transmission. The energy crisis demanding 10^9 KVA more power by 1990 requires not only more sources of energy and materials for generating but also additional means for transporting such power. The Boston Edison Company Study mentioned earlier indicates that only superconducting transmission lines could carry sufficient current densities to satisfy the projected requirements. A study performed by Linde finds that the original capital cost of 300 to 350 billion dollars to construct conventional systems would be reduced by 50% if superconducting lines were used ***and*** powers greater than about 10^9 KVA were involved. Consequently, programs have been started in the United

States, Germany, Britain, France and the U.S.S.R. to develop superconducting transmission lines. Sections of superconducting transmission line have been tested in the United States, Germany, and England. In England both a.c. and d.c. lines were tested by BICC at 33 KV and 750 MVA. With a serious development effort 10 to 12 years would be required to put a full-scale system into operation.

There are many problems associated with superconducting transmission of power, some of them quite serious. Among these are: the high heat leaks at terminations leading to normal metal conventional user systems, ever-present a.c. losses in a.c. lines, the large-scale dewars and refrigeration systems (such projects as the Stanford superconducting accelerator will help in testing design concepts), a need for expensive redundancy when one system conducts so much power, the thermal expansion and contraction at joints, and the problem of what dielectric to use at the low operating temperatures of the system.

Magnetically Suspended Trains. The need for rapid inter-city ground transportation as an alternative to air travel has prompted a number of developments of train systems with alternates to the usual "wheel-on-rail" suspension system. The need for this is the requirement of relatively high speed (>150 mph) for effective time and cost of such travel. At speeds over 150 mph, wheel systems are unsafe and air cushion or magnetic field support are preferred.

Two types of superconducting magnetic suspension systems are under investigation: the so-called null flux, and conducting sheet designs. Both require a superconducting magnet on the train. In addition, the null flux design requires a (conventional) magnet in the track, so arranged that the repulsion of the fields produces a stable suspension. Track tolerances are a bit severe in this design, but there are relatively few losses at high speeds. On the other hand, the conducting sheet design has no magnet coils in the track (which may be somewhat cheaper) and has lower tolerances for track dimensions. The "opposing" field is produced by the eddy currents in the track caused by the moving train magnet.

An obvious disadvantage is the resulting Joule heating loss of energy in the track.

In terms of cost, the larger payload allowed by the conducting sheet design somewhat favors this method. Even if a 300 mph system were used (half speed vs. jet aircraft) the cost/passenger would be one quarter of air travel for a savings of greater than two in cost/hour of travel.

At present, working models of superconducting magnet suspended trains of both designs have been tried in both Germany and Japan. Design studies are in progress at several places in the United States. The cost/year of the overseas programs is about $\$4 \times 10^6$. A full test at the United States test site at Pueblo, Colorado, would involve an $\$80 \times 10^6$ program.

11-5. Small-Scale Uses of Superconductivity

The discovery of flux quantization and the verification of the predictions of Brian Josephson concerning the behavior of weakly linked superconducting systems have produced one of the most intense spurts in basic research in recent years. The result has been a wealth of superconducting devices capable of measuring extremely small physical quantities: magnetic fields as small as 10^{-16} gauss, voltages as small as 10^{-14} V; and RF fields with detector noise equivalent powers of 10^{-15} watts/$\sqrt{\text{Hz}}$. Devices with these capabilities and more can now be purchased in the United States from about four suppliers with a backlog of orders of about $\$10^6$. The long-range commercial prospects of these devices is not that interesting if they are restricted ***only*** to laboratory use, since that market will soon saturate. However, if such uses as the superconducting gradiometer for magneto-cardiograms or for naval surveillance develop into viable techniques, the dollar value of the market could be considerably greater.

11-6. Discussion and Evaluation

A summary of our findings in terms of dollars to be spent in superconductivity in the next decade is shown in Figure 11-1. We note that this estimate is quite tenuous and the extension of

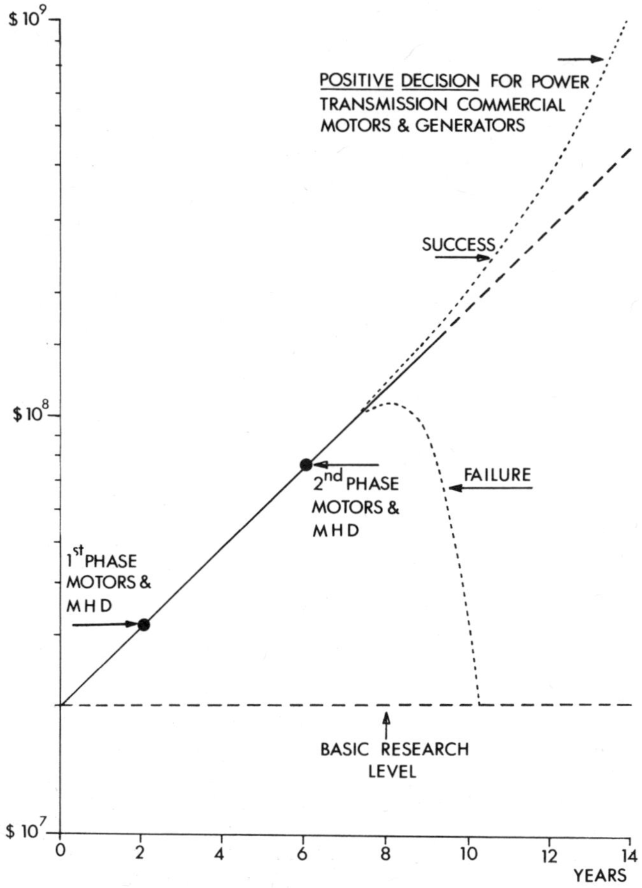

Fig. 11-1. Projected U.S. spending on superconducting systems for the next decade. The year 0 is late 1971. Success and failure as indicated in the figure refer to the fate of the large motor and MHD projects.

the curve beyond the next few years depends on the success of many projects, the largest of which are indicated on the graph. Perhaps more substantial are the qualitative conclusions we reached, as follows:

1. Almost no commercial uses of superconductivity presently exist.
2. There is a small market (somewhat depressed in late Spring 1973) for superconducting instruments.

3. There is a modest materials market (R&D).
4. There is a modest magnet market (closely linked with materials).
5. Despite many scattered "Engineering" programs, truly commercial prototypes are some years away.
6. If everyone obtains and keeps the funding they want, and if programs are on schedule, then the money involved in R&D represents a sizeable precommercial "market" for the next decade.
7. Unless developmental programs are accelerated, we expect that mainly materials companies, small instrument companies, magnet companies, and firms with engineering R&D capacities for large-scale uses will attract superconductivity money in the next decade. Only after this coming decade, and only if feasibility and sociopolitical factors are removed, will a true commercial market for superconductivity exist analagous to the semiconductor market.

It is particularly important that the "sociopolitical" factors are not ignored. For example, it is quite nice to postulate a high-speed train with a superconducting magnet suspension, but quite another thing to obtain the "right of way" required to make it a reality. In the opinion of the authors, the levitated train is the large-scale use of superconductivity most likely to succeed technically, but unfortunately it has a low probability of success because of these "sociopolitical" factors. The future of superconductivity, then, depends as much, and more, on nontechnical developments as on the true technical feasibility.

References

1. W. D. Gregory, W. N. Mathews, Jr., and E. A. Edelsack, eds., *The Science and Technology of Superconductivity* (2 vols.), Plenum Press, New York, 1973.

12

RESEARCH FOR NEW SUPERCONDUCTING MATERIALS

F. J. Di Salvo

The past history of superconducting materials is reviewed to provide an overview of the "chemical-metallurgy" approach to finding useful superconductors. While this approach has been entirely empirical, recent work is being based on more solid "theoretical" grounds. Recent theories elucidate the properties of the normal state metals that are desirable for the production of high transition temperature superconductors. While this is a large step forward, prediction of the transition temperature for a specific material is still almost impossible. Although these advances have caused some excitement in the research field, there is currently nothing to suggest that the standard "workhorse" materials, such as Nb, Nb-Ti, Nb_3Sn, will be replaced by newer ones in the near future.

There has been much controversy about the possible existence of new mechanisms for superconductivity. The electron phonon interaction appears to be responsible for all the known superconductors; the other proposed mechanisms have been generally called "exciton mechanisms," and it has been proposed that this mechanism could lead to high transition temperatures. In the last few years a class of layered materials has been discovered which may be suitable for an experimental test. Although to date no evidence for this new mechanism in organic-layered material complexes has been found, experiments in this field are continuing.

This paper is divided into two parts. The first part concerns superconductors as we presently know them—more or less from a historical perspective. The second part deals with current thoughts about the possibility of new mechanisms for superconductivity, higher transition temperatures, and materials that might show these new mechanisms.

PART I

12-1. Needs and History

Certainly the reader is aware that materials with higher transition temperatures would make possible a number of engineering advances. For instance, being able to operate a superconducting transmission line at 15° or 20°K in cold helium gas or perhaps liquid hydrogen, would considerably reduce the cooling costs compared to those at 4.2°K. Another possible advance that could occur even in presently known materials is to raise the operating critical current density in the superconductor to its theoretical limit (i.e., above 10^7 amp/cm^2 in high T_c materials). Indeed, even with current technology, superconductors appear to be breaking into the field of power generation and may also be used in power transmission and in transportation. Of course the metallurgical and engineering problems in these ventures, or for that matter in putting new superconductors to commercial use, are formidable; but it is assumed (perhaps naively) that these problems can be handled if new materials are found.

A quick sketch of the progress in the field of superconductivity, particularly in superconducting materials, is shown in Table 12-1 where the year of discovery, the material and its transition temperature where appropriate, are shown. In 1911 Kamerlingh Onnes discovered superconductivity in a sample of mercury at 4.15°K. In 1929 Meissner discovered that the compound CuS, made from two nonsuperconducting elements, is superconducting, in this case at 1.5°K. Many more examples of nonsuperconducting elements combined to form superconducting compounds are now known. The other compounds shown are intermetallic compounds that contain both a transition element and a nontransition element. Notice that in the last twenty years the maximum observed transition temperature has increased about 3°K to 21°K. This peaking out in T_c is particularly striking when one considers that most of the research for superconducting materials, as indicated by publications or dollars spent, has occurred in the last ten or fifteen years.

Table 12-1. History of Superconductivity

Year	*Material and* T_c	*Discoverer*
1911	Hg 4.15 °K	K. Onnes[18]
1929	CuS 1.5 °K	W. Meissner[19]
1930	NbC 10.5 °K	Meissner & Franz[20]
1941	NbN 15 °K	Ascherman et al.[21]
1953	V_3Si 17 °K	Hardy & Hulm[22]
1954	Nb_3Sn 18 °K	Matthias, Geballe et al.[23]
1954	Phenomenological theories	Ginzburg & Landau[24]
1957	BCS microscopic theory of superconductivity	
1961	Nb_3Sn superconducting at high fields	J. E. Kunzler et al.[25]
1967	$Nb_3(Al, Ge)$ ~21 °K	Matthias, Geballe et al.[26]
1970	Beginnings of large-scale industrial uses	
?	Higher T_c's?	?

12-2. Classes of Superconductors

Superconducting materials are conveniently classed into four groups: elemental materials, nontransition metal alloys or compounds, transition metal alloys or compounds, and finally compounds containing both transition elements and nontransition elements. Each of these groups shall be treated in turn.

Elements. In Figure 12-1 the known superconducting elements are shown along with their transition temperatures. Those elements which are superconducting under high pressure are indicated with horizontal hatching. The various groups of metals previously mentioned are shown here: the transition metals occupy the middle of the periodic table, columns IIIB through VIIIB; the rest are nontransition elements. In the polyvalent nontransition element series, columns IIIA through VIIA, there are a number of semiconductors that become superconducting at high pressure. At these high pressures the materials may undergo several phase changes, and if the new state is metallic it is always found to be

TABLE OF PERIODIC PROPERTIES OF THE ELEMENTS

IA	IIA	IIIB	IVB	VB	VIB	VIIB	VIII	VIII	VIII	IB	IIB	IIIA	IVA	VA	VIA	VIIA	O
Li	Be — 6.0			H (IA)					He (O)			B	C	N	O	F	Ne
Na	Mg											Al 1.19	Si — 7.1	P — 5.4	S	Cl	Ar
K	Ca	Sc	Ti .40	V 5.30	Cr	Mn	Fe	Co	Ni	Cu	Zn .875	Ga 1.08	Ge — 5.0	As — .5	Se — 6.8	Br	Kr
Rb	Sr	Y — 3.0	Zr .546	Nb 9.26	Mo .92	Tc 8.0	Ru .49	Rh	Pd	Ag	Cd .54	In 3.4	Sn 3.7	Sb — 2.6	Te — 3.3	I	Xe
Cs — 1.5	Ba — 5.0	La 4.8 6.0	Hf .16	Ta 4.48	W .012	Re 1.7	Os .66	Ir .14	Pt	Au	Hg 4.15	Tl 2.38	Pb 7.2	Bi — 7.25	Po	At	Rn
Fr	Ra	Ac															

Ce — 1.7	Pr	Nd	Pm	Sn	Eu	Gd	Tb	Dy	Ho	Er	Tm	Yb	Lu — 0.7
Th 1.37	Pa 1.4	U .68	Np	Am	Pu	Cm	Bk	Cf	Es	Fm	Md	No	Lw

Fig. 12-1. Elements of the Periodic Table—T_c is given for the superconductors in °K. Those elements superconducting at high pressure only are shown with horizontal hatching.

superconducting. The element with the highest transition temperature is Nb at 9.2°K. Including those under pressure, almost half of the elements are known to be superconducting, and probably more will be discovered at higher pressures and/or very low temperatures.

Nontransition Metals. The nontransition metal alloys and compounds seem to be well understood in terms of a modified BCS theory proposed by McMillan.[1,2] The highest transition temperature achieved for these alloys is approximately 9°K for Pb-Bi as predicted from McMillan's theory.[3]

Transition Metals. Alloys and compounds formed only from transition metals are moderately well understood in that empirical relationships between the electron per atom ratio and the transition temperature are usually obeyed. However, all the parameters necessary to compare with McMillan's theory have not been experimentally determined. The maximum transition temperature observed for these systems is about 15°K (for $Tc_{.75}Mo_{.25}$).[3] An example of the electron per atom rules proposed by B. T. Matthias[4] is shown in Figure 12-2, kindly supplied by J. Hulm of Westinghouse Laboratories. This is a three-dimensional plot of the transition temperature vs. alloy composition for the body-centered cubic alloys of group IVb, Vb, VIb, and VIIb. Ridges or peaks in T_c occur at 4.7 and 6.7 valence electrons per atom.

Several authors have pointed out that superconductivity is prevalent in those materials having certain crystal structures, most of them cubic.[5] This, however, may be due to the fact that almost all metallic binary alloys or compounds form in structures of high symmetry. Metals of low symmetry, say triclinic or monoclinic, are now frequently being discovered among ternary or higher order compounds, and their properties are just being studied.

Transition-Nontransition Compounds. All the known high temperature compounds ($T_c \gtrsim 15$°K) contain both transition elements and nontransition elements. Until recently, all known high T_c compounds had cubic crystal structures (the NaCl, β-W and Th_2C_3 structures). In this last year, however, B. T. Matthias and coworkers found high temperature superconductivity in a class of ternary compounds discovered by Cheverel et al. These have a complex rhombohedral structure.[6,7] The highest T_c reported for this new class is somewhat above 15°K for $PbMo_6S_8$. It is not understood why both transition and nontransition elements seem to be necessary in high T_c compounds, and again there is only empirical evidence, not solidly based theoretical evidence, that this is so.

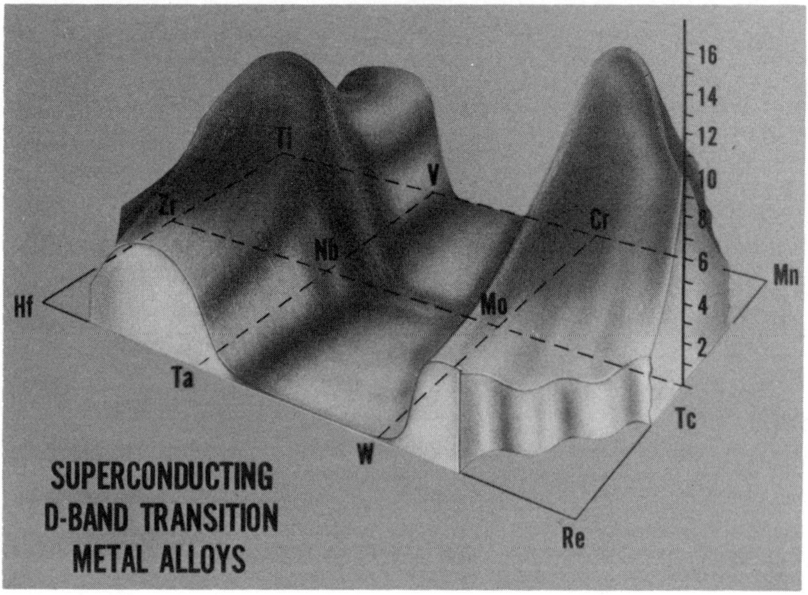

Fig. 12-2. T_c vs. alloy composition for the body-centered cubic materials of group IVb, Vb, VIb, and VIIb.

12-3. Electron-Phonon Mechanism of Superconductivity

In all superconductors, it would appear that the superconductivity is caused by the electron-phonon mechanism. The free electrons in the metal become weakly bound to each other in pairs, called Cooper pairs. A very simplified picture of how this mechanism works is illustrated in Figure 12-3, kindly supplied by Professor W. Little of Stanford University. An electron traveling through a lattice of positive ions attracts the ions to itself by electrostatic interactions. These ions respond slowly, so that by the time the ions have moved together the electron has moved away.

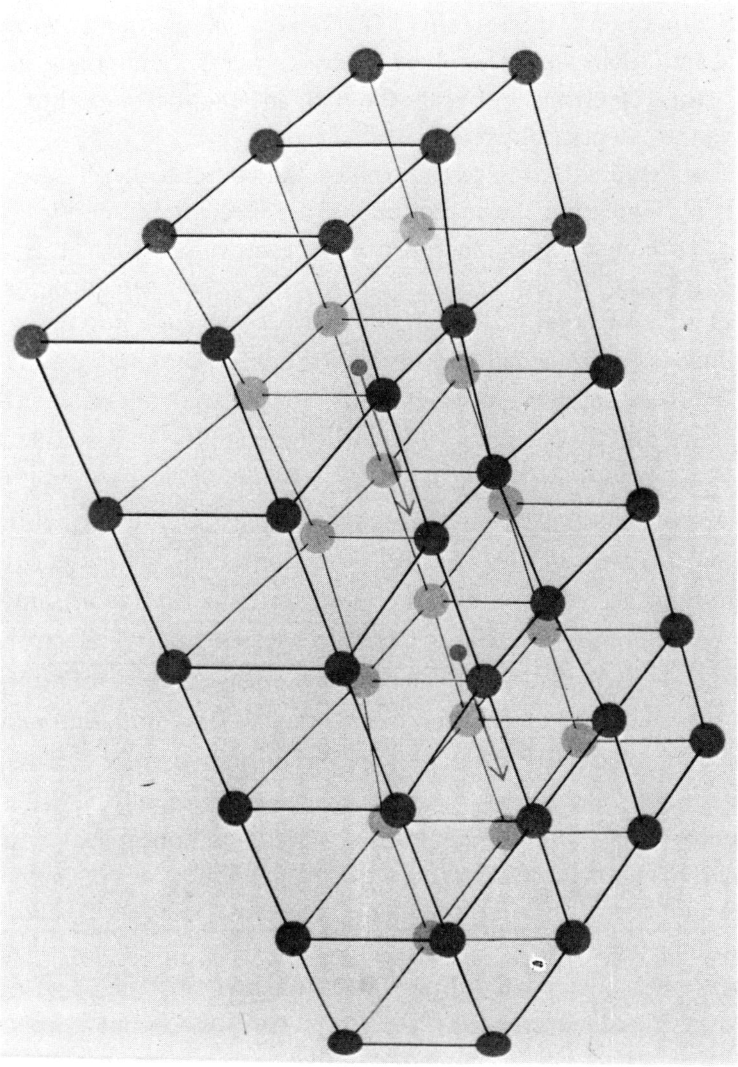

Fig. 12-3. A schematic picture of an electron interacting with a second electron via a lattice distortion—this is the basis of the electron-phonon coupling mechanism.

A second electron sees this distorted positive charge and is attracted to it. In this way the first and second electron are coupled together. The lattice distortion produced can be built up (or Fourier synthesized) from a series of sound waves or phonons (thus "electron-phonon" interaction). Obviously the picture is more complicated than indicated here since in a real metal there are many other electrons between the two interacting ones, but it conveys the essential physics.

Now we might attempt to produce higher and higher T_c compounds by increasing the magnitude of the electron-phonon interaction. (Of course experimentalists do not know how to do this in the general case.) In order to increase this interaction the ions must be freer to move; that is, the lattice must become soft. But as such a soft state is approached, the crystal structure is not stable and it spontaneously changes, or transforms, to a new structure that is not so soft, and T_c decreases! This lattice instability, as it is called, appears to be present in most, if not all, high T_c compounds.[8,9] An example that may be familiar is the martensitic transformation observed in many of the high T_c β-W compounds, where at low temperatures the usual cubic structure distorts to a tetragonal one. A particular compound that is unstable because of large electron-phonon effects is usually difficult to prepare in the laboratory. Special annealing processes may be necessary; the compound may form off stoichiometry, or may not form in the structure that was expected at all. Present work in these systems is usually aimed at controlling or "fooling" the instabilities. It is hoped that with more complex structures, perhaps of low symmetry or containing a large number of different elements, some increases in T_c above 12°K will be achieved.

Although these instabilities are being studied, I know of no great successes in overcoming their effects. Thus I would expect that the materials you are already familiar with, such as Nb, Nb-Ti alloys, Nb_3Sn and perhaps Nb_3Ga, will continue to be the standard working materials in applied superconductivity, at least for the foreseeable future.

PART II

12-4. Alternative Mechanism for Superconductivity?

For the past ten years, there has been talk about organic or other exotic superconductors that might show a new mechanism of superconductivity other than the electron-phonon mechanism, and that it might lead to very high T_c's.[10,11,12] While these theories have been rather hotly disputed and defended, there have been few experimental searches for these new mechanisms, because no convenient experimental system or material has been available to test them. The objections to these new mechanisms, known as exiton mechanisms, involve the occurrence of new kinds of instabilities.[13] For example, the one-dimensional organic molecule chains proposed by Professor W. Little may be unstable toward forming insulating ferroelectrics. Since the details of new mechanisms or possible instabilities are not understood well enough to predict accurately maximum possible transition temperatures, and since no other mechanism other than the electron-phonon mechanism is known, it would be interesting to determine if these new mechanisms exist at all.

12.5. Transition Metal Dichalcogenides

A class of compounds that may be suitable for this search has been discovered recently.[14,15] What has been done with these materials, and what is planned for the future is discussed in the remainder of this paper.

Materials. The new materials are called transition metal dichalcogenides; a chalcogenide is one of the elements S, Se, or Te. These compounds form in a layered structure as is illustrated in Figure 12-4, where an edge view of the layers is shown. The layers are three atoms thick, the top and bottom sheets of the layer are chalcogenide atoms indicated by X's, and the middle sheet is occupied by transition metal atoms indicated by M's. The layers are very weakly bonded to each other by van der Waals forces. The bonding is similar to that in graphite or mica—thus these crystals easily cleave parallel to the layers. At least 50 of these compounds have been prepared and studied, and their electrical properties range from insulating materials to superconducting metals.

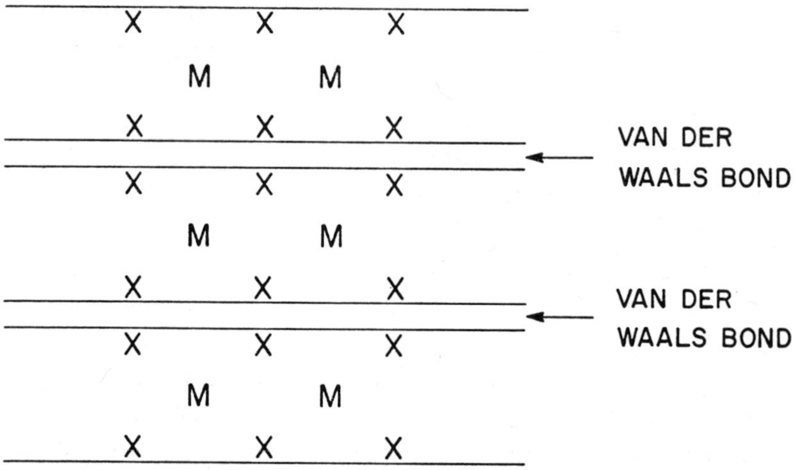

Fig. 12-4. An edge view of the layered structure of the transition metal dichalcogenides. The layers are three atoms thick and held together only by van der Waals forces.

Properties and Preparation. To illustrate the physical and chemical properties of these compounds, only one of them, tantalum disulfide (TaS_2) is discussed. The resistivity (ρ) of the metallic TaS_2 hexagonal crystals is quite anisotropic. ρ is ten times smaller parallel to the layers than perpendicular to them [ρ (parallel)=1.5 x 10^{-4} ohm cm], also TaS_2 is superconducting (T=0.8 °K).

Although the physical properties of TaS_2 itself are quite interesting, the chemical properties provide us with a way to make complex metal-organic structures. Since the forces holding the layers together are so weak, the layers can be forced apart chemically to accept guest atoms or molecules. This process, called intercalation, produces an alternating structure of TaS_2 layers and organic layers as illustrated in Figure 12-5. The 6Å thick layers of TaS_2 remain continuous and highly ordered. These layers are considerably thinner than continuous films produced by evaporation or other techniques. In Figure 12-6, an intercalation compound of TaS_2 with a long chain molecule is shown, the TaS_2 layers are separated by over 50Å. Although we have not attempted to produce larger interlayer separations, we foresee no experimental difficulties in producing virtually any interlayer separation.

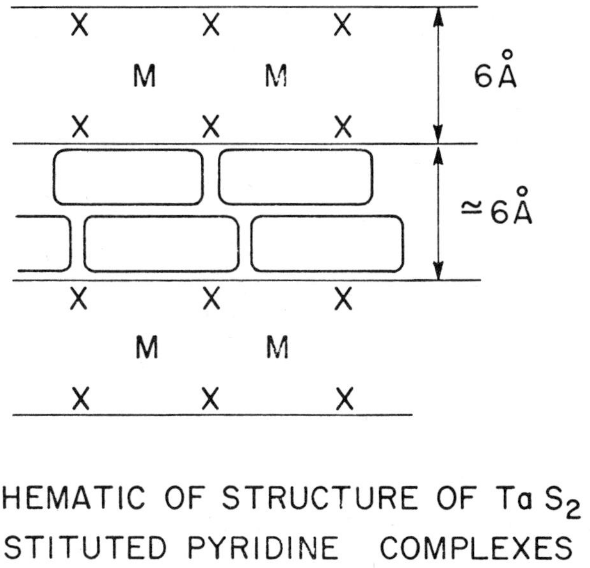

Fig. 12-5. A schematic of the intercalation compounds of TaS_2 with a ring-like molecule such as pyridine.

It is rather easy to prepare these intercalation compounds, as illustrated for TaS_2 (pyridine)$_{1/2}$ in Figure 12-7. The structure of pyridine is shown. It is a flat ring molecule similar to benzene, but a C-H group of benzene has been replaced by a nitrogen. When TaS_2 is sealed in a tube with an excess of pyridine and heated to near 200°C, the TaS_2 will quickly abosrb enough pyridine to form the compound TaS_2 (pyridine)$_{1/2}$. The structure of this compound is similar to that shown in Figure 12-5, where the flat ring molecules lie flat against the TaS_2 layers and a bilayer of pyridine forms. Even in an excess of organic, no more pyridine is absorbed by the TaS_2. In general the organic complexes have chemical formulas that can be written TaS_2 (molecule)$_{1/n}$, where n=an integer, indicating that stoichiometric compounds are formed.

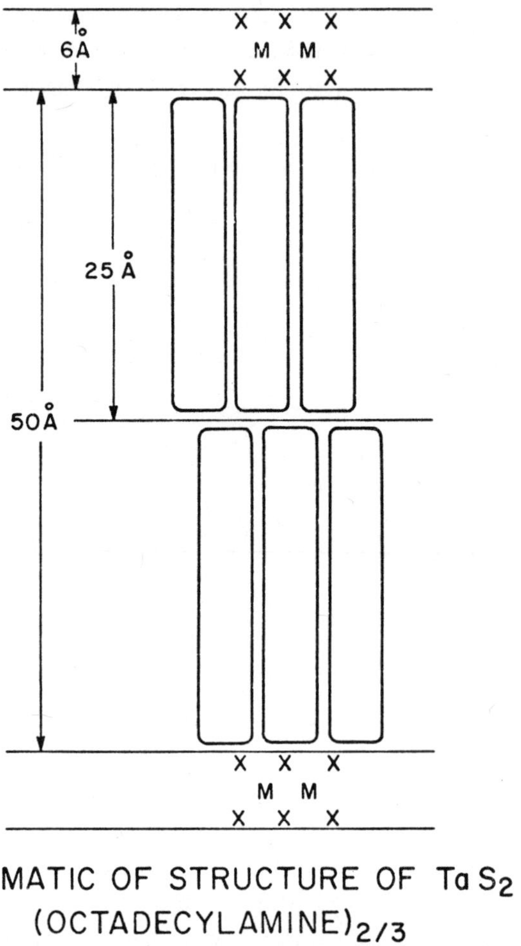

Fig. 12-6. The structure of TaS_2 (octadecylamine) showing the large interlayer spacings already produced.

Superconducting Behavior. The organic intercalation compounds of TaS_2 are superconducting, as is the starting TaS_2 material, with T_c's on the order of 3° to 4°K depending upon the molecule intercalated. Heat capacity measurements in Figure 12-8 for TaS_2 (pyridine)$_{1/2}$ show that the superconductivity is a bulk effect.[15] Also shown is the inductive transition, and it is apparent that the anomaly in the heat capacity and susceptibility occur over the same broad temperature interval.

$$\text{Ta S}_2 + \text{PYRIDINE (liq)} \xrightarrow[\approx 1/2 \text{ HR}]{200\,^\circ\text{C}} \text{Ta S}_2 \text{ (PYRIDINE)}_{\frac{1}{2}}$$

$$\text{TaS}_2 \text{ (MOLECULE)}_{\frac{1}{n}}$$

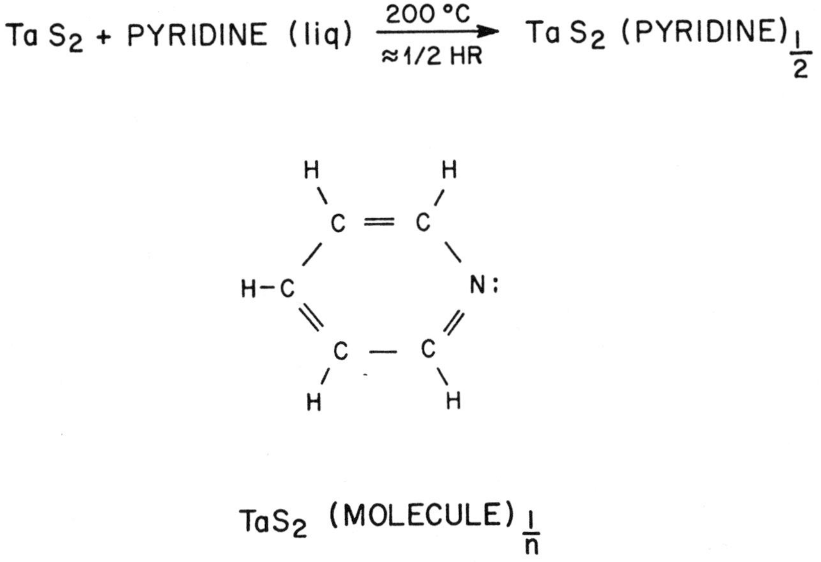

Fig. 12-7. The structure of the molecule pyridine is shown along with the conditions used to prepare the compound TaS_2 (pyridine)$_{1/2}$.

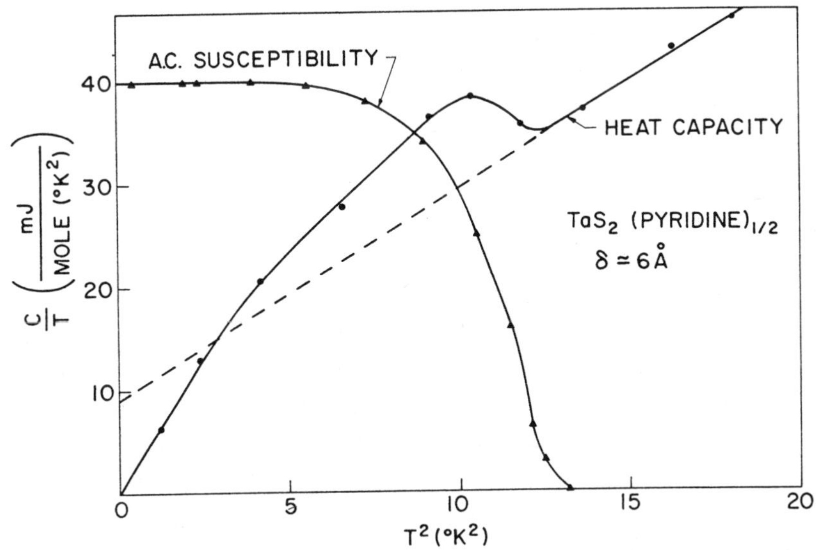

Fig. 12-8. The superconducting anomaly seen in the susceptibility is also seen in the heat capacity over the same temperature interval in TaS_2 (pyridine)$_{1/2}$, showing that the superconductivity observed is bulk and not due to minor impurity phases.

The critical field properties of these compounds are anisotropic as expected from the structure, but the critical fields are quite high, especially considering their low T_c's. Data on TaS_2 (pyridine)$_{1/2}$ with the field parallel and perpendicular to the layers have been obtained by S. Foner at the National Magnet Lab at MIT. Parallel to the layers the critical fields exceed 100 KG just a few degrees below T_c.[17]

12-6. Other Intercalation Compounds

Although several hundred of these compounds have been prepared, we have no reason at present to believe that the superconductivity in these compounds is caused by a new mechanism. Although there are a large number of organic molecules between the layers, their major role appears to be just to separate the layers. Most proponents of the exiton theories, however, would have stated previous to these studies, that not just any organic molecule would do; that exotic molecules such as dyes were needed. Perhaps it is appropriate that exotic molecules are necessary for exotic mechanisms.

At present attempts are being made to intercalate molecular systems that have low lying excited electronic states, such as dyes, to study the coupling of the electrons to these excitations. It may be that these systems will not lead to high T_c's, but they should lead to an understanding of the effects of metals upon organic systems. Also interest is directed to studying the properties of two-dimensional metallic systems and how the effects of interlayer coupling perturb the two-dimensional properties.

12-7. Summary

Hopefully, this paper has indicated the current trends in research for new superconductors, instabilities, new mechanisms and higher transition temperatures. Again it should be emphasized that, although much effort and hope characterize superconductor research, higher T_c's by new mechanisms or otherwise are not presently forthcoming from the laboratory. For one involved in this research it is easier to be somewhat pessimistic about predicting higher T_c's, while still hoping for them; for if the author is wrong in his pessimism it will be a pleasure to admit it. However, being overoptimistic and wrong is a sin much harder to live down.

References

1. W. L. McMillan, *Phys. Rev.* **167**, 331, 1968.
2. R. C. Dynes, *Solid State Comm.* **10**, 615, 1972.
3. B. W. Roberts, Nat. Bureau Stds. Tech. Note 482, 1969.
4. B. T. Matthias, *Prog. in Low Temp. Phys.,* North-Holland Pub. Co., Amsterdam 2, 138, 1957.
5. B. T. Matthias, T. H. Geballe and V. B. Compton, *Rev. Mod. Phys.* **35**, 1, 1963.
6. R. Chevrel, M. Sergent, and J. Prigent, *J. Solid State Chem.* **3**, 515, 1971.
7. B. T. Matthias, M. Marezio, E. Corenzwit, A. S. Cooper, and H. E. Barz, *Science* **175**, 1465, 1972.
8. L. R. Testardi, *Phys. Rev.* **5**, 4342, 1972.
9. A. C. Lawson, *Mat. Res. Bull.* **7**, 773, 1972.
10. W. A. Little, *Phys. Rev.* **134**, A1416, 1964.
11. V. L. Ginzburg, *Soviet Phys. JETP* **47**, 1964.
12. D. Allender, J. Bray and J. Bardeen, to be published *Phys. Rev.,* 1972.
13. M. L. Cohen and P. W. Anderson, *Proc. of Conf. on Superconductivity in d- and f-Band Metals,* Ed., D. H. Douglass, AIP Pub., 1972.
14. F. R. Gamble, J. H. Osiecki and F. J. Di Salvo, *J. Chem. Phys.* **55**, 3525, 1971.
15. F. R. Gamble, J. H. Osiecki, M. Cais, R. Pisharody, F. J. Di Salvo and T. H. Geballe, *Science* **174**, 493, 1971.
16. F. J. Di Salvo, R. Schwall, T. H. Geballe, F. R. Gamble, and J. H. Osiecki, *Phys. Rev. Letts.* **27**, 310, 1971.
17. S. Foner, E. J. McNiff, Jr., A. H. Thompson, F. R. Gamble, T. H. Geballe and F. J. Di Salvo, *Proc. of Conf. on Low Temp. Physics* **13**, *1972, to be published.*
18. H. K. Onnes, *Akad. van Wetenschappen* **14**, 113 and 818, Amsterdam, 1911.
19. W. Meissner, *Z. Phys.* **58**, 570, 1929.
20. W. Meissner and H. Franz, *Z. Phys.* **65**, 30, 1930.
21. G. Aschermann, E. Friederich, E. Justi and J. Kramer, *Z. Phys.* **42**, 349, 1941.
22. G. F. Hardy and J. K. Hulm, *Phys. Rev.* **89**, 439, 1953.
23. B. T. Matthias, T. H. Geballe, T. H. Geller and E. Corenzwit, *Phys. Rev.* **95**, 1435, 1954.
24. P. G. De Gennes, An excellent review of the results of the Ginzburg-Landau theory in *Superconductivity of Metals and Alloys,* pub. W. A. Benjamin, Inc., 1966.

25. J. E. Kunzler, E. Buehler, F. S. L. Hsu and J. H. Wernick, *Phys. Rev. Letts.* **6**, 89, 1961.
26. B. T. Matthias, T. H. Geballe, L. D. Longinotti, E. Corenzwit, G. W. Hull, R. H. Willens and J. P. Maita, *Science* **156**, 645, 1967.

13

ENGINEERING FEASIBILITY OF A FUSION REACTOR

Roger W. Boom, Gerald L. Kulcinski,
Charles W. Maynard, and William F. Vogelsang

Potentialities and design problems of a Tokamak fusion reactor are discussed, using a 1000 MW plant as an example. Immense superconducting magnets are required to confine the reacting plasma to its toroidal shape. New structural materials are needed, especially to contain the neutron moderator in a severe temperature and radiation environment. Maintenance requirements must be minimal, since the structure will be radioactive for years, perhaps decades after shutdown.

13-1. Introduction

Preliminary design studies aimed at assessing the feasibility of producing power from fusion reactions have made considerable progress in the past few years. A number of groups have pursued the evaluation of various proposed devices. The authors are members of a team organized at the University of Wisconsin at Madison to study a system known as a Tokamak as a possible power reactor. While many of the problems are common to all approaches to achieving fusion power, this paper is restricted to Tokamaks since these are the systems with which the authors are most familiar.*

Rather than attempt a comprehensive discussion of all aspects of a power Tokamak, this paper presents a brief overall description of the system, followed by a more detailed discussion of four main problems of particular interest or importance for a Tokamak.

*For a different approach, see paper No. 2 in this book; and for a more general introduction to fusion power experiments, see ref. 9. See also paper No. 22 of Volume 4.

A Tokamak is a device which contains the fusion fuel in the form of a high-temperature plasma in a toroidal region. The plasma acts as a single turn secondary of a transformer. A Tokamak is a low beta device, which means that the kinetic pressure of the particles making up the hot plasma is small compared to the pressure associated with the confining magnetic field produced by suitable coils. Many necessary laboratory studies of plasma containment may be made without the reacting fuels needed for power generation.

The plasma in the power device is expected to be a mixture of deuterium and tritium, since the fusion reaction sought is ${}_1H^2 + {}_1H^3 \rightarrow {}_2He^4 + n$. (The neutron leaves with about 14 MeV energy; the *a* particle, about 3.5 MeV.) While deuterium occurs as a small fraction of the hydrogen in water, tritium does not appear in nature and must be produced by nuclear reactions. The only practical reactions for this purpose are neutron reactions with the two isotopes of lithium, viz:

$$ {}_3Li^6 + n \rightarrow {}_2He^4 + {}_1H^3 $$

$$ Q + {}_3Li^7 + n \rightarrow {}_2He^4 + {}_1H^3 + n $$

Since neutrons are produced in the fusion reaction between deuterium and tritium, the possibility exists of producing as much or more tritium than is consumed. This tritium production is usually referred to as the breeding of new fuel, and calculations indicate that it should be possible to produce the replacement fuel required to operate the system. Tritium breeding requires that lithium in some form appear in the region surrounding the plasma in order to utilize the neutrons from the fusion process. The energy produced by the fusion reactions is partly absorbed in the plasma but most of it is carried to the first material wall of the system by the neutrons and by electromagnetic radiation. This energy heats the walls and the lithium regions so that cooling is necessary. Cooling can be provided by the lithium itself either in the form of a liquid metal or as a constituent of a molten Li-Be-F salt. It is also possible to use helium as the coolant, in which case the lithium need be circulated only enough to remove from it the tritium that is being produced. The lithium bearing region in which most of the energy is deposited is referred to as the blanket. In order to keep the plant efficiency at a satisfactory level, it is necessary to use superconducting magnets to achieve the high magnetic fields required. This in turn demands good radiation shields to prevent excessive energy deposition in the magnets from cuasing either damage to the magnets or an excessive heat load for the cryogenic refrigeration system.

The power reactor will need also a number of auxiliary systems to start up operation, provide fuel, handle the exhaust, and recover the tritium. Figure 13-1 shows schematically some of the regions and components described above. This is a cross-sectional view (see also Figure 13-8). The plasma is located in the central region of the "doughnut," confined to a part of the total space within the first structural wall. A very low pressure sheath known as the divertor region occupies the volume between the plasma and the first wall. Collision processes in the plasma allow some ions and electrons to diffuse into the divertor region where they follow the magnetic field lines through the slots and into a wetted lithium wall from which both the energy and unburned fuel can be recovered. In the system shown here the coolant is liquid lithium, which serves also as the fuel breeding medium. Behind this blanket is a shield, which in a representative case might consist primarily of steel with some boron carbide. The large "D"-shaped region contains one of the superconducting magnets and its dewar. Other features shown in Figure 13-1 are vacuum pumps at the ends of the divertor slots and the divertor coils which produce fields to guide the particles through the divertor slots. The ends of these windings are shown between the magnet "D" and the shield.

Operation of the system begins with a gaseous fuel mixture in the confinement chamber. Windings in the central core of the torus are energized to induce a voltage in the fuel region which will ionize the gas and cause a current flow in the fuel. The current heats the fuel and completes the ionization.

The plasma may be heated further by injecting more fuel with particle energies that correspond to temperatures well above

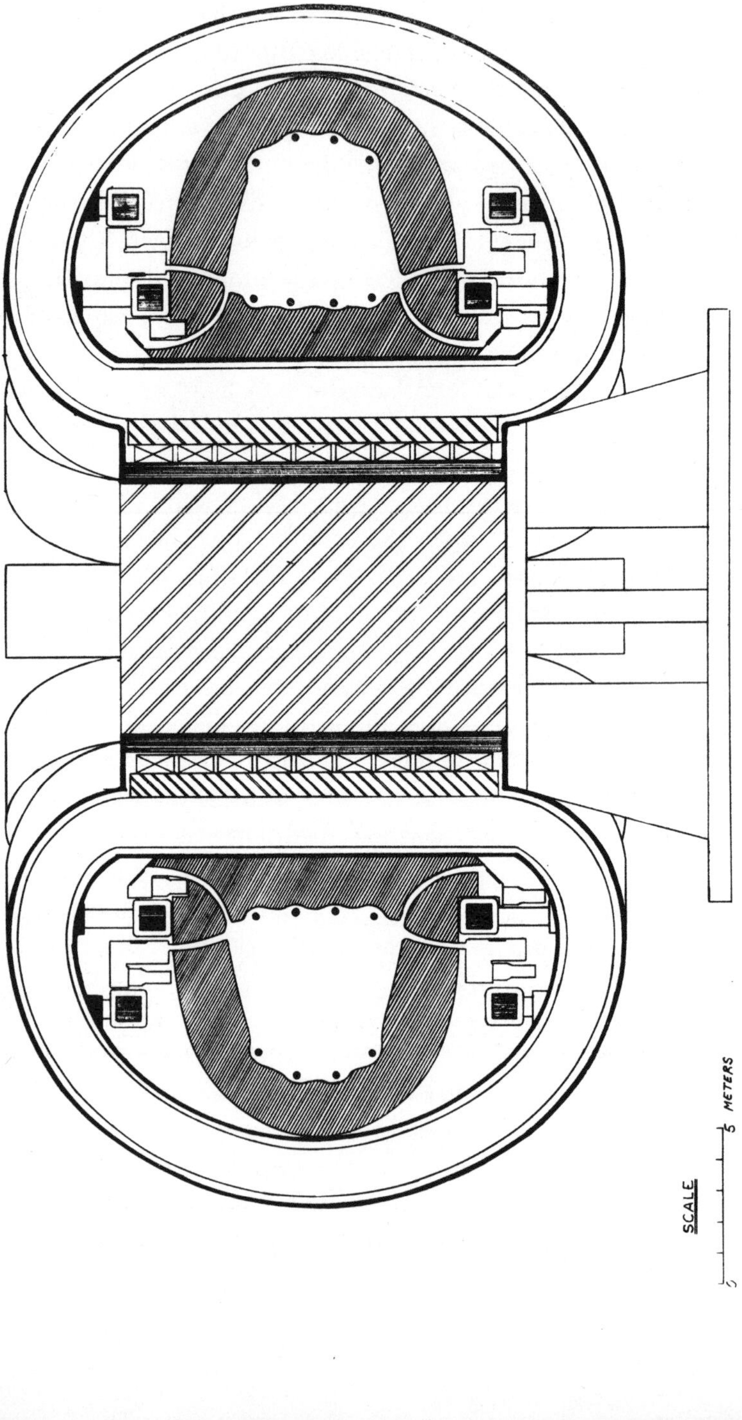

Fig. 13-1. Schematic View of a Portion of a Tokamak Fusion Reactor

the ignition temperatures, that is, well above the temperature at which heating of the plasma by fusion reactions exceeds the cooling by radiation losses and particle leakage. Since this injected fuel used for heating comes from outside the plasma, the fuel atoms cannot be ionized if they are to cross magnetic field lines and reach the plasma zone. Therefore, outside the magnetic field, a fuel mixture is ionized and accelerated to high energy (>25 kev). This fuel beam is then passed through a charge exchange cell to convert it to a neutral beam which can cross the magnetic fields and reach the plasma zone. The beam is, of course, ionized in the plasma by collisions, which also distribute the energy of the beam.

Beyond the ignition temperature of around 6 kev (a plasma temperature close to 50×10^6 degrees Kelvin), the alpha particles from the fusion reactions are sufficient to heat the plasma to a suitable operating point. The fuel and fusion product helium nuclei leak out into the divertor and are removed from the system, while the corresponding fresh fuel is continuously injected to replace the losses. Adjustments must be made in the system to create a steady-state operating condition.

The four main problems to be discussed here are the following: (1) Those aspects of the plasma which are of particular interest and importance in a power reactor assuming that confinement of the plasma is assured from the results of the experiments currently being planned. (2) The severe materials problem anticipated in the first wall, where the radiation and thermal environment is possibly the worst ever seriously envisioned for a real material. This poses the most serious challenge to the engineering feasibility of fusion power. (3) Induced radioactivity in the reactor components. As this is by no means negligible, it is important to have an early knowledge of the main features of the problem. (4) Finally, the superconducting magnets, which make the largest contribution to the costs of a fusion plant. The distinctive features of these units which are two hundred times larger than any now in existence, are discussed.

13-2. Plasma Considerations

The important problems in plasma physics up to the present have been concerned with confinement of the plasma, elimination of unexpected instabilities, and establishment of basic loss mechanisms and rates. From our present point of view, plasma stability will be assumed and loss rates will be extrapolated on the basis of data presently available.

If a magnetic field level and confinement geometry and size are chosen which seem suitable and attainable, then one can work out the temperature and fuel density for a steady state operating point. This is obtained by finding the operating parameters for which energy insertion into the plasma fusion reactions from alpha particles balances the losses by radiation, neutron transport, and charged particle leakage. With no impurities and with confinement according to present scaling laws this balance occurs at either very low temperature (<10 kev ion temperature) or very high temperature (>50 kev), and at relatively low fuel density. The resultant fusion power density is so low as to be totally unacceptable. The power is limited because a fraction of the fusion energy goes into the plasma, and only as many fusion reactions can be allowed as can be balanced by the loss mechanisms of the plasma.

To obtain acceptable power levels, the loss mechanisms for removing energy from the plasma much be enhanced. This is most readily achieved by adding impurity atoms having a relatively high atomic number to the plasma. This increases the electromagnetic radiation from the plasma and allows the plasma temperature and density to increase. The result is an increase in the fusion rate needed to attain a new energy balance. The operating temperatures move toward a more acceptable range, but this scheme gives rise to another problem. The fractional burn up of the fuel and the helium content of the plasma become too high (the large He content limits the amount of fuel which can remain in the plasma). To allow more fresh fuel to be added and to exhaust the system properly confinement time must be reduced; that is, the mean confinement time of the particles must be reduced relative to that which is predicted by present plasma scaling laws. It may be possible to do this by introducing intentional flaws in the magnetic field. Thus by controlling the impurity concentration and the fields, the power level for steady-state operation can be adjusted.

A knowledge of the necessary adjustments to obtain a desired operating point is only a beginning in controlling the system. One must have information about the thermal stability of the reactor. In the language of the reactor engineer, one must know the temperature coefficient of the power. A positive temperature coefficient results in an unstable operating point and will require a feedback control system for operation. A negative coefficient minimizes the control problem since any small perturbation of the power tends to be compensated for automatically. Our studies to date indicate that it is possible to obtain in many situations two operating points, one stable and the other unstable. There are, however, some drawbacks to the stable operating point.

The unstable operating point typically occurs at an ion temperature of around 14 kev and is rather insensitive to the system geometry. The stable equilibrium point is usually at about 30-40 kev but it is more sensitive to other reactor parameters. Generally, the cost per unit power produced will be greater for the stable than for the unstable point, often by a factor of two. In addition, it is not at this time evident that a negative power coefficient would make the control system appreciably simpler than for the unstable case because the negative temperature coefficient is small. Thus, the choice of an operating condition is still open.

To minimize the cost per unit power produced one must carry out an optimization study subject to a number of constraints imposed by the plasma and other parts of the system. For example, a Tokamak is a low-β device. Requiring β to be lower than a specific value requires large (>40 kilogauss) magnetic fields on axis. But other considerations of the system, such as the maximum radiation load to the magnets, set limits on the maximum field attainable. Preliminary results indicate that the ratio of the major radius to the minor radius of the torus should be between 2 and 2.5, or as close to these values as possible, subject to being able to put start up transformer coils and the main magnet windings in the central hole of the torus. Generally, one should use the largest magnetic field possible subject to the constraint that the power load on the first wall not exceed materials limitations. Fortunately, the cost per unit power is relatively insensitive to the trade-off between maximum field and system size, and this is the main degree of freedom available for optimizing. The choice of magnetic field and major radius of the torus can therefore be based on materials more than economic considerations.

13-3. Requirements for Materials

Once the technical problems of confinement are solved, it will be necessary to design safe, efficient and economical reactor structures to convert kinetic energy of the plasma reaction products into heat. There have been many conceptual designs to accomplish these goals, and the materials requirements for all of these proposals can generally be categorized into the following five areas:

1. Satisfactory fabricability and high temperature mechanical properties.
2. Acceptable compatibility with the coolant.
3. Reasonable neutronic characteristics.
4. Satisfactory radiation damage resistance.
5. Acceptable costs and availability to U.S. markets.

Obviously, one cannot treat all of these areas in depth here and the reader is referred to reviews of these areas.[1-3] A few points of each of the above areas are discussed below.

The shear size of the neutron moderation and absorption region in fusion reactors requires that very careful attention be paid to minimizing the amount of structural material surrounding this zone. Also, the desire to lower the thermal discharge rates and to increase thermal efficiencies drives the operating temperature upward. The object is then to find materials that have the best mechanical properties at the desired temperatures in order to reduce the wall thicknesses and hence the total amount of material. Figure 13-2 shows some typical stress rupture values for materials that have been mentioned for controlled thermonuclear reactor

(CTR) application. Typical stress levels of 10-20,000 psi will limit the useful operating temperatures to less than 650°C for 316 stainless steel, 800°C for vanadium and its alloys, 1000°C for niobium and its alloys, and 1200°C for molybdenum and its alloys. Unfortunately, the best material from the standpoint of high temperature strength, molybdenum, has the worst fabrication characteristics. The best material from a fabrication standpoint is stainless steel. Both of these factors have to be weighed when contemplating the construction of a vacuum chamber which has an inner diameter on the order of the diameter of the fuselage of a 747 aircraft.

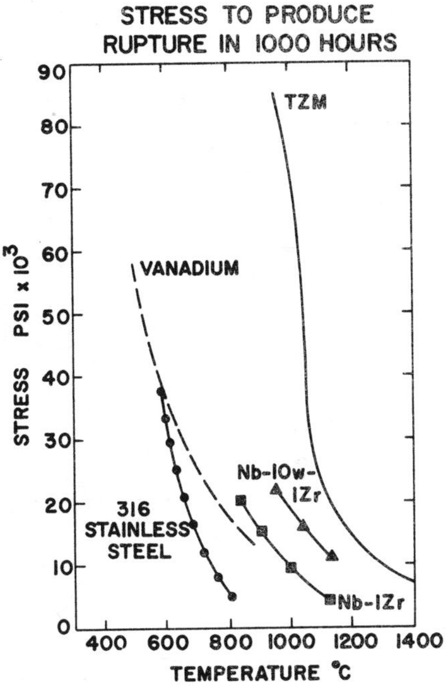

Fig. 13-2. Stress Rupture Values for Possible Fusion Reactor Materials

The chemical compatibility of the aforementioned materials with potential CTR coolants (i.e. lithium, helium or Li-Be-F Salts) is an important consideration. For example, Figure 13-3 shows the useful temperature ranges in which CTR materials can be used with respect to CTR coolants. The maximum temperature from a mechanical property standpoint is also given as a dashed line. Note that 316 stainless steel can be used up to its mechanical property temperature limit with lithium salts and helium, but that it must be limited to less than 500°C in dynamic (circulating) lithium. Both niobium and vanadium are extremely susceptible to interstitial impurity pick up at high temperatures. Thus unless the oxygen level can be kept below one atomic part per million in a helium coolant, neither vanadium nor niobium can be used above 600°C.

Molybdenum is not subject to this limit because oxygen is relatively insoluble in this metal. The situation improves when Li-Be-F salts are used. The limits become 700°C in vanadium, 800°C in niobium, and 1000°C in molybdenum. Finally, all three of the refractory metals can be used close to their mechanical property temperature limits when lithium is used as a coolant.

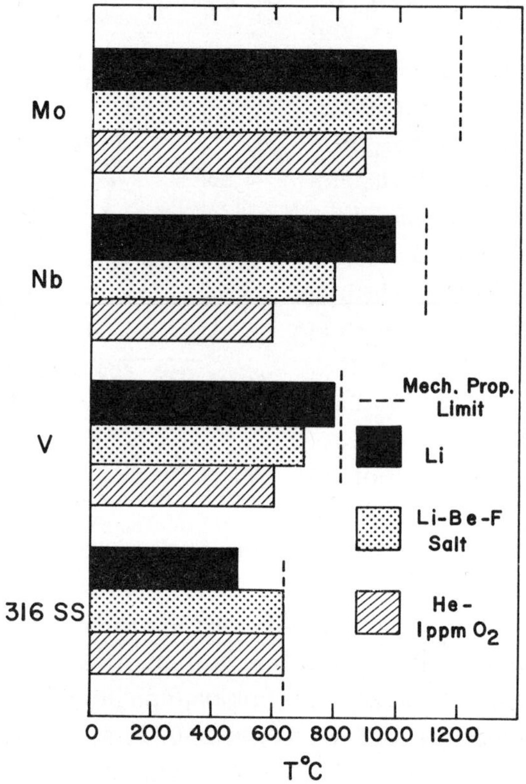

Fig. 13-3. Useful Temperature Ranges of Fusion Reactor Materials in the Presence of Possible Coolants

The necessity for tritium breeding requires that the amount of parasitic neutron capture by the structural materials be minimized. It would also be advantageous if the structural material has a high (n,2n) reaction cross section to compensate for the nonproductive absorption of neutrons. These considerations tend to favor, in increasing order, stainless steel, vanadium, niobium and molybdenum. However, thus far the neutronic differences between the various materials have not been great enough to clearly favor any one material. The exceptions to this statement are radioactivity and afterheat problems which will be considered later.

The most severe problems envisioned with CTR materials, namely radiation damage, are basically connected to the process of slowing down the 14 Mev neutrons. These neutrons lose energy to the atoms in the structure and coolant, and these atoms in turn are propelled through the material displacing other atoms. In metals the vacant lattice sites remaining after the atoms are displaced have enough energy to migrate and collect into three-dimensional aggregates called voids. These voids can cause the metal to expand, and large volume changes may take place. The amount of swelling anticipated in stainless steel[4] is plotted in Figure 13-4 as a function of time, for a reactor which has been recently proposed. Note that data are available for only five years of projected operation. Various theories predict either a saturation of swelling at 15% or continued swelling that may approach 100% in 10 years. Not only will such expansions be difficult to control by themselves, the swelling gradients set up in the blanket between regions of different temperatures and neutron exposure will induce tremendous stresses. It is expected that all materials will show somewhat the same behavior, but there has been some recent

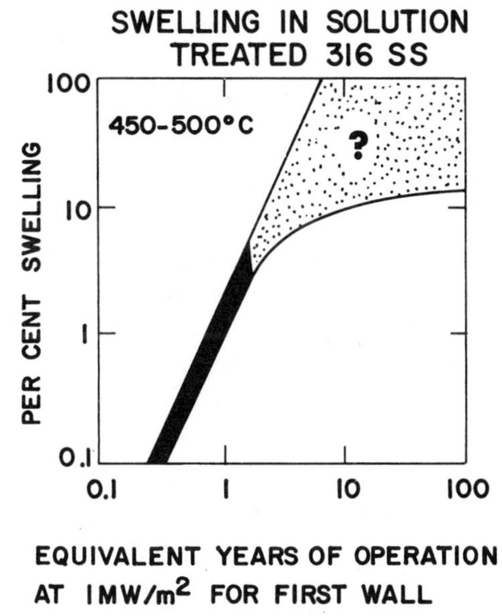

Fig. 13-4. Anticipated Swelling of 316 Stainless Steel with Time in a Fusion Reactor. Dark area represents available data.

evidence[4] that metallurgical treatments such as cold working could alter the final result. Vanadium-titanium alloys have shown some swelling resistance.

Embrittlement is perhaps an even more serious problem from a radiation damage standpoint than is swelling. Figure 13-5 shows, again for 316 stainless steel in a particular reactor design, the variation of uniform ductility and yield strength expected during operation of a CTR. Unfortunately, experimental data[4] are again available only for the first two to three years of projected operation. It does indicate, however, that less than 1% uniform elongation can be tolerated after that time. The increase in yield strength is also quite dramatic in the first few years, but it will probably saturate thereafter at four times the unirradiated value. The main cause of the embrittlement is felt to be due to the helium generated by (n,α) reactions in the metals. The situation is particularly aggravated in a CTR (as opposed to fission reactors) because of the much larger gas production cross sections for 14 MeV neutrons emitted from a D-T reaction. Niobium and vanadium have the lowest helium production cross sections of the materials considered for CTR application; stainless steel has the highest. Much more work is needed to clarify the role of helium in embrittlement before one can be sure that a CTR vessel will retain enough ductility over its lifetime to insure safe start-up and shut-down of the reactor.

The problem of material cost can be appreciated when it is realized that approximately 5000 metric tonnes of structural material alone will be required for the construction of a 1000 MW plant (excluding the magnets). A large portion of this mass can be stainless steel because it will be at relatively low temperatures and in low irradiation fields. However, as much as 1000 metric tonnes per plant will be in the critical areas of the CTR where refractory metals might be used. Multiplying this by a thousand plants (to produce 33% of the projected electrical power required by the year 2020) predicts requirements on the order of 10^6 metric tonnes.

Such numbers represent only 10% of the world's reserves of niobium, vanadium or molybdenum, but they also represent 10 times the known U.S. reserves of niobium and vanadium.

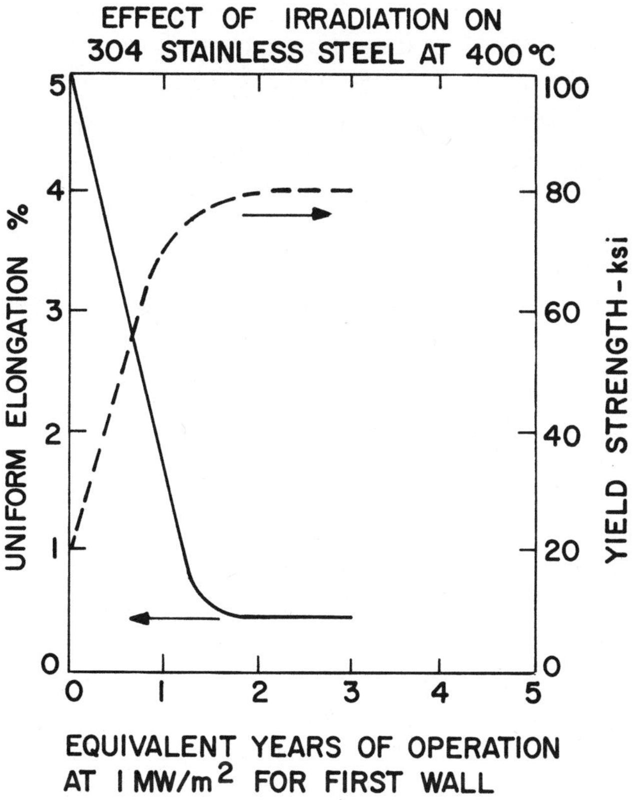

Fig. 13-5. Variation of Uniform Ductility and Yield Strength of 304 Stainless Steel with Time in a Fusion Reactor

Fortunately, only 20% of the known U.S. supply of molybdenum would be needed. There are also concerns about the amount of chromium required for stainless steels, since the U.S. has essentially no chromium reserves at its disposal.

It should be obvious from these figures that dependence on foreign suppliers for the structural material for energy sources is not much different from depending on them for fuel (i.e. oil). The associated national security and balance of payments problems should be closely examined. Several problems not discussed here also present possible limitations on the fusion reactor. These areas are: charged particle and neutron sputtering of the wall facing the plasma, blistering due to helium injection from the plasma, degradation of electrical insulators in the blanket, swelling of nonmetallic blanket and shield components such as graphite and B_4C, and radiation degradation of organic thermal and electrical insulation in the magnet, to name just a few. While the list is formidable, it is expected that it can be solved by properly conducted research programs.

In conclusion, there are many factors to consider in choosing materials for fusion reactors. Stainless steel, vanadium, niobium and molybdenum or their alloys are top candidates for structural materials, and lithium, helium or Li-Be-F salts are the most favored coolant materials. No particular metal or coolant combination has been shown to be far superior to any other in all of the required areas.

13-4. Radioactivity and Afterheat

In addition to the changes in physical properties of the reactor structure, neutrons from the plasma may also induce radioactivity. Furthermore, the tritium produced from the lithium is radioactive. The magnitude of this radioactivity must be assessed since it bears upon the accessibility of the plant for maintenance, presents a potential hazard should it be released, and if present in any quantity leads to an afterheat which must be dissipated by the plant after the plasma is extinguished.

Regardless of the material chosen for the structure, some general comments may be made. The energy of the primary neutrons is quite high, approximately 14 MeV and nearly monoenergetic. (By comparison, neutrons from fission have a broad energy distribution with an average energy of approximately 2 MeV.) Neutrons of this energy are above the threshold for many of the charged particle reactions, e.g. (n,α), $(n,2n)$ (n,p). While the fuel is changed in a fission system about every three years, the structure in a fusion system may remain in use for up to twenty years. Thus the isotopes produced by successive capture of neutrons may also be important. These isotopes may not make a large contribution to the initial radioactivity, but if they are long-lived, they may be important in terms of long-term storage and disposal of used structures. Fortunately for the disposal process, the volume of a fusion plant will be quite large so that while the power from radioactive afterheat may be significant, the power density may be low. An appreciation of the complexity of the situation is obtained if a stainless steel system is considered. Even with all minor alloying elements and impurities ignored and only captures in the isotopes originally present considered, the transmutation cross sections for a minimum of four reactions and the associated decay schemes as a function of neutron energy for each of fourteen different isotopes leading to approximately three times as many daughter nuclei must be known.

Steiner has considered the problem for the fusion system developed at Oak Ridge National Laboratories.[5] Since the calculations are dependent on the specific reactor design they were repeated for the system proposed at Wisconsin and extended to include stainless steel as a structural material. This proposed reactor, designed with a 316 stainless steel structure, operates at a power of 1000 MW thermal. As mentioned previously, the volume of the system is quite large. There are 1000 metric tonnes of structure in the blanket alone. The calculations reported here are for an operating time of 10 years, and the radioactive decay is followed for 200 years after shutdown. Since cross-section information is not available for most of the product nuclei, neutron reactions with the radioactive product nuclei were not considered. Thus these calculations represent a lower limit on the radioactivity but should still yield representative values.

The results are shown in Figure 13-6 in terms of the energy deposited by gamma and beta radiation as a percentage of the power prior to shutdown. The decay heat at shutdown is approximately 0.7% dropping rather slowly to a factor of ten lower in about two years. At long times after shutdown the residual activity is due to ^{63}Ni, which has a half life of 92 years. However, at these long times this activity has decayed sufficiently that long half-lived isotopes from neutron capture in impurities could be important and must be considered in developing the complete picture. In looking further into the source of the decay heat it is found that approximately 50% comes from the wall between the plasma and blanket regions and that this fraction remains roughly constant with time.

By comparison, a fission reactor would have an initial decay heat about one order of magnitude larger. This would fall off somewhat more rapidly with time, but because of the long-lived

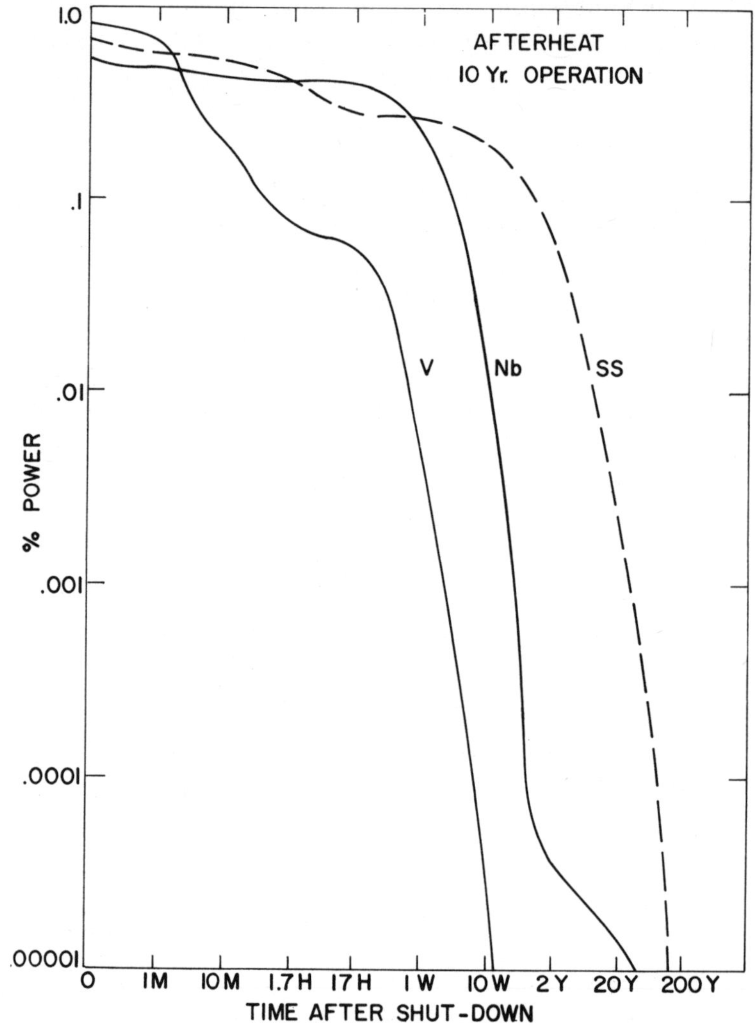

Fig. 13-6. Afterheat Following Shutdown of a Tokamak Fusion Reactor

unstable fission products, the activity would be greater than for a comparable fusion reactor even at long times. The results may also be expressed in terms of Curies of radioactivity per megawatt prior to shutdown. These results are shown in Figure 13-7 (1 Curie represents 3.7 x 10^{10} disintegrations/sec.). The initial activity is about 1 M Curies/MW with the first wall contributing about 50%. After 200 years the activity is due to ^{63}Ni and has dropped to

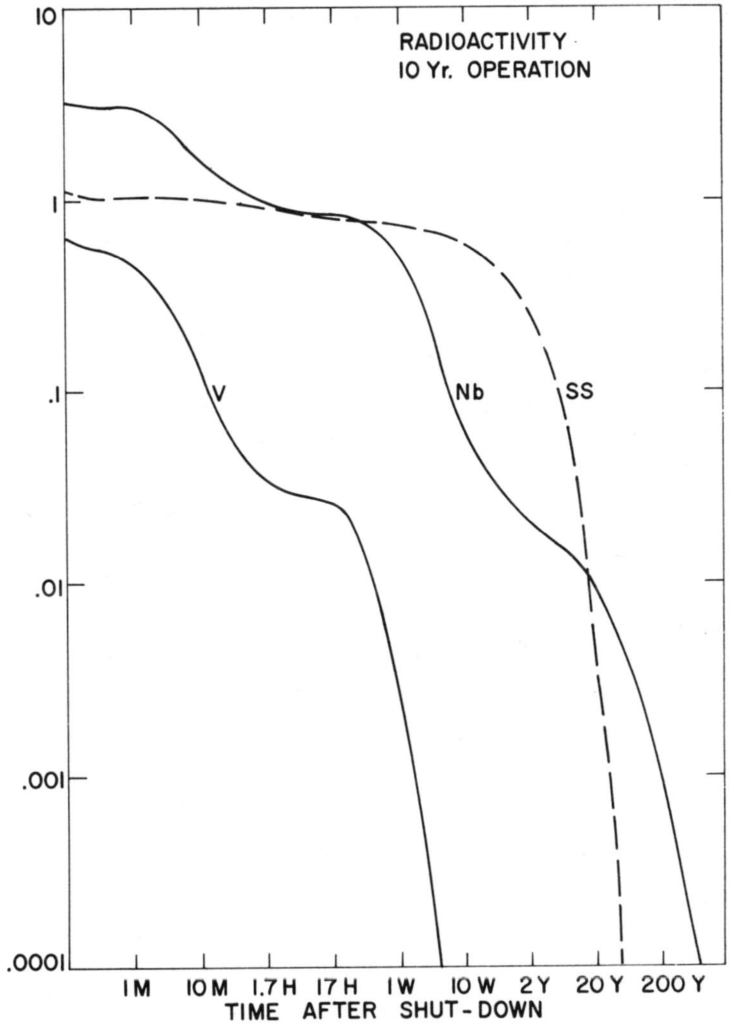

Fig. 13-7. Radioactivity Following Shutdown of a Tokamak Fusion Reactor

0.217 Curies/MW. Because of the large mass involved this corresponds to a specific activity in the first wall of only 0.017 milli-Curies/cm^3 for this particular design.

Stainless steel is not the only material which has been suggested. In particular both niobium and vanadium are potential candidates offering the possibility of higher operating temperatures than stainless steel. The afterheat and radioactivity were calculated for both these systems assuming a direct substitution of materials. The afterheat and radioactivity for a niobium system are shown in Figures 13-6 and 13-7. It is seen that the initial afterheat is about the same as for stainless steel but the initial drop-off is somewhat more rapid, coming down an order of magnitude in about 20 weeks. The long-lived activity in niobium is dominated by ^{94}Nb with a half life of 2 x 10^4 years, and thus niobium will have greater afterheat than stainless steel at very long times. The radioactivity curves show somewhat the same behavior.

The afterheat and radioactivity curves for vanadium are also shown in Figures 13-6 and 13-7. Vanadium shows a somewhat higher afterheat but drops off very rapidly coming down an order of magnitude in about two hours. The long-lived activity in this case is ^{51}Cr with a half life of 28 days so that after only 10 weeks the decay heat is down four orders of magnitude. Certainly in this case the afterheat at long times will be dominated by radioactive products of alloying elements and impurities. The radioactivity curve shows a similar behavior.

In summary it is found that because of the high energy neutrons produced in the fusion process the structure in the immediate vicinity of the plasma will become radioactive. The decay heat will not be excessive and should present no particular problems such as loss of coolant. The radioactivity levels are quite significant and may be rather long lasting although the half lives for stainless steel and vanadium appear to be shorter than those of residual fission products. Radioactivity will present problems in shielding and access to the reactor, and maintenance will have to be remote. Provision will have to be made for the long-term storage of used material and for preventing the dispersion of radioactivity released through corrosion.

13-5 Magnets

The confinement of plasmas by magnetic fields requires the use of superconducting magnets. Such use is far from routine since the magnets envisioned involve the new technology of superconductivity in which the conductor has zero resistance at temperatures near absolute zero. For the fusion reactor extrapolations from a 100-ton magnet, the largest in present use, to 20,000 tons, i.e. by more than two orders of magnitude, are necessary. Below are considered some of the engineering problems which cause these huge magnets to be the most expensive part of the reactor system.

An electromagnet is a pressure vessel with the force on each segment of current equal to I x B, where I is the current and B the magnetic field. Since B is caused by I, the internal pressure is proportional to B^2 and can be scaled in relation to 5800 psi at 100 kilogauss. The magnet for the reactor discussed earlier is of the constant tension design in which the local radius of curvature of the conductor and support structure is varied so that the tension in the structure remains the same.[6] The average tensile stress is

$$\sigma = p \frac{R}{t} \frac{1}{\lambda}$$

where p is the local magnetic pressure $B^2/2\mu_o$, R is the radius of curvature, t the structural thickness and λ is the structural space factor.

For example, in Figure 13-8 at r = 17.5 m, if the average stress for low-temperature stainless steel is limited to σ = 24,000 psi, p = 794 psi (37 kilogauss), λ = 0.23 and R = 8.75 m, then t/R = 0.14. Thus the steel thickness is 1.25 m as shown in Figure 13-8, and the steel therefore weighs 7,600 metric tonnes. For magnets this size or larger the structural material completely dominates the weight and cost.

It can be shown[7] that the structural mass required to contain a magnetic field must satisfy

$$M \geq \rho \frac{E}{\sigma}$$

where E is the total energy stored in a magnetic field, ρ is the structure density and σ is the average working stress of the structure. For stainless steel at 48,000 psi this amounts to about

$$0.18 \text{ lbs/watt hr. magnetic energy}$$

as the unavoidable minimum structure. Stainless steel is used here because it is a face-centered cubic metal which does not become brittle at low temperature. If the support structure could be at room temperature, then ordinary carbon steel could be used at less cost.

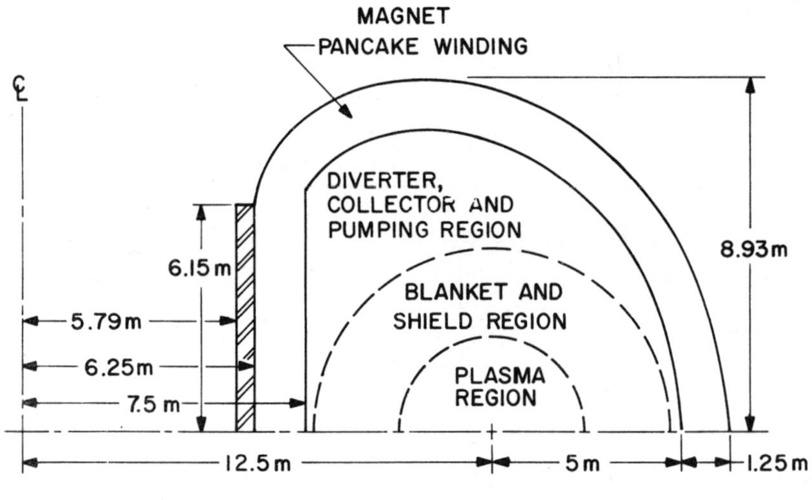

Fig. 13-8. Toroidal Magnet Cross Section

The second most important design consideration is the conductor. A composite superconductor-normal metal conductor supplies three functions for the operation of a magnet. First, the superconductor carries the transport current in a lossless fashion. The alloys of Ti-Nb are the best choice for use up to 86 kilogauss at 4.2°K and their use can be extended to 100 kilogauss by operation at lower temperatures. Second, the normal metal part of the conductor must be able to carry the current even when the superconductor has been driven normal. Copper, which is usually used, must be sized to carry this current without allowing the temperature to increase above the superconducting transition temperature for the Ti-Nb. In this way recovery to the superconducting state is possible. The third function of the conductor is to carry its unavoidable fraction of the mechanical load without deleterious effects on itself, the superconductors, or the insulator between the conductor and the steel support frame.

The composite conductor is mechanically in parallel with the steel structure and must experience the same strain. The copper strain should be kept below 0.001, the nominal yield point at 12.000 psi for OFHC copper. If copper is allowed to yield, its electrical resistance increases and mechanical hysteresis results. The latter produces additional helium cooling losses. While neither effect is disastrous it is better to avoid yielding if possible. The superconductor is usually subdivided into many small filaments inside the copper matrix and is subject to breakage if excessive strains are allowed.

The proposed design[6] uses the 304 stainless steel support to its yield point without yielding the copper. There is thus no need to consider yielding the copper, though its allowed strain could be subject to later adjustment. The steel forms would be solid forgings which can be stretched radially while the copper conductors are epoxied in grooves. Removal from the winding fixture will then relieve the tension in the steel and compress the copper to -12,000 psi. Under magnetic loading the prestressed copper can be carried from -yield to +yield by absorbing 24,000 psi load while at the same time steel will absorb 48,000 psi since its modulus is twice that of copper. Stainless steel yields at 60,000 psi at low temperature; it is prudent to design to 48,000 psi, since there is no recovery possible from structural yielding.

The 13,400 metric tonne magnet structure can be cooled in three stages, following Purcell's[8] example with the 12-ft. bubble chamber magnet at Argonne and the 15-ft. magnet at the National Accelerator Laboratory. The magnet will be divided into 12 sections or modules each with its own one kilowatt liquefier at 4°K. The following numbers are given for one section. The rate of heat extraction during cooldown must be limited to 60 KW to control differential stresses caused by contraction. The first-stage cooling lasts 16 days, requires 300,000 liters of liquefied nitrogen which exchanges heat with circulating helium gas and reduces the temperature to 115°K. The second-stage cooling is accomplished by operating the liquefiers as refrigerators. The temperature can be brought down from 115°K to 20°K within nine days. Fifteen thousand liters of liquid helium will be added to drop the temperature to 4°K and to fill each module dewar with helium. The operating capacity of each liquefier is 1200 liters per hr. of which 900 liters per hr. are available for removal of the neutron and gamma ray heat load. Compromise designs will consider the trade-off between shield thickness, radiation heat load and cool-down rate—all of which affect the choice of refrigerator size.

Following the 30-day-cool-down the current is brought up to 10,000 amperes with a 100-volt power supply. Charging requires four days for the 2600 henry inductance at 100 volts. Low voltages are desirable so that cryogenic insulation between the conductors and the steel backing can be simple. An occasional short is tolerable for low-voltage systems. Turning off the total magnet, warming up a magnet section, recooling and recharging add up to a 64-day turnaround time for repair or replacement of parts.

13-6. Summary

In the context of examining the feasibility of using a large Tokamak device in a fusion power system, four areas important to the system have been discussed. It should be emphasized that ability to design a power reactor is a long way off. In looking at these areas one finds that while the problems are real and challenging, they do not appear insoluble. The materials problem may be an exception, but even here a combination of ingenuity and additional data from a well thought out research program can be expected to resolve the issues.

References

1. Proceedings of the International Working Sessions on Fusion Reactor Technology, Oak Ridge, Tennessee, June 28-July 2, 1971, CONF-71-06.24.
2. "Fusion Reactor First Wall Materials," ed. by L. C. Ianniello, January 1972, WASH-1206.
3. G. L. Kulcinski, "Major Technological Problems for Fusion Reactor Power Stations," International Conference on Nuclear Solutions to World Energy Problems, Washington, D. C., 1972, American Nuclear Society, Inc., Hinsdale, Ill., 1973, p. 240.
4. We are indebted to scientists at Hanford Engineering Development Laboratory (HEDL) for data on the swelling and mechanical property limitations of 316 stainless steel during neutron irradiation.
5. D. Steiner and A. P. Fraas, "Preliminary Observations on the Radiological Implications of Fusion Power," *Nuclear Safety* **13**, 353, 1972.
6. W. C. Young and R. W. Boom, "Materials and Cost Analysis of Constant-Tension Magnet Windings for Tokamak Reactors," *Fourth International Conference on Magnet Technology,* Brookhaven National Laboratory, Conf.-720908, p. 244-252, 1972.
7. R. H. Levy, "Comments on Radiation Shielding of Space Vehicles by Means of Superconducting Coils," *ARS Journal,* 787, 1962.
8. J. Purcell, Argonne National Laboratory, private communication.
9. R. F. Post, "Prospects for Fusion Power," *Physics Today,* **26**, No. 4, p. 30, April 1973.

Acknowledgments

This paper was presented at the 35th annual meeting of the American Power Conference, May 8-10, 1973, sponsored by the Illinois Institute of Technology, Chicago, Illinois.

We would like to acknowledge the contributions of all the members of the University of Wisconsin design group, in particular Drs. R. W. Conna and D. E. Klein.

We would also like to acknowledge support for this work from the Wisconsin Electric Utilities Research Foundation and the United States Atomic Energy Commission.

14

CRYOPUMPING A LARGE ACCELERATOR

J. T. Tanabe and R. A. Byrns

A cryogenic vacuum pump, designed to lower the base pressure from 2×10^{-6} torr to 3×10^{-7} torr was installed in the Bevatron in February 1972. The Bevatron, a particle accelerator at the Lawrence Berkeley Laboratory, contains approximately 11,000 ft^3 of free volume and an excess of 100,000 ft^2 of outgassing surfaces pumped by 24 freon-baffled 32-inch oil diffusion pumps. The gas loads consist of metallic and organic outgassing surfaces as well as air leaks from nonrepairable radiation-hardened aged gaskets and mechanical actuators.

Nine cryopanels were installed, fed by 20 °K recirculating helium refrigerators and 80 °K boiling liquid nitrogen. Monte Carlo computations predict noncondensible (at 77 °K) pumping speeds of 140,000 liters/sec and condensible pumping speed of >500,000 liters/sec. Final measurements are included in this report.

14-1. Introduction

The Bevatron is a large synchrotron completed in 1954 at the Lawrence Berkeley Laboratory. Figure 14-1 shows the main ring of four straight sections and four curved sections. Each curved section or quadrant consists of a rectangular tube approximately 4 ft x 1 ft x 75 ft long. Parallel to this "beam tube" is a 2 ft x 3 ft crawl space designed for personnel access, see Figures 14-2 and 14-3. The straight sections are large steel tanks 10 ft x 8 ft x 20 ft long. The original pumping system has 24 32-inch oil diffusion pumps, six located in each of the four straight sections. Each pump is baffled with a freon-cooled Chevron-type shield maintained at

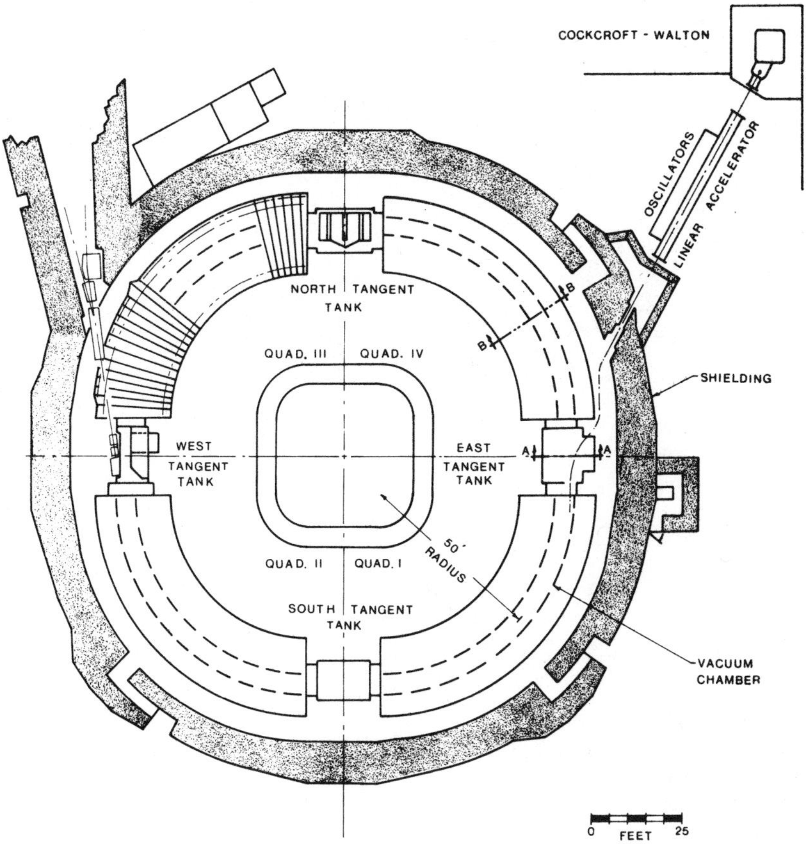

Fig. 14-1. The Bevatron

-15°F to reduce oil backstreaming to the Bevatron. Typically, 16 to 20 of the 24 pumps are operating, with the balance valved off for maintenance, repair, or to defrost the baffles. Instrumentation in the Bevatron consists of an ion gage in each of the four straight sections and two of the four curved sections. Prior to cryopumping, typical pressures were 1 to 2×10^{-6} torr in the straight section and 3×10^{-6} torr in the curved sections. These pressures were obtained only after several months of pumping after the system had been up to air. This slow pumpdown rate severely restricted internal machine access, since this access had to be balanced against operating time at reduced beam current due to poor vacuum.

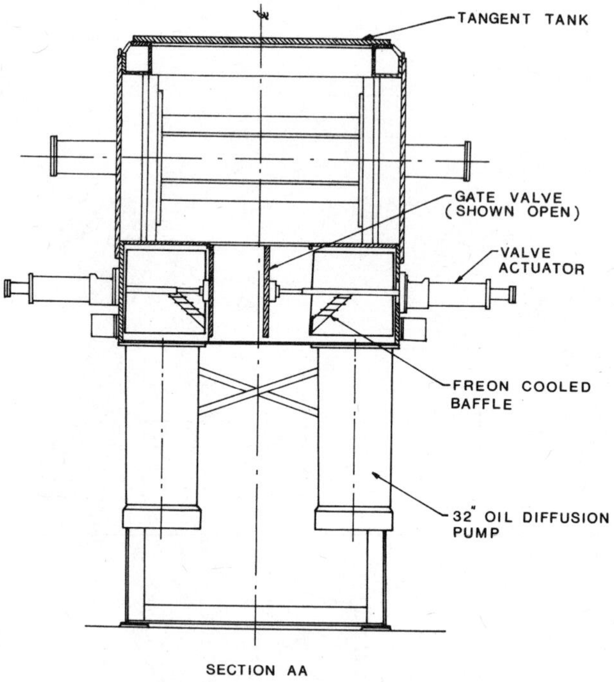

Fig. 14-2. Tangent Tank and Diffusion Pumps

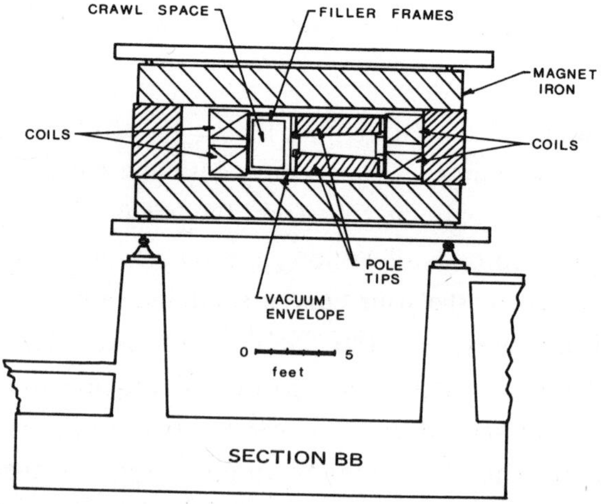

Fig. 14-3. Quadrant Cross Section

14-2. Design Objectives

It was always felt that increasing the pumping speed in the Bevatron would have a positive effect on the proton beam current. However, with the advent of the heavy-ion program at LBL the quality of the vacuum took on new importance. Pumpdown rates became an important consideration also because of the anticipated need for more frequent changes of equipment inside the Bevatron vacuum envelope. The general requirements were to decrease the Bevatron base pressure by about a decade, and in the process increase the pumpdown rate.

14-3. Gas Load

The Bevatron consists of 11,000 ft^3 of free volume with approximately 25,000 ft^2 of free outgassing area. The magnet pole tips, which consist of about 7,000 0.50-in.-thick enameled steel laminations, add approximately 80,000 ft^2 of restricted outgassing area. Radiation-hardened gaskets add to the gas load by providing small air leaks which are repaired periodically with "dux seal" (many of these gaskets are inaccessible except by major overhaul of the machine). Generous quantities of epoxy fiberglass and other plastics, as well as long-stroke reciprocating mechanical actuators, also add to the gas load problem. A large percentage of the free surfaces of the inside of the vacuum envelope are rough steel surfaces and aluminum castings which act as strongbacks; these have soaked up some quantity of the backstreamed pump oil over a period of years.

14-4. Measurements

Design of an improved vacuum system required knowledge of the quantity, quality, and distribution of the gas load. The quantity of the gas load was measured by a simple pressure rate of rise technique. This measurement compared favorably with the computation of the load by using the measured average base pressure of the vacuum envelope and estimates of the net speed of the diffusion pumps at that pressure. The data varied between 45 to 100×10^{-3} torr-liters/sec, and were rather well clustered around 60×10^{-3} torr-liters/sec.

A knowledge of the quality of the residual gases is crucial to the design of many types of vacuum systems because of the highly selective pumping characteristics of different pump types. Residual gas analyzer data showed relatively large and fairly equal peaks at mass 18 and mass 28, corresponding to large populations of air and water. Smaller peaks in the high-mass range plus a peak for mass 1 indicated the presence of hydrocarbons and the hydrogen ionized from these hydrocarbons.

Measurements at each straight section revealed that a factor of 2 pressure reduction was achieved by adding liquid nitrogen shielding to the ion gage. In general, we concluded from the above results that the residual gas was approximately half water and other condensibles and half air.

Since the distribution of the gas load would dictate the pump location, this measurement also became crucial to the design. The relative quantities of the gases in the four straight sections were simple to measure from relative base pressures of the tangent tanks. The proportion of the gas in the curved sections was a more difficult measurement: a nude ion gage was mounted on a travel target assembly, inserted through an air lock, and traversed through a quadrant. The monitored pressures were plotted against position, and the plot displayed the characteristic parabolic shape of a pipe with uniform gas generation pumped only at its ends. The data points were least-squares fitted to a parabola whose coefficient of the second-order term could be related to the gas load according to the following expression:

$$p = \frac{Q_T X^2}{2C_T \ell^2} + C_1 X + C_2,$$

where

$$C = \frac{Q_T}{2C_T \ell^2}$$

and

p = pressure (function of X),

Q_T = total gas load in the curved section (torr-liters/sec),

C_T = conductance of the rectangular pipe length ℓ (ℓ/s),

ℓ = length (units same as X in parabolic equation).

The conductance, C_T, was estimated from a standard geometric formula. Figure 14-4 shows a surprisingly close fit to a parabola of data taken on a day when system pressures were unusually high. The results of this experiment indicated that approximately half the gas load occurs in the curved sections and that a serious pressure gradient existed because of the relatively low through-put of the curved section to the straight sections where the diffusion pumps supplied the only molecular sink.

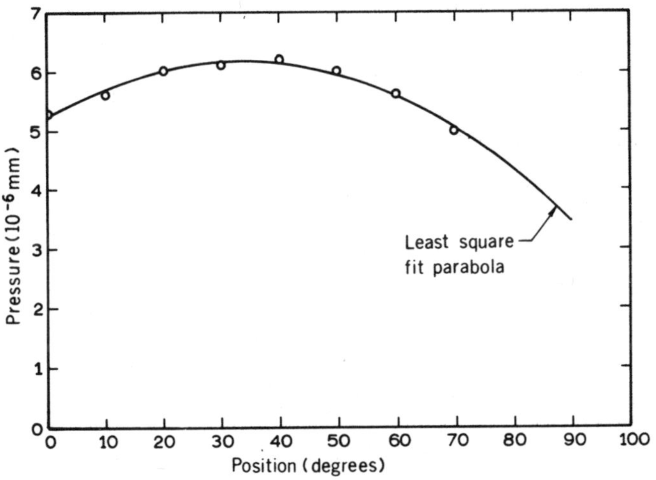

Fig. 14-4. Parabolic Pressure Distribution in Quadrant

14-5. Summary of Design Parameters

1. Gas load ≅ 0.060 torr-liters/sec.
2. Half the gas load is noncondensible at 77°K.
3. The gas load is fairly uniformly distributed about the perimeter of the Bevatron.
4. The desired base pressure is 2 to 3×10^{-7} torr.

5. Any LN_2-shielded 20°K pumping surface will necessarily have a high "condensible" pumping speed compared to its air speed. Thus, the design requires careful sizing of the air pumping speed:

$$S_{req} = \frac{\dot{Q}_{total/2}}{P_{req}} = \frac{60 x 10^{-3}/2}{3 x 10^{-7}}$$
$$= 100{,}000 \text{ liters/sec.}$$

14-6. Cryopanel Design

A two-dimensional Monte Carlo computer program was devised so that accurate predictions of the pumping speed of the final geometry could be made. Monte Carlo techniques trace the orbits of a large population of particles being emitted, reflected, lost, or absorbed in a two-dimensional array of straight lines and circles. A census is then taken of the particles lost or absorbed to determine the capture probability of a particular geometry. The Monte Carlo computation of the geometry shown on Figure 14-5 yielded a capture fraction of 0.2683.

Pumping speed computations based on this capture fraction are computed as follows:

$$S = C_p \times \ell_e \times s = 131 \text{ liters/sec-in.}$$

where

S = pumping speed per inch length of panel,

s = "blackhole" speed,

C_p = capture fraction,

ℓ_e = length of the emitting surface.

Preliminary tank measurements of a 60-in.-long test panel confirmed this capture probability within 5%. Nine cryopanels, each 10 ft long, are positioned in the Bevatron as shown in Figure 14-6.

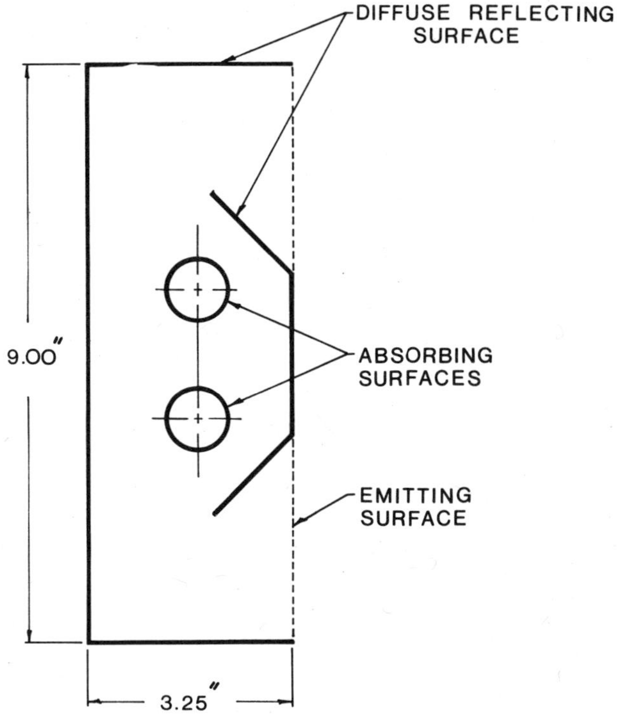

Fig. 14-5. Monte Carlo Model

14-7. Heat Loads

The nine cryopanels have a computed total "air" pumping speed of 140,000 liters/sec. The convex perimeter of the cryopanel is 24.5 inches. Estimates of the heat load assumed 90 ft of cryopanel and a curve for the emissivity of a polished surface as a function of water cryodeposit. Computations representing two to three days of operation at the fairly high vacuum load levels anticipated yielded emissivities of 0.2 to 0.4. Because hydrocarbons make up a fair fraction of the gas load, and since periodic regeneration of the cold surfaces by temperature cycling becomes problematical because of the low vapor pressure of these hydrocarbons, a conservative value for emissivity of 0.5 was used in heat computations.

$$\text{Heat load} = A\,\epsilon h,$$

where h = black-body radiation from 77 to 310°K
= 0.384 watts/in.^{2}.

Heat load = 90 x 12 x 24.5 x 0.5 x 0.384
= 5080 watts ≅ 115 liters/hour LN_2 consumption.

Operating cost at this rate for a machine which operates continuously would be too high. Therefore, six layers of multilayer insulation were wrapped on all the external surfaces. The LN_2-cooled perimeter which looks directly at the warm vacuum chamber was thus reduced to 4.72 inch/inch of panel. The inside surface of the LN_2 shield was painted a dull black to prevent reflection to the 20°K surface. The black-body heat load was reduced to approximately 2 kW or LN_2 consumption rate of 45 liters/hour. For this geometry, the "condensible" pumping speed is about 400,000 liters/sec. Thus, the partial pressure of condensibles in the vacuum system should be approximately

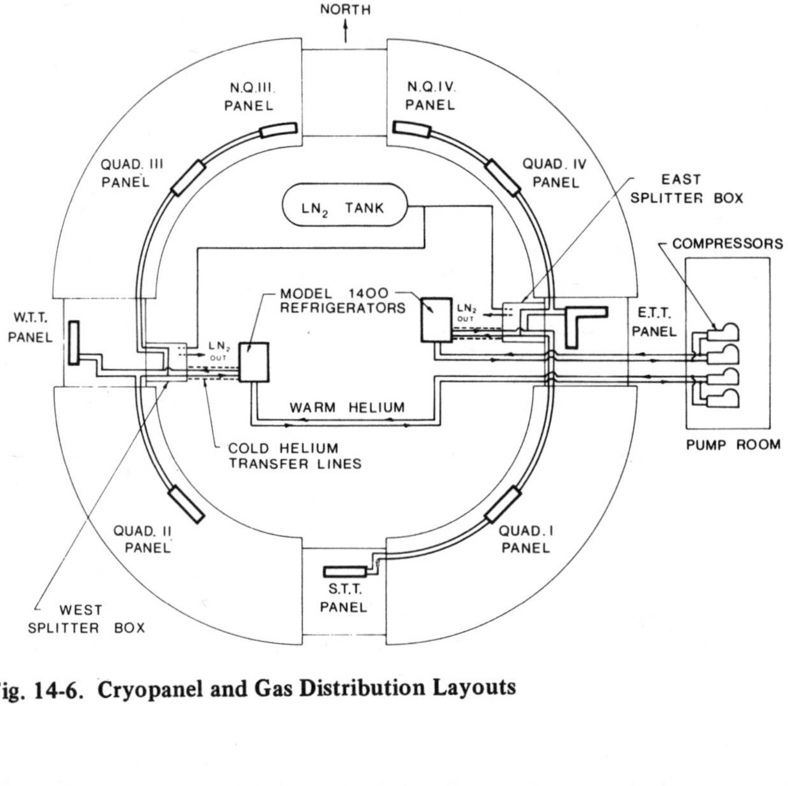

Fig. 14-6. Cryopanel and Gas Distribution Layouts

7.5×10^{-8} torr. In addition to the 90 ft of cryopanel, there are 264 ft of transfer lines inside the vacuum space and 70 ft of external transfer lines. The internal helium lines are LN_2-shielded, and the external lines are insulated with multilayer insulation against radiant heat transfer.

14-8. Helium Refrigerators

There are four circuits in the cryopanel system. Three circuits feed two cryopanels, and the fourth circuit feeds three. Two CTI 1400 refrigerators each feed two circuits. Two refrigerators were chosen although a single refrigerator would have had enough capacity. Reasons for deciding on two refrigerators follow:

1. Uncertainties existed in the calculated refrigeration loads due to eddy current heating and lack of precise pressure drop data. The reserve capacity was built in to take care of unanticipated contingencies.
2. A finely tuned feedback control system would be required to balance the flow if a single refrigerator fed four branching circuits. The precision of this balancing procedure becomes very critical if the load approaches the full capacity of the refrigerator, especially since the vapor pressures of N_2 and CO are a sensitive function of temperature in the 20 to 25°K range.
3. Redundancy in the circuit allowed the operation of half the cryopump system in the event of refrigerator outage.
4. Operational experience will establish accurate refrigeration load values. Expansion of the system then becomes a possibility, since in any vacuum system the vacuum is seldom "good enough."

The CTI units offered a range of refrigeration capacities between 100 and 350 watts at 20°K, depending on choice of options (LN_2 pre-cool and an extra compressor). Since the refrigerator circuit is not separate from the circulating fluid, the compressor provides the driving force in circuits with up to 220 ft of 0.310-i.d. line. These and other features made this unit attractive, especially in view of the possible expansion of this system.

14-9. Design Constraints

The constraints imposed by the physical characteristics of the Bevatron and the operating schedule presented some unique challenges.

1. A three-week period was set aside for the Spring of 1972. Design, fabrication, and inspection of all components to be installed in the vacuum envelope had to be completed before this shutdown. All installation and leak checking had to be completed in that period.
2. Residual radiation limited the amount of access time for any installer.
3. The extreme congestion of some of the areas required some fairly complex designs.
4. The maximum circuit length is 105 ft. Rather generous expansion joints were required in the transfer line between fixed points (usually the cryopanels) to handle the thermal shrinkage.
5. Space, clearance, and access problems required a large number of joints in the vacuum space. The time factor and lack of working space made the prospect of soldering or welding extremely unattractive. Thus, a reliable mechanical joint technique had to be selected which could withstand numerous temperature excursions.

14-10. Mechanical Design

Figure 14-6 shows the location of the cryopanels in the Bevatron. Six cryopanels are located in the quadrants and three are mounted in the tangent tanks. A cryopanel was omitted in the north because of the physical clearances required by the high-voltage r.f. accelerating electrodes mounted in this tank. The extra panels in quads III and IV near the tank handle the pumping required in the north. Cross sections of the quad panel and the tangent tank panel are shown in Figure 14-7. Their installation is shown in Figure 14-8.

The time-varying magnetic field dictated both the radial location and the materials of construction of the cryopanels mounted in the quadrants. The quad panels were mounted in the crawl space to avoid the high-field areas. These panels were made from stainless steel to minimize the eddy current heat loads generated by the time-varying magnetic field in the quadrants. The use of stainless steel required close spacing of the LN_2 circulating tubes because of low thermal conductivity. Copper was used in the tangent tank panels where eddy current heating was negligible. This allowed simpler construction because of more generous spacing of the LN_2 tubes. The deeper panels provided a slight increase of the capture fraction.

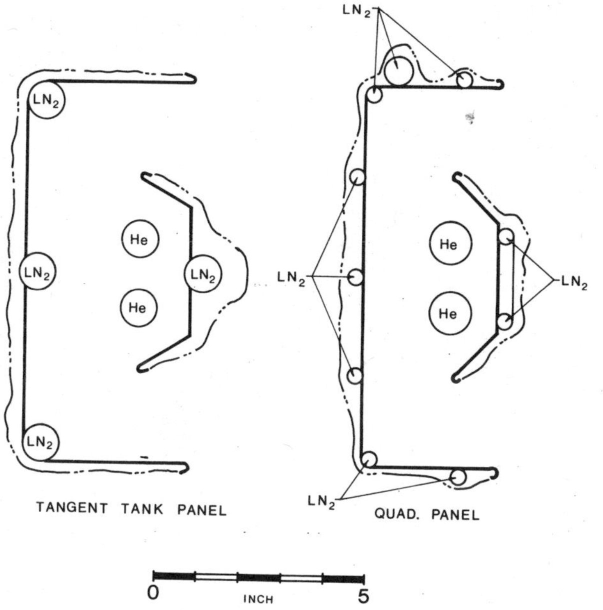

Fig. 13-7. Cryopanel Cross Sections

Inside the vacuum envelope there are 16 sections of transfer line of two basic types as shown in Figure 14-9. The helium lines in the quad transfer lines were wrapped with alternate layers of aluminized Mylar and nylon mesh. The quad lines were comparatively long and straight and could easily be wrapped with the superinsulating material. This material acted as a physical spacer

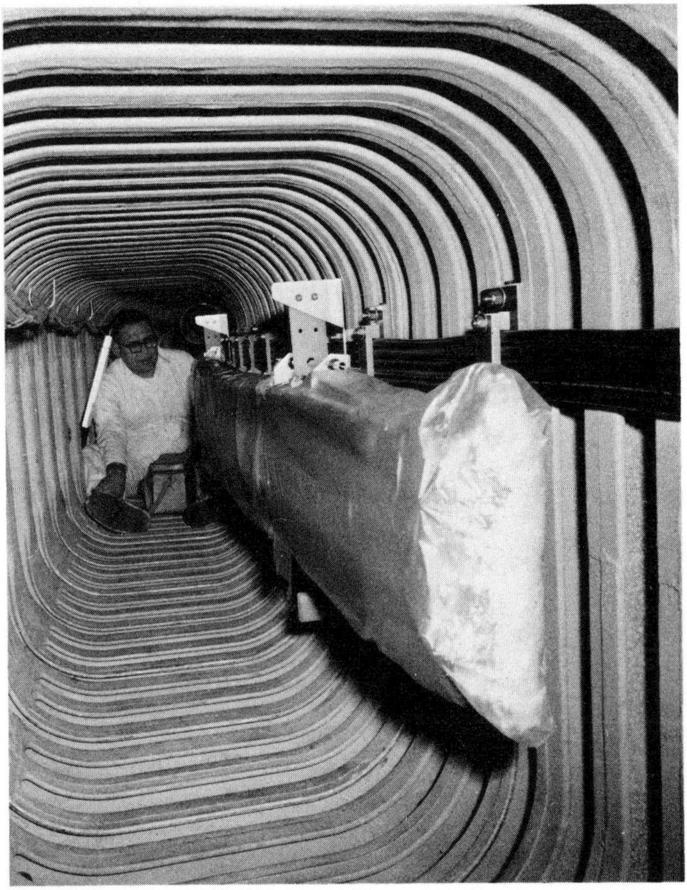

Fig. 14-8. Installed Quadrant Cryopanel

between the 80°K surface and the 20°K surface, as well as a radiant insulator.

Most of the expansion joints and all of the severe bends occurred in the tangent tank transfer lines. Here, physical separation between the LN_2 lines and 20°K He lines was achieved with plastic spacers and plastic grommet inserts. These transfer line sections vary from 30 to 2.5 ft long, depending on their ease of installation. There are 20 joint sections connecting the various lengths of transfer line and cryopanels inside the vacuum envelope. Each joint

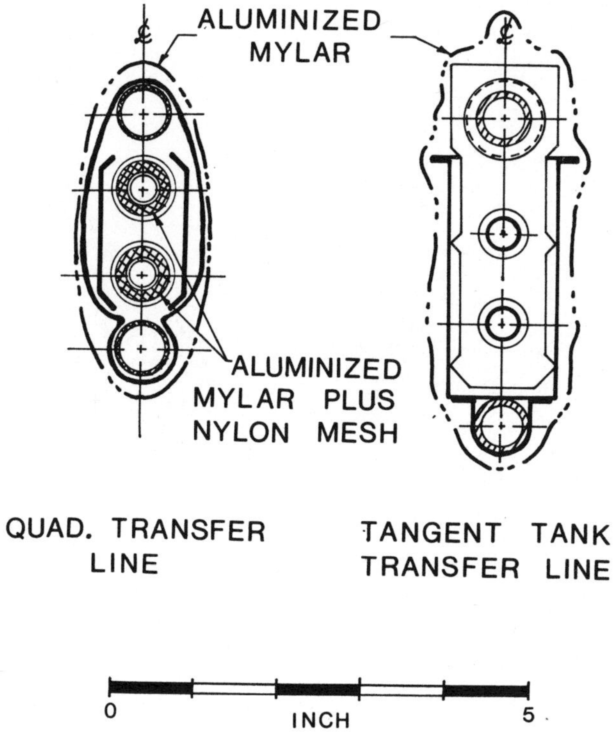

Fig. 14-9. Transfer Line Cross Section

section contains four unions: a feed and return for both LN_2 and helium. The "Cajon" union was selected because it is almost foolproof in installation and extremely reliable during the numerous temperature cycles required during operation. This union features a nickel gasket compressed between two stainless steel glands with a smooth raised ridge. Relatively high torques are required in installation areas with very limited working space. Therefore, a special tool was designed with a 10:1 mechanical advantage in torque which prevented relative rotation of the glands during the torquing (Figure 14-10).

14-11. Cooling Circuits

The helium and nitrogen circuits are shown in Figures 14-6 and 14-11. Splitter boxes are mounted on the east and west tangent tanks where flows for the helium circuit are controlled with

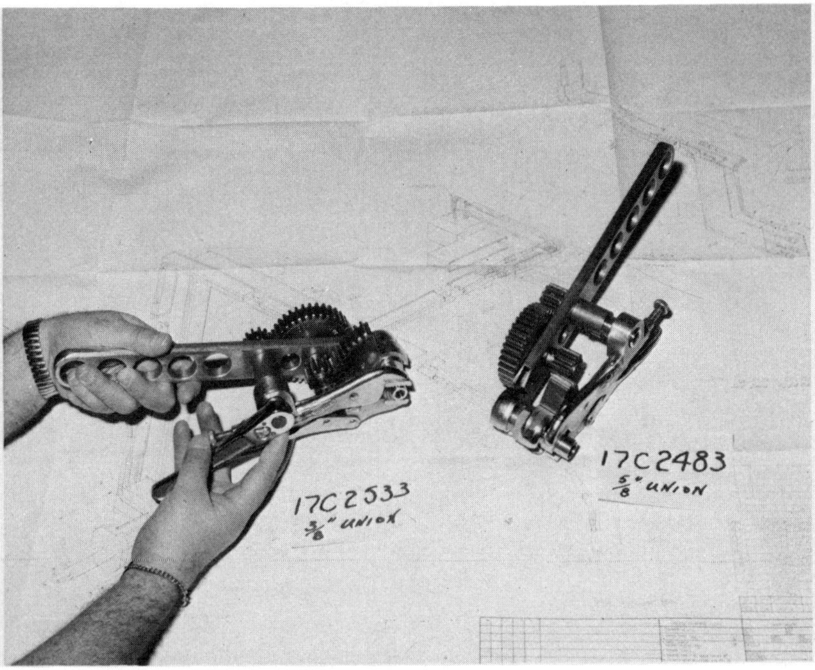

Fig. 14-10. Torquing Tool for the Compression Fitting

broken-stem valves. Gas bulb thermometers for each of the input and output lines are provided for balancing the flows.

14-12. Performance

The cryopanels were installed in the Bevatron in February 1972. Liquid nitrogen was circulated in the LN_2 circuits for the first time on February 25, 1972. The external helium transfer lines were installed during a two-week shutdown in June 1972. The refrigerators were installed in July, and cold gas was circulated in the helium circuits for the first time on August 2, 1972. Figure 14-12 illustrates the average pressure history in the Bevatron during this initial period. The Bevatron has been up to air once since the initial run with 18°K helium. In addition, the nitrogen and helium circuits have been warmed up several times to exhaust their accumulation through the diffusion pumps.

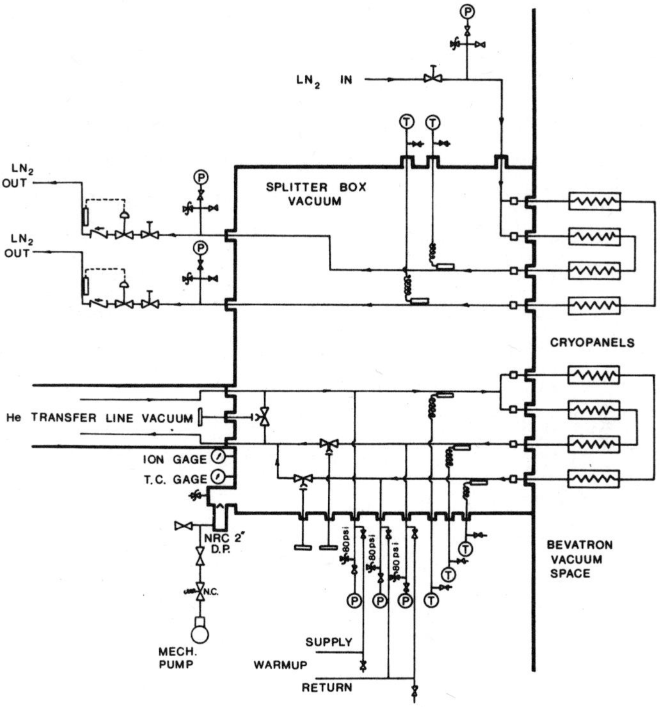

Fig. 14-11. Splitter Box Circuit Detail

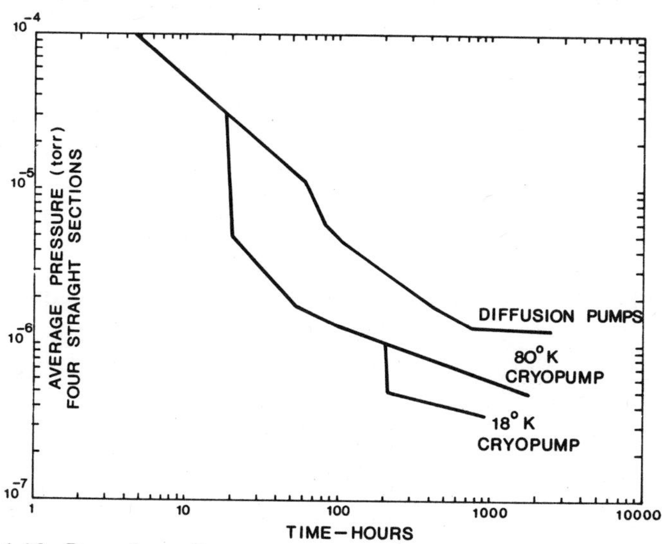

Fig. 14-12. Pumpdown Curves

In general, good operating pressures (2 to 4×10^{-6} torr) have been achieved within a few hours after introduction of LN_2. Previously, several weeks were required to achieve equivalent pressure after the vacuum tank had been up to air. Introduction of 18°K helium reduced the pressure by a factor of 2 below the base pressure achieved by LN_2 alone. Thus far, the best average pressure achieved with the full system has been 3.0 to 5×10^{-7} torr.

It should be noted that the vacuum performance described above does not reflect the full capability of the cryopump system. There has not been any extended period under vacuum to allow the outgassing rate to base out. Previous data have indicated that two months is required for the tank to dry out. Also, there are several leaks in the vacuum tank system which have not been adequately repaired.

Thus far, we have not been successful in holding a base pressure in the 10^{-7} torr range when we close the gate valves on the diffusion pumps. At these pressures, the net pumping speed of the diffusion pumps should be between 5% and 10% of the cryopanel pumping speed. A slow pressure rate of rise with the D.P. gates closed is interpreted as a verification of the existence of air leaks and the introduction of the fractions of air (24 ppm neon, helium, and hydrogen) which are not pumped at 20°K.

The press of heavy-ion and high-energy physics experiments in the Bevatron has not allowed time to perform the obvious residual gas analyzer experiments to supplement the rather crude experiments related above.

The refrigeration load has been close to predicted design calculation. Eighty watts is the estimated load on each of the two helium refrigerators operating in the circuit, based on return gas temperature of 15°K and supply temperature of 12°K. The total measured consumption rate of LN_2 is 70 liters/hour. Presently, the LN_2 circuit is a once-through circuit which operates on a temperature sensor demand feedback control system. The consumption rate thus depends on how closely this system can be adjusted to most efficient setting. A recircualting system is planned which should reduce the LN_2 consumption.

14-13. Conclusion

Given sufficient time under vacuum to allow the substantial outgassing area of the Bevatron to base out, and assuming that future effort will be successful in pinpointing and repairing some of the more obvious leaks, the design pressure of 2 to 3×10^{-7} torr should be achieved.

Refrigeration loads fairly close to design prediction allow us a fair degree of confidence in our ability to expand this system or design new systems. There is nothing quite like an empirical datum point to extrapolate in the design process.

References

1. R. G. Blackely, R. W. Moore, Jr., Vivienne J. Harwood, "TRIUMF, The Conceptual Design of the Vacuum Pumping System for a 500 MeV Cyclotron," University of British Columbia, August 1969.
2. R. L. Merrian and R. Viskanta, "Radiative Characteristics of Cryodeposits for Room Temperature Black Body Radiation," in *Advances in Cryogenic Technology*, **14**, 1968, Plenum Press, New York, 1969.
3. F. S. Reinath, "Proposed Modification of Bevatron Vacuum System," Lawrence Berkeley Laboratory Engineering Note M4318, July 1970.
4. J. T. Tanabe, "Proposed Cryogenic Pumping System," Lawrence Berkeley Laboratory Engineering Note M4282, April 1970.
5. J. T. Tanabe, "Bevatron Vacuum Measurements," Lawrence Berkeley Laboratory Engineering Note M4302, June 1970.
6. J. T. Tanabe, "Proposed Cryopanel Cross Section," Lawrence Berkeley Laboratory Engineering Note M4326, September 1970.
7. J. T. Tanabe, "Monte Carlo Program," Lawrence Berkeley Laboratory Engineering Note M4338, October 1970.
8. F. S. Reinath, R. M. Richter, J. T. Tanabe, and E. Zajec, "A Cryogenic Pumping System Proposed for the Bevatron," presented at the 1971 Particle Accelerator Conference, Chicago, Ill. Lawrence Berkeley Laboratory Report UCRL-20197, February 1971.

Acknowledgments

The work described in this paper was done under the auspices of the U.S. Atomic Energy Commission.

We would like to acknowledge the leadership provided on this project by Walt Hartsough, Hermann Grunder, Ken Lou, and

Bob Richter. It would be difficult to acknowledge all the effort spent in design, procurement, fabrication, and installation. However, we would like to express our gratitude to the large group of individuals who have expended extra effort to bring in a good job on time.

15

CRYOPUMPED HYDROGEN JET TARGET

V. Bartenev, J. Klen, A. Kuznetsov, E. Malamud, B. Morozov, V. Nikitin, Y.Pilipenko, V. Popov, B. Strauss, D. Sutter and L. Zolin

The hydrogen jet target, a product of joint USSR-USA research, replaces conventional foil targets for proton-proton interactions in the main accelerator of the National Accelerator Laboratory. Cryogenic systems are necessary both to prepare the jet target and to maintain the accelerator vacuum against the introduction of hydrogen gas pulses.

15-1. Introduction

In order to carry out proton-proton interaction experiments with the internal circulating beam of the NAL main accelerator, a jet target was designed. This jet is required so that the high backgrounds inherent with polymer foil targets are eliminated. This target produces a pulsed jet of condensed hydrogen with the following parameters: diameter, ~10 mm; duration, 300 msec; density, $\sim 10^{-6}$ g/cm^3.[1-5] The jet is pumped from the vacuum chamber with a helium-cooled cryopump. The target has been in operation for a year and has serviced five experiments including the first USSR/USA high-energy physics collaboration.

15-2. Target Description

A schematic of the target is shown in Figure 15-1. Boiloff helium is used to cool the main heat exchanger (3). When this is done, a measured amount of compressed hydrogen, which has been gathered in the buffer volume between valves 1 and 2, is released

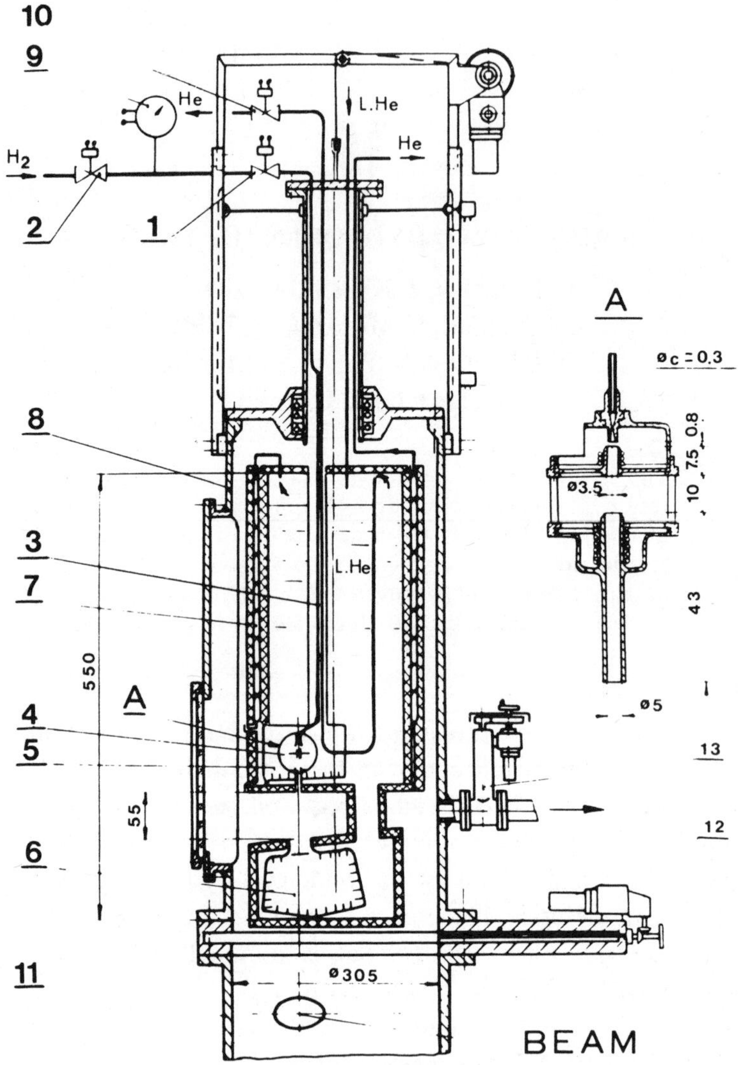

Fig. 15-1. Cross section of the target for the production of a condensed hydrogen jet (dimensions are in mm). Legend: (1, 2) supply hydrogen valves; (3) heat exchanger; (4) nozzle and collimator system; (5) upper cryogenic pump; (6) lower cryogenic pump; (7) shield cooled by vapor helium; (8) target isolation chamber; (9) pressure gauge; (10) heat-exchanger control valve; (11) accelerator chamber; (12) vacuum gate; and (13) valve for evacuation of sublimated hydrogen.

into the heat exchanger train. Here it is cooled, condensed, and injected into the target. Injection is perpendicular to the proton beam of the accelerator. The jet crosses the accelerator chamber at the beam line and is captured by the helium-cooled cryopump. This cryopump has two portions: a main trap with a round throat of 19.6 cm^2 and an additional trap with a throat of 36.6 cm^2. The additional trap helps to scavenge gas which has diffused into the chamber during injection. The target is periodically withdrawn from the accelerator chamber (11) to an isolation chamber (8) where it is defrosted, and where the solid hydrogen clustered in the traps is sublimated.

Because of working conditions, such as high radiation levels in the accelerator tunnel, the target must be operated remotely. Through remote control panels, operators can shift the target in a vertical direction, as well as open and close vacuum gates and carry out all cryogenic operations, such as cooling and filling with liquid helium. All main parameters are registered on recorders and a computer interface is provided.

15-3. Precooling and Hydrogen Condensation

A "see-saw" heat exchanger is used to cool the hydrogen in the jet. Basically it consists of two capillaries attached along their length. The heat exchanger is first cooled by a measured amount of boiloff gas from the liquid helium cooling the cryopump. When this flow is stopped, a premeasured amount of LN_2-cooled hydrogen is bled into the other side of the heat exchanger, where the sensible heat of the copper cools the gas to the region of 20°K, and then it is delivered to the jet nozzle. The timing of these gas pulses is controlled by a special timing circuit which is triggered with respect to cycles in the main accelerator. Thus we do not have a standard heat exchanger but rather a short-term sensible heat storage device.

15-4. Jet Collimation

In order to minimize the local vacuum deterioration in the accelerator, the divergence of the hydrogen jet must be minimized.

This is done with a series of collimators. These are scrapers with small cryopumps at their side. Even with this, there is a large local variation of vacuum in the vicinity of the jet, as shown in Figure 15-2. Oil diffusion pumps are used in a scavenger system for 10 meters upstream and downstream of the jet.

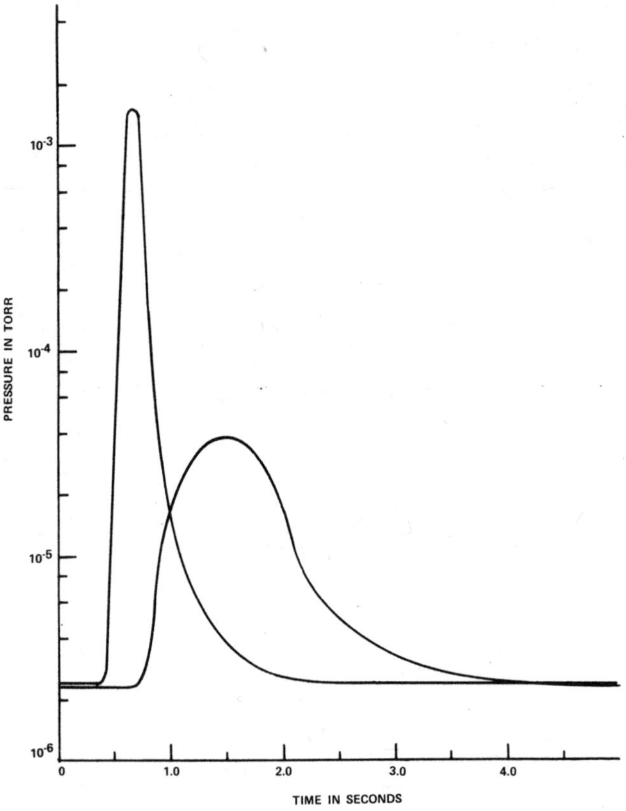

Fig. 15-2. Pressure variation in the accelerator vacuum chamber during injection of the jet. The base pressure is 2.4 x 10^{-6} Torr. The first peak shows the pressure rise in the vacuum chamber near the jet. The peak pressure is 1.5 x 10^{-3} Torr and each major time division is equal to 500 msec. The second peak shows the vacuum change 10 m upstream of the jet. Peak pressure is 3.9 x 10^{-5} Torr. The delay is due to diffusion through the proton beam tube. Notice the rapid pumping speed of the cryopump as evidenced by the recovery leg of the first peak.

To protect synchrotron operation from the effects of excessive local vacuum pressure rise due to the hydrogen jet, two predetermined pressure and safety levels are continuously monitored. One is on overall pressure rise. If the local vacuum pressure rises above this level, then jet operation is electronically vetoed until recovery occurs. The second safety level is applied to multiple jet operation during one synchrotron acceleration cycle. Here the vacuum pressure rise associated with each jet firing must recover below the safety level before the next jet is due, or that jet will be electronically vetoed. In this way the average vacuum near the jet can be maintained on the average at a quality good enough to cause only minimal interference with the circulating proton beam.

15-5. Helium Supply System

The system for providing liquid helium to the target is based around a CTI Model 1400 liquefier. This machine makes liquid continuously into a 1000 liter dewar; the liquid is then batch transferred to a 500 liter dewar near the target in the accelerator tunnel. The transfer line which is 35m long and nitrogen-shielded has a steady state loss of about 10 watts including that from six bayonet connectors. A remotely-controlled system transfers liquid in small batches from the downstairs dewar to the target itself. The Model 1400 liquefier is used also for recovery of the boiloff helium gas. The boiloff line from the target is connected to the suction of the compressor system. Upon a rise in suction pressure, the system can shunt the full output of the compressors to the recovery tank. The system has recently been brought up to a capacity of 30 liters/hour by the addition of a third compressor.

References

1. V. Bartenev, A. Valevich, Y. Pilipenko, and V. Smelanski, JINR Preprint P13-6058, Dubna, 1971.
2. K. D. Tolstov, JINR Reprint, 1698, Dubna, 1964.
3. K. D. Tolstov, JINR Preprint, 1-4103, Dubna, 1968.
4. V. Bartenev, et al., JINR Preprint, P13-6323, Dubna, 1972.

5. V. Bartenev, et al., "Cryopumped, Condensed Hydrogen Jet Target for the NAL Main Accelerator," *Advances in Cryogenic Engineering,* **18**, 1973.

Acknowledgments

The authors would like to thank the many people who contributed to this paper; in particular, A. Belushkina, A. Valevich, V. Smelanski, and others of the staff of the Cryogenic Laboratory at the Joint Institute for Nuclear Research, Dubna, USSR, who did the early development work on this apparatus. They would also like to thank the following at NAL: the Internal Target technicians and mechanical design group, and the NAL machine shop.

16

HYDROGEN SORPTION PUMPING DYNAMICS OF CRYOFROSTS

A. M. Smith

Recent experiments on the dynamic vacuum pumping of H_2 by cryosorption on CO_2 cryofrosts are reviewed. The effects of frost thickness, formation rate, and temperature on the H_2 pumping speed and capacity are presented. In addition, the effect of warming the frost is discussed. Results are given also for dynamic sorption pumping of H_2 by SO_2 and CH_3Cl cryofrosts.

16-1. Introduction

Cryodeposited frosts formed by condensing certain gases (such as CO_2, A, O_2, and N_2) on surfaces at temperatures between 10K and 20 K have been shown to provide an effective means of pumping hydrogen gas.[1-9] This sorption pumping technique is basically a dynamic process during which the pumping speed decreases from some initial value and approaches zero as the frost cryosorbent is saturated by the H_2 sorbate at some equilibrium condition. The equilibrium sorption properties of several frosts for H_2 have been measured in some of the previously referenced investigations. However, only a meager amount of dynamic cryosorption data are available for H_2 pumping by cryofrosts.[3,10] This paper presents the results of a recent experimental investigation on the sorption pumping dynamics of H_2 by CO_2, SO_2 and CH_3Cl cryofrosts.[1]

16-2. Experimental Apparatus

The vacuum chamber used for the hydrogen cryosorption studies is shown schematically in Figure 16-1. Its pumping system consisted of a mechanical roughing pump and a 6-in. oil diffusion pump equipped with an LN_2-cooled cold trap. A 6-in. high-vacuum, air-operated gate valve was used to isolate the chamber from the pumping system. The cryosurface upon which the frost sorbent was deposited was a 6-in.-diam. stainless steel sphere (Figure 16-1) with an available pumping area of 970 cm^2. It was attached to a helium refrigerator via vacuum jacketed transfer lines. The refrigerator supplied gaseous helium in the temperature range from 12K to 70K.

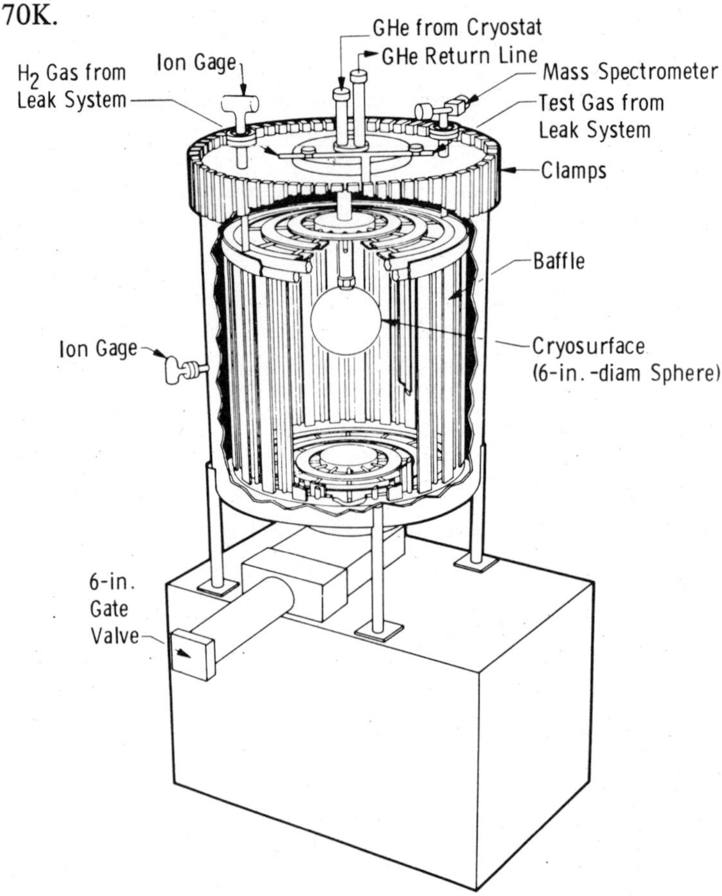

Fig. 16-1. Experimental apparatus used for H_2 cryosorption studies

Hydrogen and the various gases predeposited to form the sorbents were obtained from commercially available bottles. These 300K gases were introduced into the top of the chamber in the region between an optically-tight baffle and the chamber wall by means of two leak systems. Since the gas molecules would experience many collisions with the 300K chamber walls and baffles before reaching the cryosorption surface, it is believed that the molecules struck the pumping surface randomly from every direction. After the sorbent had been added on the pumping surface at temperatures below 20K and before the H_2 addition was started, the base pressure decreased to about 1×10^{-8} to 4×10^{-8} torr.

An ion gage and a magnetic-deflection type mass spectrometer were used to measure the pressures in the chamber (see Figure 16-1). Tubulation inside the chamber was employed to prevent the gages from directly sensing the pumping surface, and oriented so that the gages would sense about the same flux of particles as the pumping surface. For higher pressure tests the ion gage was replaced with an Alphatron. The sensitivities of the ion gage and mass spectrometer for H_2 were determined prior to each series of tests. A hydrogen vapor-pressure thermometer was used to determine the temperature of the gaseous helium cooled surfaces at temperatures between 10K and 25K. At higher temperatures a helium gas thermometer was employed. The bulbs of both of these devices were located inside of the sorption pump and heliarc welded into the pumping surface. It is shown in Ref. 8 that the surface temperature of the frost is adequately indicated by the bulb temperature.

16-3. Experimental Procedure

All runs were started by pumping the chamber to its base pressure level with the diffusion pump. Next, the cryosurface was cooled to the desired temperature by circulating cold gaseous helium from the cryostat through it. When this temperature was reached, the chamber was isolated from its pumping system and ready for admission of the sorbent gas. Most of the sorbents were formed at different pressure levels or strike rates (number of molecules/sec/cm^2) by varying the leak forepressure. However, a few were formed by pumping the chamber to its base pressure and then backfilling it with helium to pressures as high as 10^{-1} torr. The sorbent gas was then introduced into the chamber and the sorbent formed at relatively high helium partial pressure levels. Both the leak forepressure and chamber pressures were recorded continuously and held constant during the prescribed addition time. Then, the helium was removed from the chamber by the diffusion pump.

Once the frost was deposited, the H_2 sorbate gas was admitted to the chamber, which remained valved-off from its pumping system. Two different sorbate addition techniques were used: (1) constant sorbate flow rate, and (2) variable sorbate flow rate. In the first technique, the leak forepressure was held constant during the addition of sorbate to maintain the addition rate constant at some desired value. Typical time histories of the chamber pressure and sorbate forepressure for this mode of operation are given in Figure 16-2. When the chamber pressure had increased to about 3 x 10^{-4} torr, the frost sorbent was essentially saturated with the sorbate and its pumping speed was reduced to a few liters per sec. The test runs were terminated at this point by shutting off the sorbate flow. In the second technique, the forepressure in the sorbate leak system was initially set at some value and the

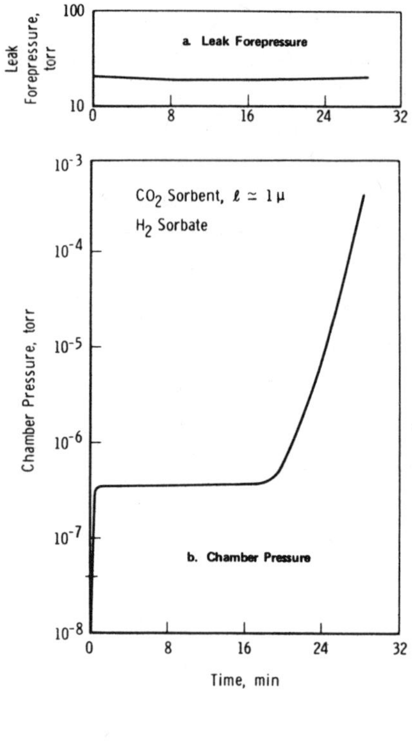

Fig. 16-2. Typical pressure history for experiment with constant sorbate flow rate (Ref. 2).

flow started which resulted in some corresponding value of the chamber pressure. Then, as the pumping speed of the frost characteristically decreased as it sorbed hydrogen, the forepressure on the leak and, hence, the sorbate flow rate, were steadily decreased in a manner to maintain a constant chamber pressure. The test was completed when the sorbate forepressure was reduced to zero. Forepressure and chamber pressure time histories for this mode of operation are presented in Figure 16-3.

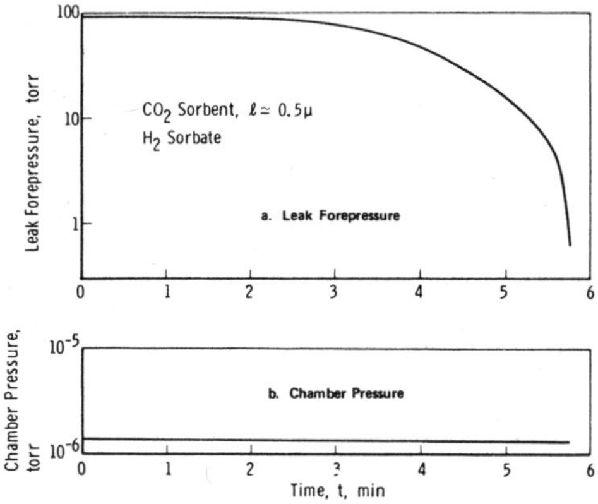

Fig. 16-3. Typical pressure history for experiment with variable sorbate flow rate (Ref. 2).

16-4. Data Reduction Procedures

The amount of gas added to the chamber to form the frost sorbent was computed from

$$Q_{sorbent} = t[P_a K]_{sorbent}$$

where t is the time interval of sorbent addition, P_a is the forepressure in the leak system, and K is the measured conductance of the leak. Assuming that all the sorbent gas added to the chamber was deposited on the cryosurface, the thickness ℓ of the sorbent frost was calculated by

$$\ell = Q_{sorbent}\, m_g/(A\rho_f k T_g)$$

where m_g is the mass of the gas molecule, T_g is the gas temperature, A is the cryosurface area, k is Boltzmann's constant, and ρ_f is the frost density. In this calculation, estimated values were used for the frost densities. These values were 1.60, 1.80, and 1.20 gm/cm^3 for CO_2, SO_2 and CH_3Cl, respectively.[2]

The rate at which the H_2 sorbate gas was added to the chamber was determined from

$$\dot{Q}_{sorbate} = [P_a K]_{sorbate}$$

When the sorbate addition rate was constant, the amount added was calculated by

$$Q_{sorbate}]_{P_a = const} = t[P_a K]_{sorbate}$$

For experiments where the sorbate addition rate was varied to hold the chamber pressure P_c constant, the amount of sorbate gas added was computed from

$$Q_{sorbate}]_{P_c = const} = [K \int_0^t P_a\, (t')dt']_{sorbate}$$

with the integral being evaluated from figures such as Figure 16-3A.

The pumping speed S of the sorbent frost for H_2 gas was determined from the relation

$$S = \frac{\dot{Q}_{sorbate} - V\, dP_c/dt}{P_c - P_u}$$

where V is the free volume of the chamber, 300 liters, and P_u is the ultimate pressure of the chamber prior to adding the sorbate. Values of P_c and dP_c/dt were obtained from pressure-time plots such as shown in Figure 16-2b. Note that the pumping speed definition used here accounts for the fact that in a closed chamber the frost sorbent must also pump an additional gas load due to chamber outgassing and leakage. The unit pumping speed of the frost was calculated from s = S/A.

The sorption capacity C of the frost sorbents for H_2 sorbate was computed from

$$C = Q_{sorbate}/Q_{sorbent}$$

This quantity C has also been referred to as the "mole ratio" by some investigators[5] and as the "concentration" by others.[6] In effect, C specifies how many sorbate molecules may be sorbed by each molecule predeposited in the frost sorbent.

16-5. Sorption Pumping Results

Figure 16-4a shows the dynamic H_2 pumping speed curves for three thicknesses of 12.4K CO_2 frost formed at a chamber pressure of 2 x 10^{-5} torr on the 12.4K cryosurface. These results are for operation in the constant H_2 addition rate mode. It is seen that the curves are displaced by time intervals which are proportional to the frost thickness. Consequently, if these data are replotted in terms of the dimensionless capacity parameter, C, they produce essentially a single pumping curve as shown in Figure 16-4b.

A similar series of sorption pumping curves is given in Figure 16-5a in which the H_2 sorbate flow rate was varied in order to maintain a constant chamber pressure during sorption. Because the sorbate flow rate is continuously decreasing, the pumping speed is shown as a function of the amount of H_2 sorbed. Again, on the basis of the dimensionless frost capacity, the pumping curves collapse into essentially a single curve as seen in Figure 16-5b. However, the scatter is somewhat greater than in Figure 16-4b, particularly in the vicinity of the knee in the curve.

The pumping speed of the "bare" cryosurface at a temperature of 12.4K is also given in Figures 16-4a and 16-5a. It appears that the stainless steel surface has a high pumping speed for several seconds and sorbs appreciable quantities of H_2. However, this is not the correct interpretation; the surface was actually covered with a thin but unknown amount of cryodeposit from the residual and desorbed gases in the chamber. This was witnessed by the fact that the chamber pressure decreased about a decade as the cryosurface was cooled to 12.4K. Consequently, the bare surface pumping effect is due to cryosorption.

It is observed from Figures 16-4 and 16-5 that the initial unit pumping speed for H_2 on CO_2 frost is about 30 liters/sec-cm^2. Since the theoretical strike rate of 300K H_2 on a surface corresponds to an equivalent maximum pumping speed of 44 liters/sec-cm^2, this implies that the maximum effective capture coefficient of 12.4K CO_2 frost for H_2 is approximately 0.7. It is also seen in Figures 16-4 and 16-5 that the pumping speed of CO_2

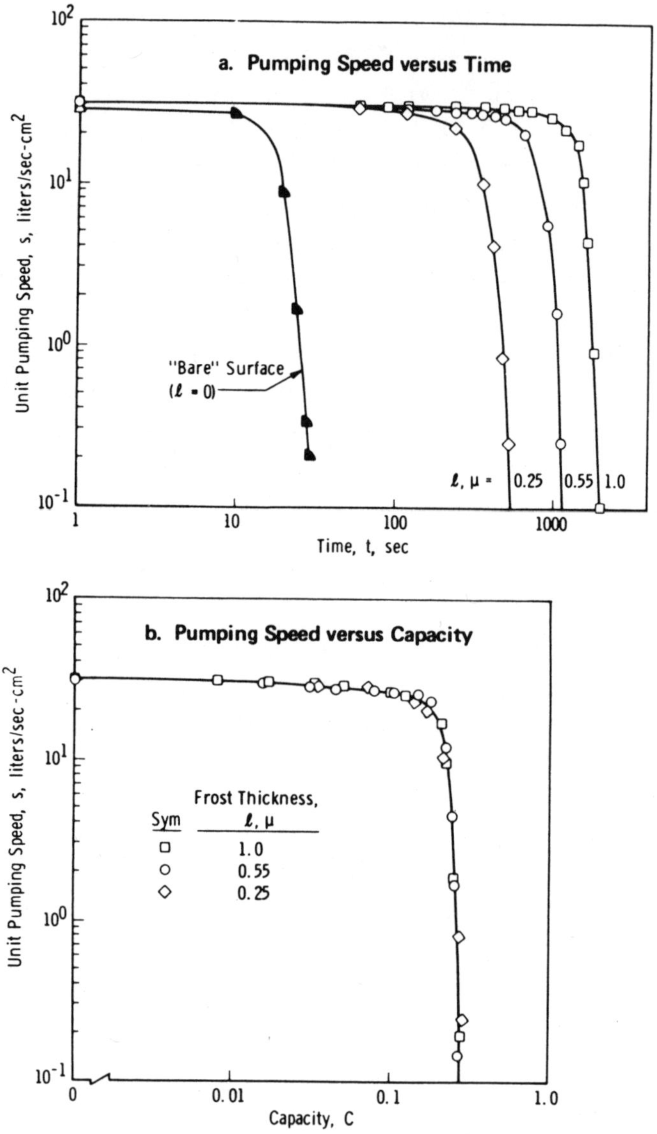

Fig. 16-4. Dynamic pumping speed curves for various thicknesses of 12.4K CO_2 frost formed at a chamber pressure of 2×10^{-5} torr on 12.4K surface—Constant H_2 sorbate addition rate (Ref. 2).

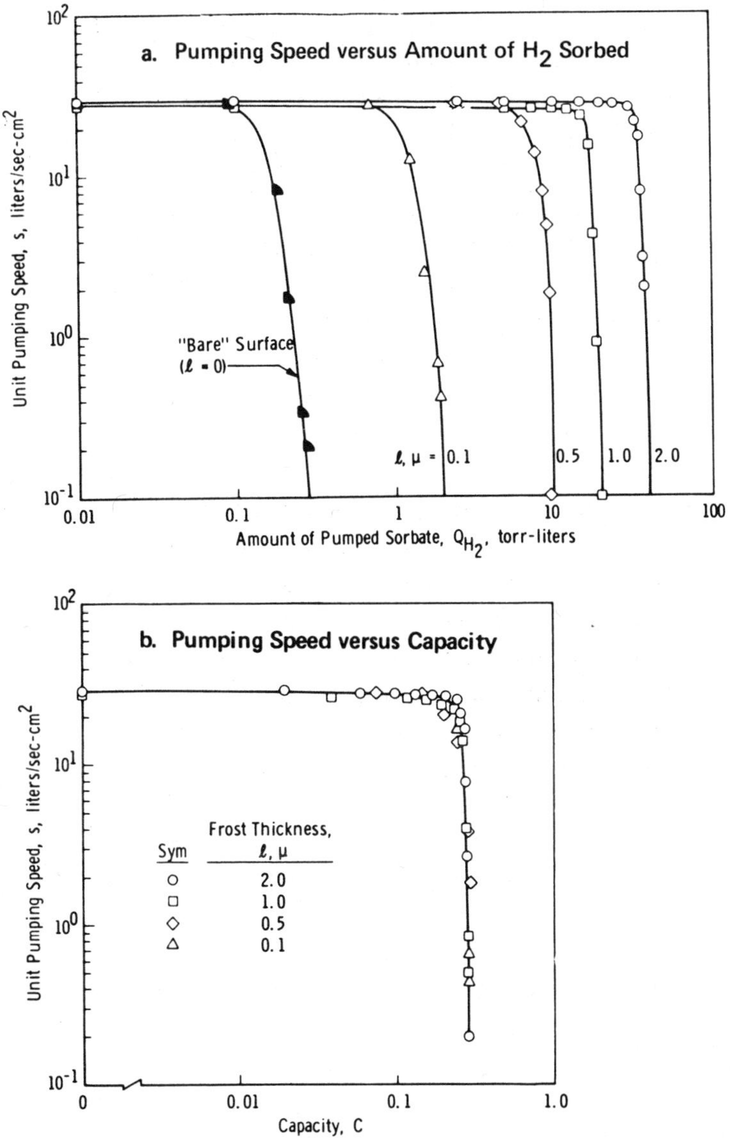

Fig. 16-5. Dynamic pumping speed curves for various thicknesses of 12.4K CO_2 frost formed at a chamber pressure of 2 x 10^{-5} torr on 12.4K surface–Variable H_2 addition rate at constant chamber pressure (Ref. 2).

frost always decreases rather slowly from its initial value as H_2 is sorbed. Although it appears on the log plots that the pumping speed is constant for some time, it actually always decreases steadily as the capacity increases. Then it falls off very rapidly as saturation is approached. Figures 16-4 and 16-5 further illustrate two important points. First, the initial unit pumping speed of the frost is completely independent of the frost volume or, for a constant cross-sectional area, the thickness. Second, the maximum amount of H_2 sorbed by a CO_2 frost is directly related to the frost volume or thickness. Consequently, a pumping speed curve measured for a frost at one thickness can be used to estimate the dynamic pumping characteristics of frosts formed at the same conditions but having different thicknesses.

Figure 16-6 shows the effect of frost temperature T_f on the H_2 pumping speed curves for CO_2 frost deposited at the indicated temperatures and two different formation pressures, P_{form}. The results shown are for a constant H_2 addition rate. It is seen that higher temperature frosts have lower initial pumping speeds. Also, the capacity range, or time interval over which the pumping speed is constant, decreases as the frost temperature increases. It is also noted in Figure 16-6 that for a given frost temperature the frosts formed at the higher pressure have lower initial pumping speeds and higher maximum sorption capacities.

Figure 16-7 shows the dynamic H_2 pumping speed curves for 12.4K CO_2 frosts deposited at various formation or strike rates, $\dot{n}_{form}$. Note that linear coordinates are employed here to better illustrate the differences in pumping characteristics of the different CO_2 frosts. It is seen that the CO_2 frosts formed at larger strike rates have smaller initial pumping speeds but greater maximum sorption capacities. In particular, frosts formed at strike rates greater than about 10^{15} molecules/cm^2-sec exhibit slowly decreasing pumping speeds over a wide range of sorption capacity. As the saturation point is approached, the pumping speed falls off abruptly. In contrast, the frost formed at the lower strike rate has a pumping speed which decreases rather rapidly during the initial stages of pumping. It is apparent in Figure 16-7 that the advantage

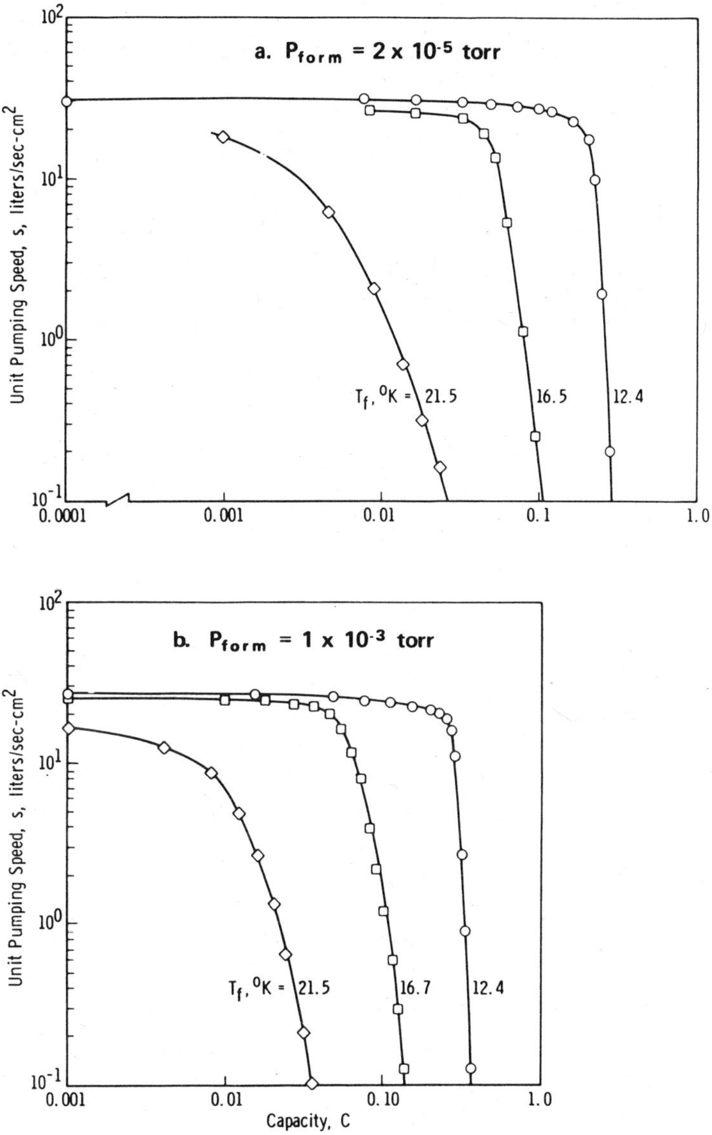

Fig. 16-6. Effect of frost temperature on dynamic H_2 pumping speed curves for CO_2 frosts deposited at two different formation pressures—Constant H_2 addition rate (Ref. 1).

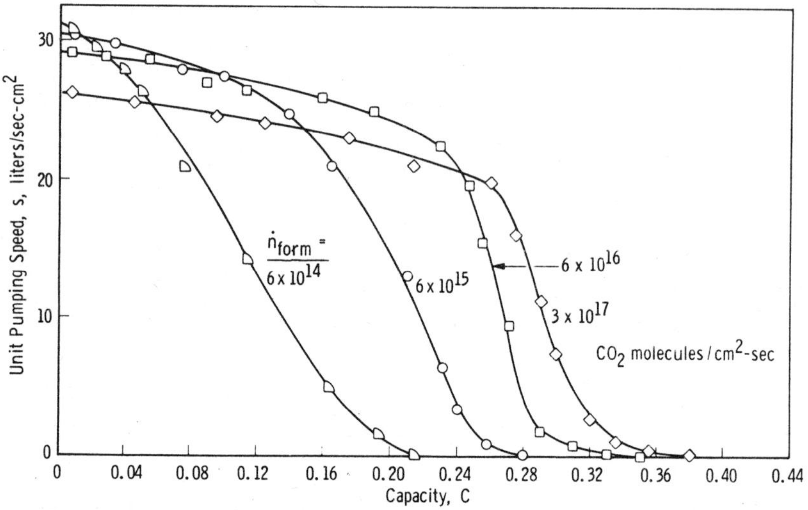

Fig. 16-7. Dynamic H_2 pumping speed curves for 12.4K CO_2 frosts formed at different strike rates (Ref. 2).

gained in increased sorption capacity by forming the CO_2 frost at high strike rates greatly outweighs the modest decrease in initial pumping speed.

Figure 16-8 shows how intermediate warming of the CO_2 frost affects the dynamic H_2 pumping characteristics. It is seen that when the frost is cycled through some temperature range its dynamic pumping speed curves are shifted to lower capacity. Figure 16-8 also shows that there is large loss in maximum sorption capacity if the frost is cycled to a temperature as high as 35K. In addition, it is noted from the curve through the filled circles that a CO_2 frost formed at 21K but cooled to 12.4K before sorption has a significantly smaller capacity than an unwarmed frost formed at 12.4K.

Figure 16-9 presents the dynamic H_2 pumping speed curves for 12.4K SO_2, CH_3Cl, and CO_2 frost sorbents. These frosts were all formed at essentially identical conditions and have equal thickness. All the curves have similar shapes but somewhat different values of initial pumping speed and saturation capacity. The CH_3Cl frost exhibited the highest initial pumping speed but the lowest saturation capacity.

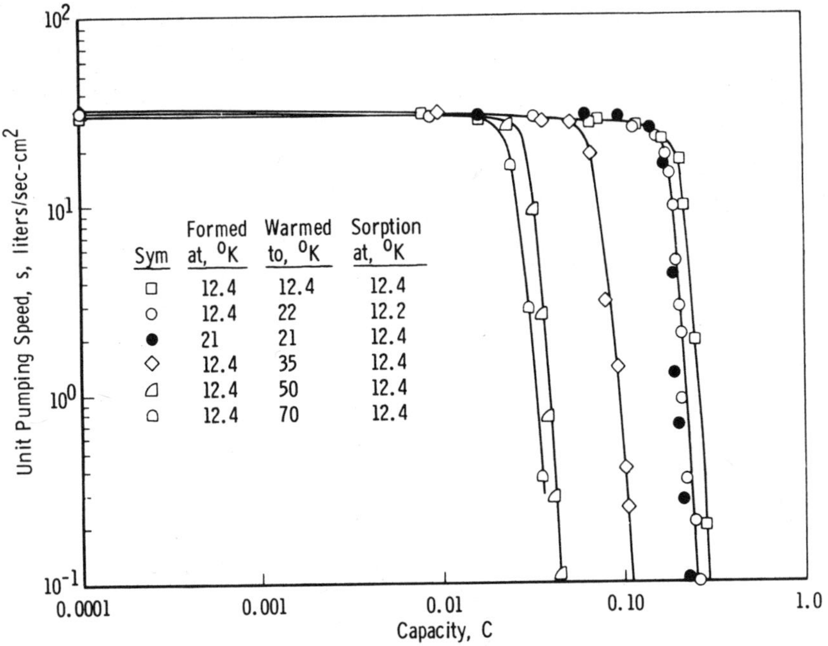

Fig. 16-8. Effect of cycling frost temperature on the dynamic H_2 pumping speed curves for CO_2 frosts formed at strike rate of 6.3 x 10^{15} molecules/sec-cm^2 (Ref. 2).

16-6. Summary

From the dynamic sorption pumping results, it is found that the initial unit pumping speed of H_2 by 12.4K CO_2 frost is 30 liters/sec-cm^2, which implies that the maximum effective capture coefficient is about 0.7. As the amount of H_2 sorbed by the CO_2 frost increases, the H_2 pumping speed decreases rather slowly at first and then very rapidly as saturation is approached. It is observed that initial unit pumping speed is independent of the frost thickness and the maximum amount of H_2 sorbed by the frost is directly proportional to the frost thickness. It is further observed that higher temperature CO_2 frosts have lower initial pumping speeds and lower maximum sorption capacities while CO_2 frosts formed at larger strike rates have somewhat smaller initial pumping speeds but much greater maximum sorption capacities. In addition, cycling the temperature of the CO_2 frosts through some intermediate range shifts the dynamic pumping speed curves to lower capacity. Finally, SO_2 and CH_3Cl cryofrosts have dynamic H_2 pumping speeds comparable to CO_2 frosts but somewhat lower saturation capacities.

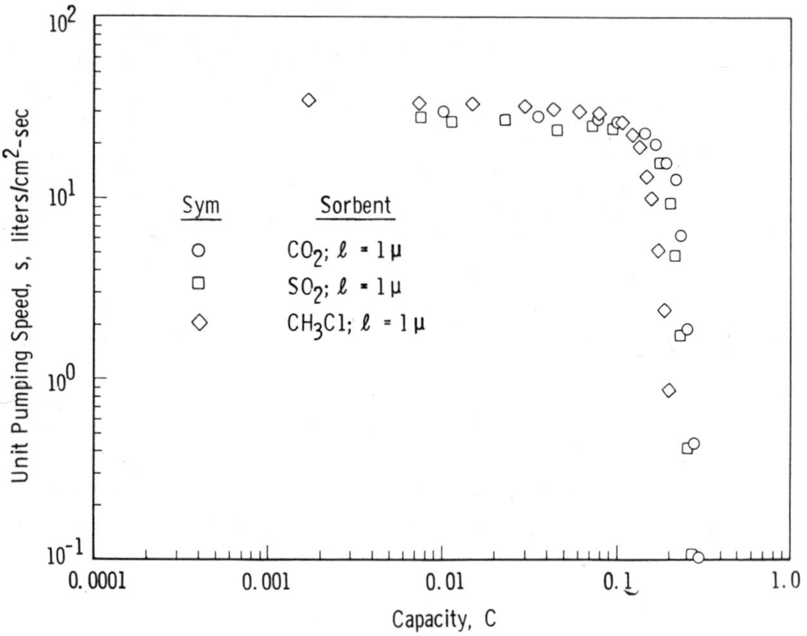

Fig. 16-9. Dynamic H_2 pumping speed curves of 12.4K CO_2, SO_2, and CH_3Cl frosts formed at the same conditions (Ref. 2).

References

1. R. T. Brackmann and W. L. Fite, "Condensation of Atomic and Molecular Hydrogen at Low Temperatures," *Journal of Chemical Physics*, **34**, 1572-1579, 1961.
2. K. E. Tempelmeyer, "Sorption Pumping of Hydrogen by Cryodeposits—Dynamic Pumping Characteristics," AEDC-TR-70-102 (AD712373), Arnold Engineering Development Center, Arnold Air Force Station, Tenn., 1970.
3. A. L. Hunt, C. E. Taylor, and J. E. Omahundro, "Adsorption of Hydrogen on Solidified-Gas Films," *Advances in Cryogenic Engineering Volune 8*, 100-109, Plenum Press, 1963.

4. R. E. Southerlan, "10-22°K Cryosorption of Helium on Molecular Sieve 5A and Hydrogen on Condensed Vapors," AEDC-TR-65-18 (AD455375), Arnold Engineering Development Center, Arnold Air Force Station, Tenn., 1965.
5. Ronald Dawbarn, "Cryosorption of Hydrogen by 12-20°K Carbon Dioxide Cryodeposits," AEDC-TR-67-125 (AD655067), Arnold Engineering Development Center, Arnold Air Force Station, Tenn., 1967.
6. V. B. Yuferov and F. E. Busol, "Sorption of Hydrogen and Neon by Layers of Solids Formed by Vapor Condensation," *Soviet Physics–Technical Physics,* **11**, 1518-1524, 1967.
7. E. Muller, "Adsorption Isotherms on Solid Carbon Dioxide," *Cryogenics,* **6**, 242-243, 1966.
8. K. E. Tempelmeyer, R. Dawbarn, and R. L. Young, "Sorption Pumping of Hydrogen by Carbon Dioxide Cryodeposits," *Journal of Vacuum Science and Technology,* **8**, 575-581, 1971.
9. K. E. Tempelmeyer, "Correlation of the Equilibrium Adsorption Isotherms of Low Temperature Cryodeposits," *Cryogenics,* **11**, 120-127, 1971.
10. K. E. Tempelmeyer, "The Sorption of Hydrogen by Sulfur Dioxide Frosts," *Journal of Vacuum Science and Technology,* **8**, 612-613, 1971.

Acknowledgment

The research reported in this paper was sponsored by the Arnold Engineering Development Center, Air Force Systems Command, under Contract No. F40600-72-C-0003 with ARO, Inc.

17

CRYOFROST BIDIRECTIONAL REFLECTANCE

A. M. Smith

The bidirectional reflectance of CO_2 and H_2O cryofrosts formed on polished copper and black paint cryosurfaces is presented. These results are given as a function of frost thickness and the irradiance zenith angle and wavelength. The various optical phenomena that occur in the bidirectional reflection distributions are discussed.

17-1. Introduction

Traditional cryopumping studies have concentrated on determining the capture coefficients of gases on cold surfaces and the vapor pressure associated with the resulting cryodeposits or cryofrosts. However, in the recent past, there have been several investigations which dealt primarily with the cryodeposit itself. The objective of this paper is to report on one of these investigations: studies[1] of the bidirectional reflectance of H_2O and CO_2 cryodeposits on LN_2-cooled black paint and polished copper substrates.

17-2. Bidirectional Reflectance

To begin, it is necessary to define bidirectional reflectance. In Figure 17-1, let $e_i(\psi)$ denote the radiant flux of the collimated beam incident at zenith angle ψ on the cryodeposit of thickness τ. Also, let $E(\psi,\theta,\tau)$ designate the reflected radiant flux leaving the deposit through solid angle $\Delta\omega_r$ in the direction defined by the zenith reflection angle θ and the plane of incidence. The bidirectional reflectance of the deposit-substrate surface system is then defined as $\rho_{bd}(\psi, \theta, \cdot\tau) = I_r(\psi, \theta, \tau)/e_i/\pi$ where $I_r(\psi, \theta, \tau)$ is the reflected radiant intensity, related to the reflected flux by $E=I_r$ cos

$\theta\ \Delta\omega_r$. Physically, this bidirectional reflectance definition can be interpreted as the ratio of the intensity reflected by an actual surface system (at angle θ in the plane of incidence) to the intensity that would be reflected in the same direction by a surface system which was perfectly reflecting and uniformly diffusing.

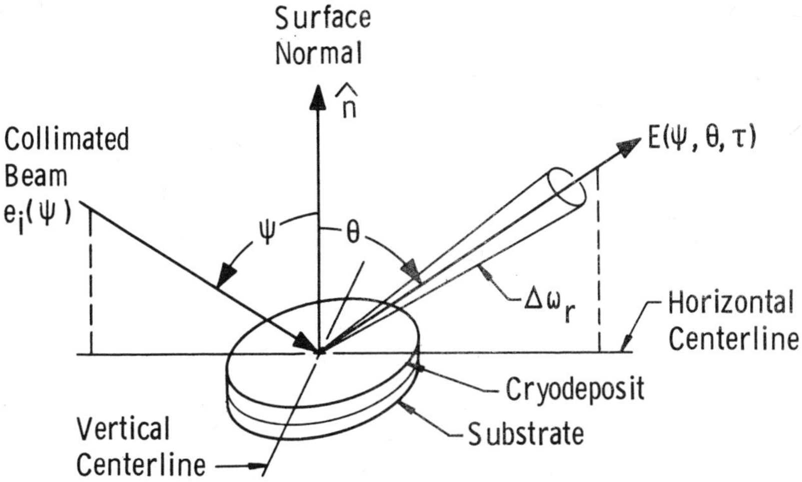

Fig. 17-1. Bidirectional reflectance measurement technique.

17-3. Experimental Apparatus

All bidirectional reflectance experiments were conducted in a spherical vacuum chamber which contained an LN_2-cooled cryoplate, as shown in Figure 17-2. To minimize internal reflections, all interior surfaces of the chamber except the cryoplate were painted flat black. The pumping system for the chamber consisted of a mechanical forepump and a 6-in. oil diffusion pump equipped with an LN_2 cold trap. Pressures of 10^{-7} torr could be obtained with this system. Chamber pressure was measured with a thermal ionization gage at pressures below 10^{-3} torr and with an alpha particle ionization gage at pressures above 10^{-3} torr. The copper cryoplate located in the center of the chamber and the LN_2 lines leading to it were completely shielded by a vacuum jacket except for the flat front face of the cryoplate where the cryodeposits were formed. This face was polished to an rms surface roughness of approximately 0.01 μ. It was used as the test surface and was either coated with black epoxy paint or was just the bare polished copper. A copper-constantan thermocouple was attached to the test surface to measure the deposition temperature. The test surface could be rotated about its vertical centerline to obtain any desired polar angle between the collimated beam of incident radiation and the test surface normal. As shown in the schematic, the collimated irradiance was obtained from a xenon arc lamp by means of a system of lenses and apertures. Monochromatic irradiance could be obtained by inserting narrow bandpass interference filters into the radiation beam. These filters typically had a bandwidth at half-maximum transmission of 0.015 μ ($\pm$ 0.007 μ). All measurements of reflected radiant flux were made using a silicon solar cell detector mounted on a remotely rotatable arm (see Figure 17-2). Carbon dioxide gas of 99.99% purity could be introduced into the vacuum chamber at a constant flow rate by means of calibrated orifice leaks. Water vapor was introduced into the chamber at a constant rate by allowing degassed distilled water to evaporate under vacuum from a reservoir and then pass through a rotameter and needle valve.

17-4. Experimental Procedure

After the vacuum chamber had been evacuated to a pressure of about 1×10^{-7} torr and the xenon lamp turned on, a transmission filter of a certain wavelength λ was inserted into the radiation beam and the test surface rotated to the desired zenith incidence angle ψ. Then, with the test surface still at 300K, the detector was rotated about it in the plane of incidence and a reference trace of the angular distribution of the reflected radiant flux was made for the surface when no cryodeposit was present (τ=0). The test surface was rotated to other values of ψ and the mapping procedure repeated for various wavelengths. Next, the cryoplate was cooled to 77K, and, after valving off the chamber from the pumping system, a cryodeposit was continuously formed on the test surface by flowing H_2O vapor or CO_2 gas into the chamber at a constant rate. The deposition rate $\dot{\tau}$ for the cryodeposit was measured in situ by recording thin film interference patterns for two different incidence angles.[2,3] With the constant deposition rate known, a cryodeposit layer of specified thickness τ was formed by flowing vapor (or gas) into the chamber for a specified time. The thickness of this layer was always chosen larger than the maximum deposit thickness at which thin film interference was observed. Throughout deposition the chamber pressure, p, was approximately 4×10^{-4} torr. After the vapor (or gas) flow was stopped, angular distributions of the reflected radiant flux were recorded for various incidence angles and wavelengths. Then, the vapor (or gas) was again added to the chamber forming another cryodeposit layer on top of the first and the above angular distribution measurements were made again. This procedure was repeated and angular distribution measurements made for each resulting deposit thickness.

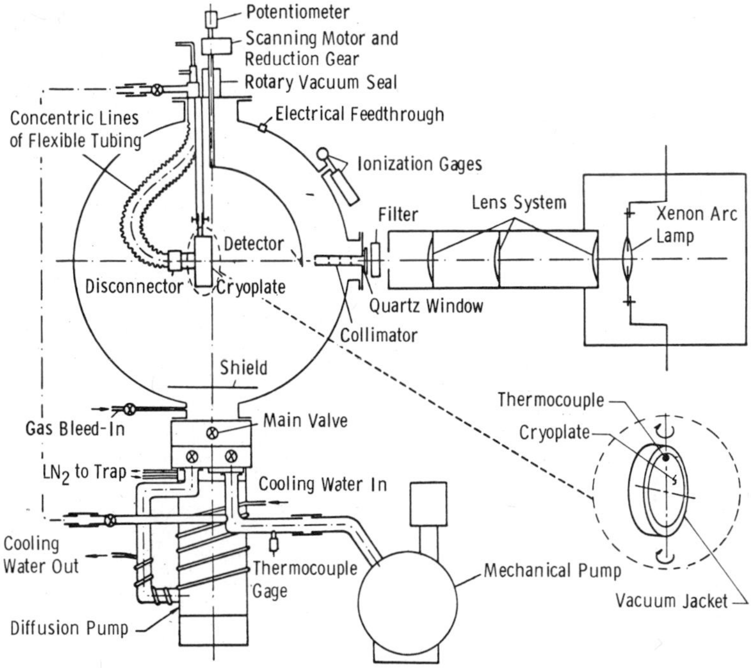

Fig. 17-2. Experimental apparatus for bidirectional reflectance measurements.

17-5. Reflectance Results for H_2O and CO_2 Deposits on Polished Copper

Figure 17-3 presents the relative bidirectional reflectance for various thicknesses of H_2O and CO_2 deposits formed on a polished copper substrate. These particular results are for an irradiance wavelength of 0.9 μ and a zenith incidence angle of 11 degrees. The reflectance values of less than 0.012 in Figure 17-3 are displayed on a greatly increased scale in Figure 17-4. It is seen from Figures 17-3 and 17-4 that the presence of CO_2 or H_2O cryodeposit on a polished copper substrate causes the bidirectional reflectance in the specular direction (θ=11 deg) to drastically decrease for deposit thicknesses of 20 μ and greater. Note that for a given thickness, the reduction in the specular peak of the bidirectional

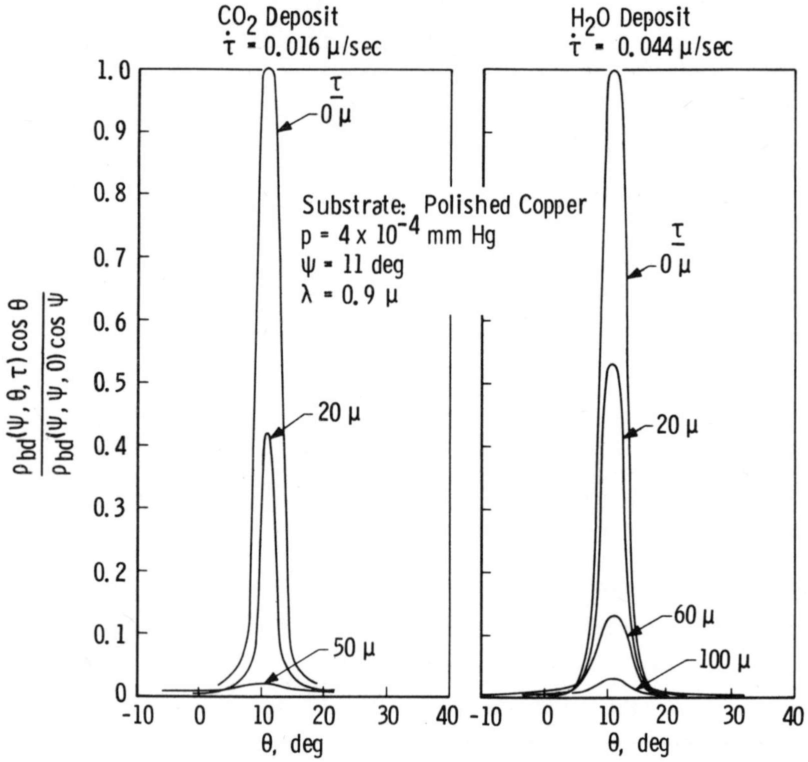

Fig. 17-3. Relative bidirectional reflectance distributions for various thicknesses of CO_2 and H_2O deposits, ψ=11 deg, λ=0.9 μ.

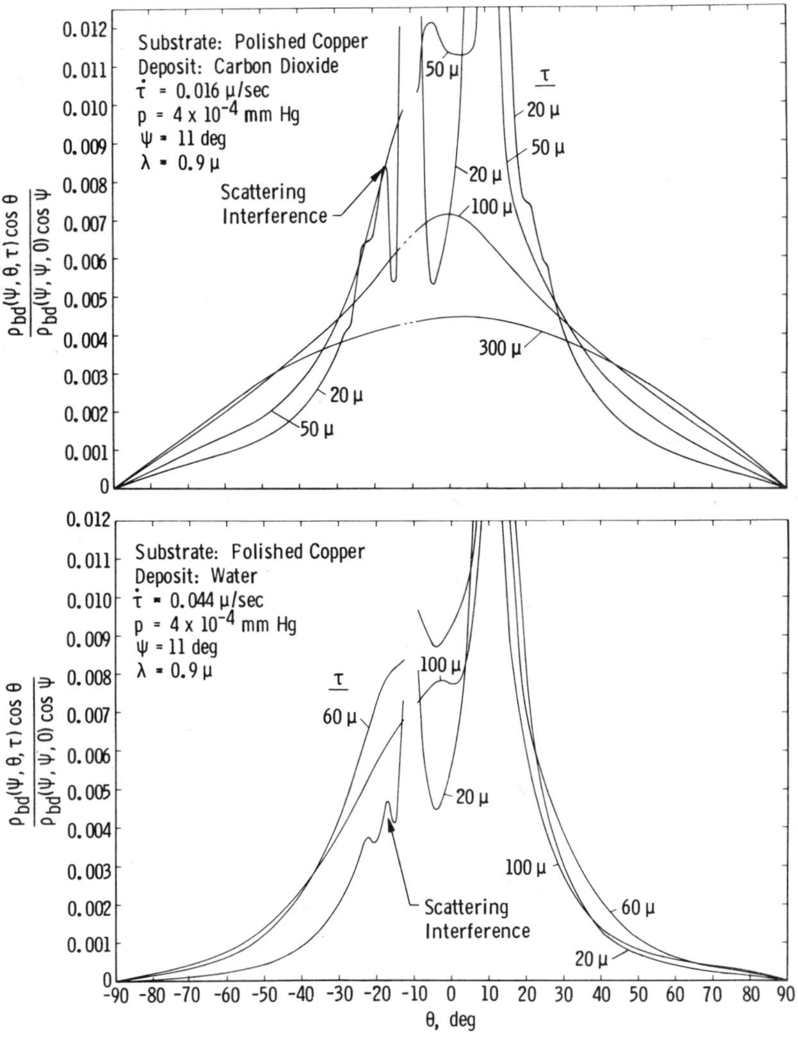

Fig. 17-4. Bidirectional reflectance distributions for various thicknesses of CO_2 and H_2O deposits on polished copper, ψ=11 deg, λ=0.9 μ (Note change in ordinate scale).

refelction distribution is greater for a CO_2 deposit than an H_2O deposit. At a thickness of 100 μ, the specular peak has disappeared in the distribution for CO_2 deposit but is still well-defined for the H_2O deposit, although relatively small. As observed in Figures 17-3 and 17-4, this large decrease in the specular peak as the cryodeposit thickens is accompanied by an increase in the bidirectional reflectance in nonspecular directions. For a CO_2 cryodeposit, the reflection distribution becomes essentially diffuse as the deposit thickness approaches 300 μ, while for an H_2O deposit, the distributions are increasingly nonspecular but do not become diffuse for the thicknesses investigated. Such behavior can be explained by noting that due to both surface and internal scattering by the cryodeposit, the incident monochromatic radiation will be reflected in directions other than just the specular direction. At first, when the deposit is relatively thin, the surface scattering will likely predominate. As the deposit thickens, the internal scattering becomes increasingly significant thereby causing the reflection distribution to become even more diffuse. It is apparent from Figure 17-4 that surface and internal scattering by the H_2O deposit is appreciably less than that for the CO_2 deposit, since the distributions for the H_2O deposit are much less diffuse than those for a CO_2 deposit of approximately equal thickness. Also observed in Figure 17-4 for a 20 μ thick deposit are the scattering interference peaks[3] which occur in the bidirectional reflection distributions for relatively thin cryodeposits formed on a specularly reflecting substrate. These peaks (and valleys) are a result of interference which is generated by scattering of the incident radiation at the vacuum-cryodeposit interface.

The bidirectional reflection distributions obtained for CO_2 and H_2O cryodeposits on polished copper using large irradiance incidence angles are shown in Figure 17-5. These results are for monochromatic irradiance of 0.9 μ wavelength and are presented in a manner similar to that of Figure 17-4. For these large incidence angles, it is observed that the specular peak decreases and the reflection distributions become more diffuse with increasing deposit thickness in a manner similar to that observed for ψ=11 deg. There are, however, several phenomena observed in the distributions for large incidence angles that were not readily apparent in the distributions for small incidence angles. One of these is the "backscattering" of reflected radiation into the quadrant of incidence ($\theta < 0$ deg) in the general direction of the irradiance, $\theta \cong -\psi$. As seen in Figure 17-5, this phenomenon is primarily observed for relatively thin CO_2 and H_2O deposits. Another

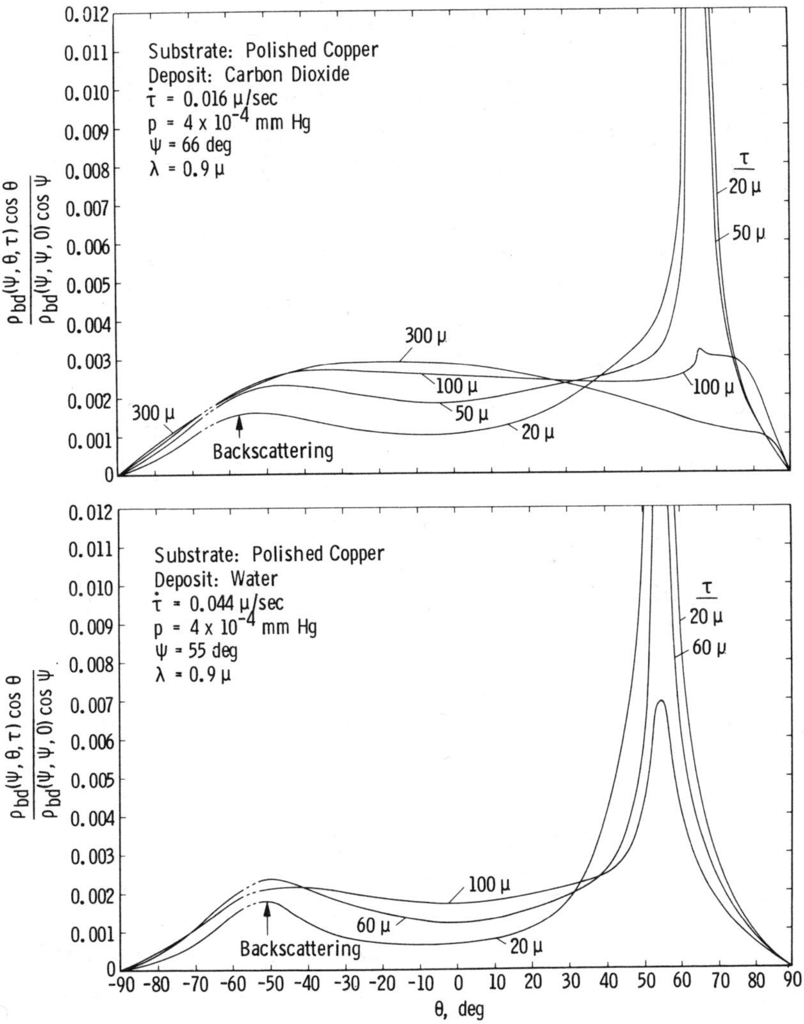

Fig. 17-5. Bidirectional distributions of λ=0.9 μ radiation reflected from various thicknesses of CO_2 and H_2O deposits on polished copper for large incidence angles.

interesting difference between the distributions for large and small incidence angles is seen in comparing the results of Figures 17-4 and 17-5 for a CO_2 deposit of 100 μ thickness. Note in Figure 17-4, which is for ψ=11 deg, that there is not a specular peak in the reflection distribution for the 100-μ-thick CO_2 deposit. However, for the same CO_2 deposit thickness and ψ=66 deg, there is a specular peak at θ=66 deg in the reflection distribution as seen from Figure 17-5. The appearance of this peak in the distribution results for ψ=66 deg is not due to specular reflection off the polished copper substrate because the extinction of the transmitted incident beam inside the deposit is greater for larger incidence angles as a result of the increased geometrical path length. Hence, the emergence of the specular peak for ψ=66 deg, when one did not appear for ψ=11 deg, must be due to specular reflection off the vacuum-deposit interface of the 100-μ-thick CO_2 deposit. This appearance of a specular peak in the reflection distribution as the incidence angle increases is in agreement with existing experimental and theoretical work[4,5] on specular reflection from a rough dielectric surface. From the results of Ref. 4 it was found that with increasing incidence angle the specular peak first becomes noticeable in the reflection distributions when the condition $\sigma \cos \psi_0 = \lambda$ is satisfied, where σ is the root mean square roughness of the surface, λ is the wavelength of the incident radiation, and ψ_0 is the incidence angle for which the specular peak first appears.

Another phenomenon observed in Figure 17-5 for the 100 μ CO_2 deposit on copper is that in addition to the presence of a specular peak in the reflection distribution, there is also an off-specular peak.[6] This off-specular peak occurs for an incidence angle of ψ=66 deg and is located at $\theta > \psi$ in the neighborhood of $\theta \cong 75$ deg. The appearance, for large incidence angles, of both a specular peak and an off-specular peak in reflection distributions has been observed previously[7] for rough dielectric surfaces with a σ/λ ratio somewhat larger than unity. An explanation of this phenomenon is presented in Ref. 7 and is based on a representation of the rough dielectric surface as an aggregate of microscopic facets with different inclinations, each of which reflects radiation in accordance with the laws of geometrical optics. It is shown in Ref. 7 that the occurrence of the specular peak and off-specular peak in the reflection distribution can be fully attributed to the specular reflection of incident radiation from the microscopic facets and is not a result of internal scattering. The off-specular peak is found to be caused by Fresnel reflection from microscopic facets that have relatively small inclinations with respect to the macroscopic surface. The specular peak is considered to be produced by interference of the radiation specularly reflected from those microscopic facets which are parallel to the macroscopic surface.

It is also seen in Figure 17-5 that for a 300 μ CO_2 deposit on polished copper and ψ=66 deg, the reflection distribution is essentially diffuse except for the off-specular "peak" occurring in the 80 deg $\leq \theta <$ 90 deg region. Note that the specular peak which was seen in the distribution for the 100 μ deposit has vanished and also the off-specular peak observed for the 100 μ deposit has shifted to larger polar reflection angles. Such behavior would result if the vacuum-deposit interface of the 300 μ deposit was considerably rougher than that of the 100 μ deposit. The specular peak would vanish due to the increased value of σ and the off-specular peak would shift to larger reflection angles due to the increased rms inclination of the microscopic facets.

It is finally observed in Figure 17-5 that for H_2O deposits on polished copper the reflection distributions for a large incidence angle (ψ=55 deg) do not approach diffuseness for any deposit thickness investigated. In addition, a quite well-defined specular peak appears in all the distributions. Also, no off-specular peaks occur in any of the distributions. This again indicates that the internal and surface scattering of the H_2O cryodeposit is appreciably less than that of the CO_2 deposit.

17-6. Reflectance Results for H_2O and CO_2 Deposits on Black Paint

Figure 17-6 presents the bidirectional distributions of total visible and near IR radiation reflected from CO_2 and H_2O deposits formed on a black epoxy paint substrate. The results shown are for an incidence angle of 33 deg and deposit thicknesses ranging from 10 to 500 μ. It is seen that for CO_2 and H_2O deposit thicknesses up to 50 μ the bidirectional reflectance in the specular direction is greatly reduced. Note that for a given thickness this reduction is larger for the CO_2 deposit. At a deposit thickness of 50 μ the peak in the specular direction has disappeared for the CO_2 deposit on black paint but is still relatively well-defined for the H_2O deposit. Just as for the CO_2 and H_2O deposit formed on polished copper, this decrease of the peak in the distribution and the accompanying increase in the bidirectional reflectance in nonspecular directions is due to surface and internal scattering. As the deposit thickens, the internal scattering becomes increasingly significant. This is vividly illustrated in Figure 17-6 by noting that an increase in CO_2 deposit thickness from 50 to 300 μ causes the bidirectional reflectance in most all directions ($\theta \neq \pm 90$ deg) to increase about an order of magnitude. The same thickness increase for H_2O causes a similar but smaller bidirectional reflectance increase indicating again that internal scattering by the H_2O deposit is less than that for the CO_2. It is noted that for a CO_2 deposit thickness of 300 μ the reflection distribution approximates a diffuse distribution. However, the bidirectional reflectance in the direction normal to the surface $\theta=0$ deg is slightly less than that for adjoining directions lying within the quadrants of incidence ($\theta < 0$ deg) and specular reflection ($\theta > 0$ deg). For an H_2O deposit on black paint, the reflection distributions do not become completely diffuse for any thickness investigated since a peak is always observed in the specular direction as seen in Figure 17-6.

The bidirectional distributions of total visible and near IR radiation reflected from CO_2 and H_2O cryodeposits on black paint for an incidence angle of 66 deg are shown in Figure 17-7. It is seen that for the CO_2 deposit the peak in the bidirectional reflectance decreases and the reflection distributions become more diffuse with increasing thickness in a manner somewhat similar to that observed for $\psi=33$ deg. Note that the peak in the distributions does not vanish at as small a deposit thickness as for $\psi=33$ deg. For the H_2O deposit, the peak in the distribution first decreases then increases as the deposit thickens and the distributions become more diffuse as seen in Figure 17-7. Note that these distributions are not as diffuse as those obtained for ψ=33 deg.

Figure 17-8 presents the effect of irradiance incidence angle on the bidirectional reflectance distributions for a 300 μ thick CO_2 deposit on a black epoxy paint substrate. The incidence angle dependence is shown for λ=0.9 μ with the distribution results for different incidence angles being normalized by $\rho_{bd}(5^o, 5^o, 0) \cos 5^o$. It is seen that for normal incidence, ψ=0 deg, the reflection distribution is essentially diffuse. However, for non-normal incidence (ψ=33, 55, and 66 deg), the distributions have two rather broad peaks which are located symmetrically about θ=0 deg. The magnitude of the peak located in the incidence quadrant appears to be slightly larger than the magnitude of the peak located in the specular reflection quadrant. Thus, for large incidence angles, thick CO_2 deposits on a black paint substrate tend to scatter more in the backward direction than the forward. It is also seen in Figure 17-8 that the bidirectional reflectance increases as the incidence angle becomes larger. In addition, it is noted that for an incidence angle of 66 deg an off-specular peak occurs in the reflection distribution at $\theta \cong 80$ deg.

Figure 17-9 presents the effect of irradiance wavelength on the reflection distribution for H_2O and CO_2 deposits on black paint. The results shown for the 300 μ CO_2 deposit are for an incidence angle of 33 deg while those shown for the 500 μ H_2O deposit are for an incidence angle of 66 deg. It is seen that for the CO_2 deposit the bidirectional reflectance decreases as irradiance wavelength increases. A similar decrease is observed for the H_2O deposit at reflection angles away from the vicinity of the specular peak. This reduction in the bidirectional reflectance is attributed to a decrease in internal scattering with increasing wavelength. It is seen in the distribution results for the H_2O deposit that the specular peak, which is due to specular reflection off the vacuum-deposit interface, increases in magnitude as the irradiance wavelength becomes larger. This behavior is consistent with the wavelength dependence predicted and observed[4,5] for roughened dielectric interfaces, $\exp[-(4\pi\sigma \cos \psi/\lambda)^2]$.

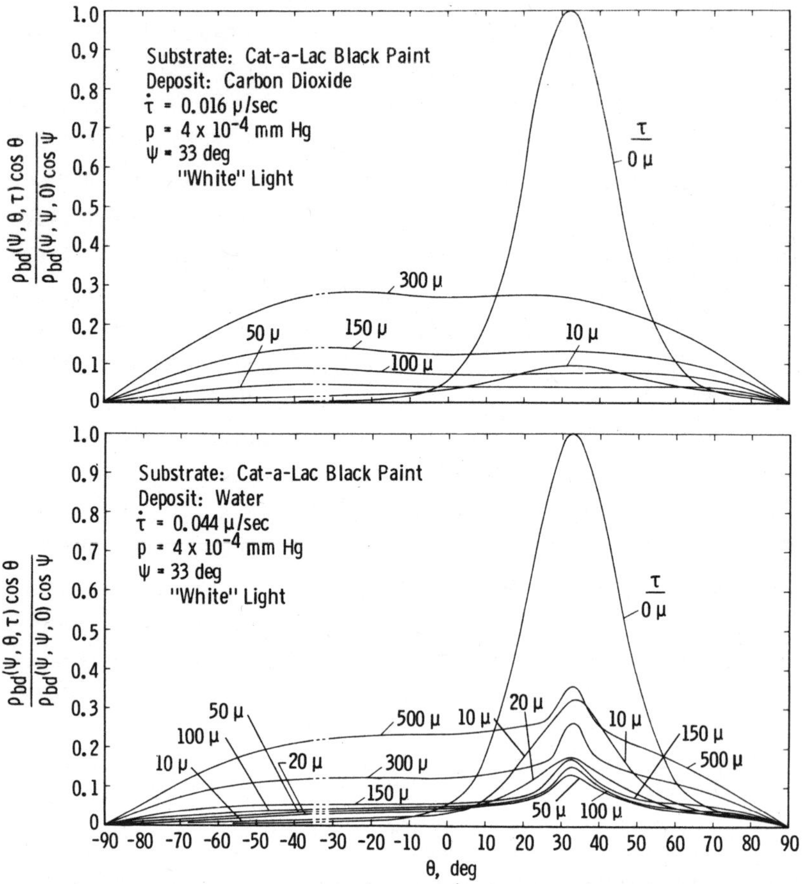

Fig. 17-6. Bidirectional distributions of visible and near IR radiation reflected from various thicknesses of CO_2 and H_2O deposits on black epoxy paint, ψ=33 deg.

17-7. Off-Specular Peaks in the Reflection Distributions

It was noted previously that for an incidence angle of ψ=66 deg off-specular peaks occur in the reflection distributions for relatively thick CO_2 deposits on polished copper (see Figure 17-5). As seen in Figure 17-7, off-specular peaks are also observed in the visible and near IR distribution results for CO_2 and H_2O deposits on black epoxy paint. These peaks occur at reflection angles θ_p which are greater than the specular reflection angle, $\theta=\psi$, and only

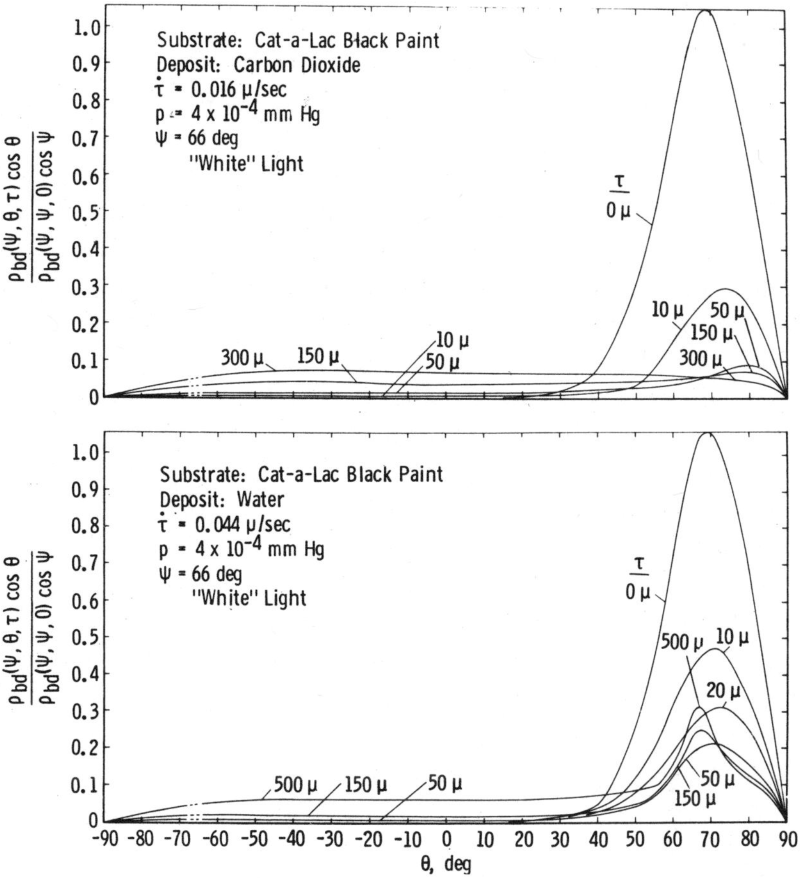

Fig. 17-7. Bidirectional distributions of visible and near IR radiation reflected from various thicknesses of CO_2 and H_2O deposits on black epoxy paint, ψ=66 deg.

appear for large incidence angles as indicated by comparison of Figure 17-7 with Figure 17-6. In Figure 17-6, which is for ψ=33 deg, no off-specular peaks are observed in the distributions. However, in Figure 17-7, which is for ψ=66 deg, off-specular peaks are present in the distributions for the bare substrates and all CO_2 deposit thicknesses and also are observed in the distributions for H_2O deposit thicknesses up to and including 50 μ. It is also noted in Figure 17-7 that the difference between the angular locations of the off-specular peak and the specular direction, $\Delta\theta=\theta_p-\psi$, depends upon the thickness of the deposit. This is better illustrated in Figure 17-10 where $\Delta\theta$ for an incidence angle of 66 deg is displayed as a function of deposit thickness for CO_2 and H_2O deposits on black paint. The results shown for the CO_2 deposits were obtained from reflection distributions for irradiance wavelengths of 0.7 and 0.9 μ while those presented for the H_2O deposits are from distributions for 0.9 μ wavelength. It is observed that for CO_2 cryodeposits the angular displacement of the off-specular peak relative to the specular direction increases from approximately 3 deg to about 14 deg as the deposit thickness increases from 0 μ to 50 μ. For CO_2 deposit thicknesses larger than 50 μ, $\Delta\theta$ remains essentially constant at approximately 14 deg.

The results shown in Figure 17-10 for H_2O deposits are taken from the reflection distributions for films formed at deposition rates of 0.012, 0.044 and 0.098 μ/sec. From these data, it appears that there is not an appreciable effect of deposition rate on the angular displacements of the off-specular peaks. It is seen in Figure 17-10 that for H_2O deposits the angular displacement of the off-specular peak relative to the specular direction increases from 3 deg up to a maximum of 7 deg as the deposit thickness increases

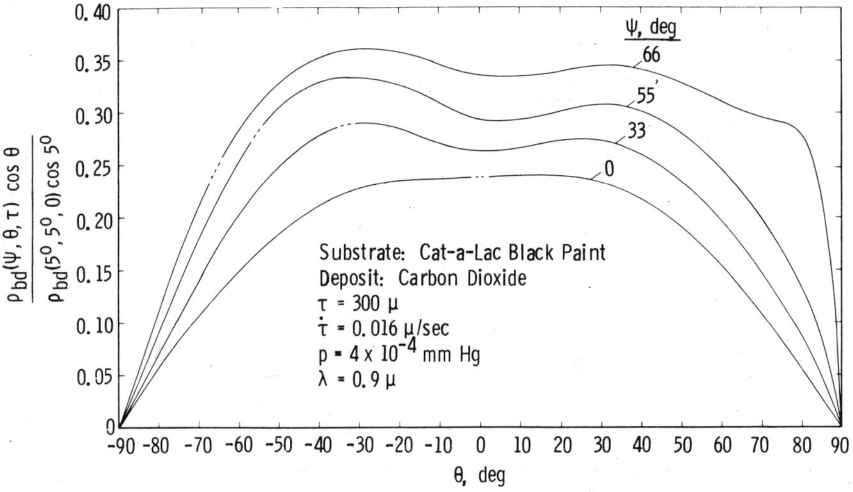

Fig. 17-8. Effect of incidence angle on bidirectional reflection distribution for a 300 μ CO_2 deposit on black epoxy paint, λ=0.9 μ.

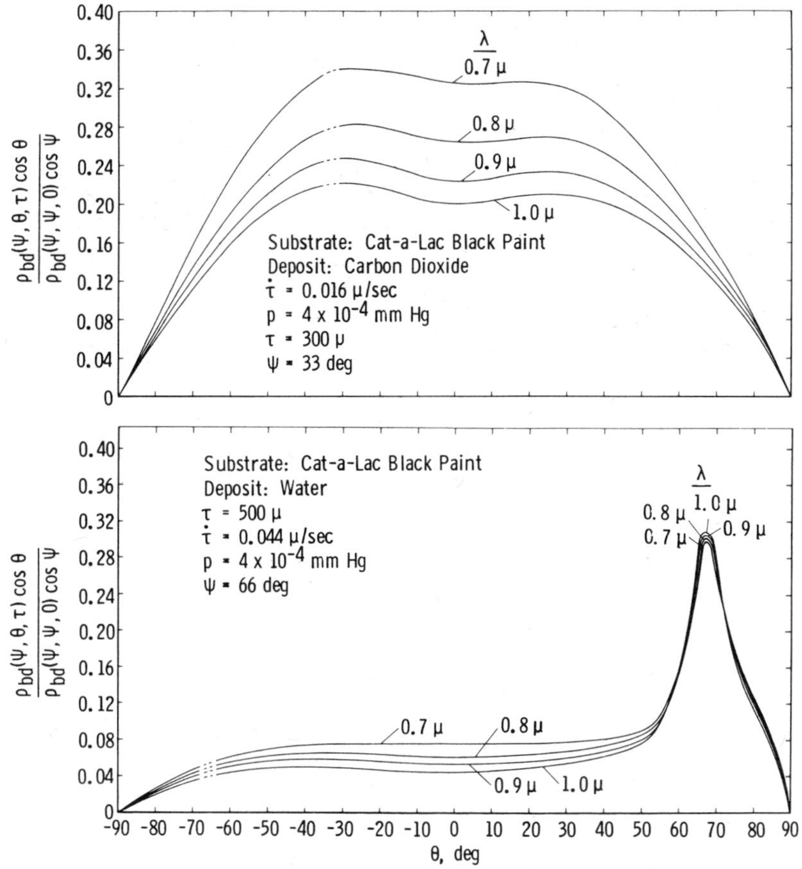

Fig. 17-9. Effect of irradiance wavelength on the bidirectional reflection distributions for thick CO_2 and H_2O deposits on black epoxy paint.

from 0 to approximately 25 μ. When the deposit thickness increases above 25 μ, the angular displacement of the off-specular peak begins to decrease. An example of this decrease may be seen in Figure 17-7 by comparing the 20 and 50 μ distributions for H_2O. It is further observed in Figure 17-10 that additional increase in the H_2O deposit thickness causes a continuing decrease in the angular displacement of the off-specular peak. As the H_2O deposit thickness approaches 150 μ, the off-specular peak merges with the arising specular peak. This can be seen, for example, in

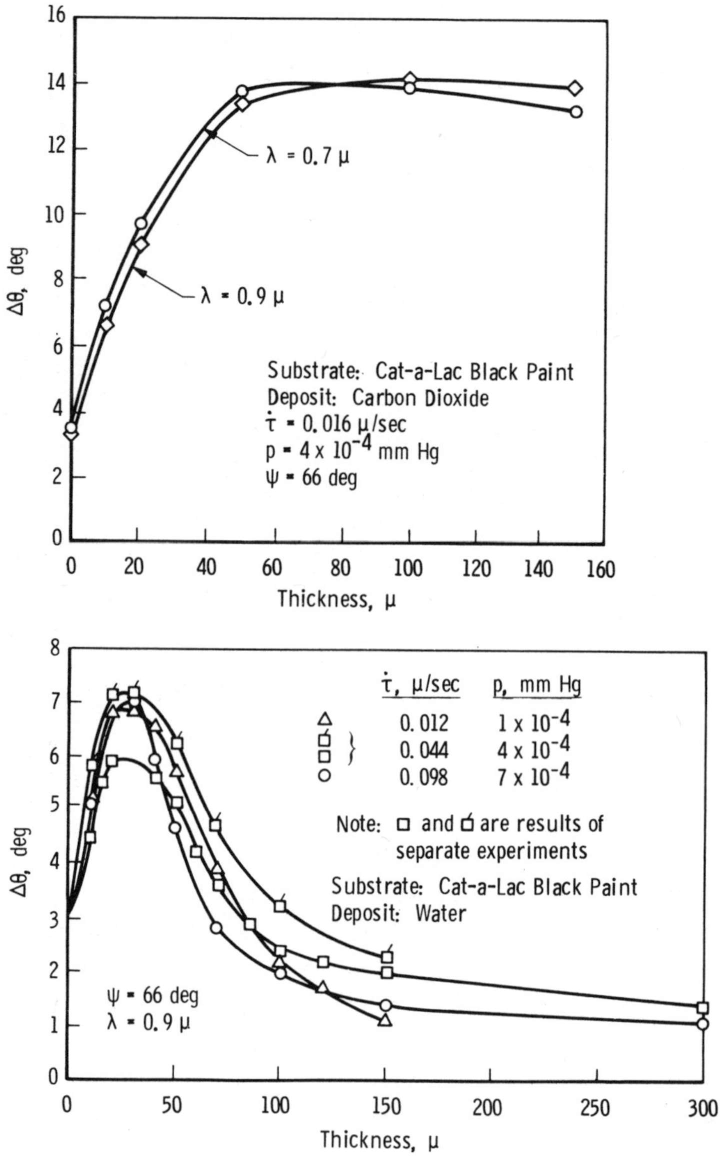

Fig. 17-10. Angular displacement of the off-specular peak relative to the specular direction for various thicknesses of CO_2 and H_2O deposits on black epoxy paint, ψ=66 deg.

the 150 μ H_2O distribution of Figure 17-7. Further increase of the H_2O deposit thickness above 150 μ causes the magnitude of the specular peak to rise in a manner similar to that observed for the 150 to 500 μ thickness increase in Figure 17-7. The resulting 500 μ H_2O distribution for λ=0.9 and $\dot{\tau}$=0.044 μ/sec is shown in Figure 17-9.

The occurrence of off-specular peaks for CO_2 and H_2O cryodeposits formed on a black epoxy paint substrate, as seen in Figure 17-7, and the dependence of their angular location on deposit thickness, as shown in Figure 17-10, can possibly be explained as follows. It is speculated that the rms roughness of the surface of the black paint substrate was comparable to or greater than the wavelength of the incident radiation and the rms inclination of the microscopic facets comprising this surface, while small, was large enough to cause the off-specular peak observed in Figure 17-7 for the bare substrate. It is further postulated that the top surface of the thin cryodeposits formed on the black paint substrate was even rougher than the substrate surface and thus the rms facet inclination for the cryodeposit surface was larger than the rms facet inclination for the bare paint surface. If this postulate is correct, the angular displacement of the off-specular peak relative to the specular direction should be larger as is observed for the 10 μ deposits in Figures 17-7 and 17-10. The increase in the angular displacement of the off-specular peak as the deposit thickened then indicated that the deposit surface became increasingly rougher and its rms facet inclination increased until apparently reaching some maximum value at a certain deposit thickness. For the rough CO_2 deposit, this maximum value of the rms facet inclination appears to be a limiting value since further increase in the deposit thickness did not change the angular displacement of the off-specular peak. For the H_2O deposit surface, the rms facet inclination, after attaining its maximum value, apparently decreased to a very low value with further increase in deposit thickness since the angular displacement of the off-specular peak decreased and became quite small as seen in Figure 17-10. This indicates that the H_2O deposit surface, after reaching a maximum roughness, became increasingly smoother with additional deposit thickness increase. Such behavior for the H_2O cryodeposit surface is further confirmed by the emergence of a specular peak in the distributions of Figure 17-7 and the increase in the magnitude of this peak as the deposit thickened.

From the above discussion and the results in Figure 17-10 for H_2O and CO_2 deposits, it appears that the surface profile characteristics of the substrate have a large effect on the surface profile characteristics of thin cryodeposits but little influence on the surface profile characteristics of thick cryodeposits. This observation is further borne out by noting that in the distribution results for the deposits formed on the polished copper substrate, no off-specular peaks ever occurred for the H_2O deposits and off-specular peaks did not appear for CO_2 until the deposit thickness approached 100 μ (see Figure 17-5). In this case, the rms surface roughness of the substrate (0.01 μ) was much less than the wavelength of the incident radiation. Hence, the top surface of a thin deposit formed on the polished copper, while likely to be rougher than the substrate surface, would probably not have an rms facet inclination large enough to cause an off-specular peak. However, as the deposit thickened, its surface would become rougher and the rms facet inclination would increase to a value which might be large enough to produce an off-specular peak. This is apparently what happened for the CO_2 deposits formed on polished copper. For the H_2O deposits formed on polished copper, it is speculated that the deposit surface became rougher and the rms facet inclination increased with thickness until reaching a limiting value which was not large enough to cause an off-specular peak. Hence, the surface of the H_2O deposit, although rougher than that of the copper substrate, was probably relatively smooth compared to the wavelength of the incident radiation.

One may ask why the surface of CO_2 cryodeposit is rough while that of the H_2O cryodeposit is smooth. This is a result of the different structure these two cryodeposit specie have when formed at 77K and low pressure. As shown by x-ray diffraction in Ref. 8, carbon dioxide deposit formed under these conditions has a polycrystalline structure and is comprised of large macro-crystals thereby causing the surface of the deposit to be rough. However, water cryodeposit formed at 77K and low pressures has the form of vitreous ice and x-ray diffraction studies of these deposits in Ref. 9 indicate that they have an amorphous structure and thus a relatively smooth surface.

17-8. Summary

From the bidirectional reflectance measurements, it is found that the presence of either H_2O or CO_2 cryodeposit on a polished copper or black paint surface causes radiation to be reflected more diffusely. For a deposit thickness of 300 μ, the bidirectional reflection distribution of a CO_2 deposit is essentially diffuse. However, for an H_2O deposit, the reflection distribution retains a significant specular peak even at a deposit thickness of 500 μ. Off-specular peaks are observed in the distributions for H_2O and CO_2 cryodeposits and the angular location of these peaks relative to the specular direction is found to be a function of deposit thickness, substrate, and irradiation incidence angle. In addition, the experimental results indicate that thin CO_2 and H_2O cryodeposits backscatter significantly for large incidence angles. Also, scattering interference patterns are seen in the bidirectional measurements for H_2O and CO_2 cryodeposits formed on polished copper.

References

1. P. R. Muller, W. Frost, and A. M. Smith, "Measurements of Refractive Index, Density, and Reflected Light Distributions for Carbon Dioxide and Water Cryodeposits," AEDC-TR-69-179 (AD692714), 1969, Arnold Engineering Development Center, Arnold Air Force Station, Tenn.
2. K. E. Tempelmeyer and D. W. Mills, Jr., "Refractive Index of Carbon Dioxide Cryodeposit," *Jour. of Applied Physics*, **39**, 2968-2969, 1968.
3. A. M. Smith, K. E. Tempelmeyer, P. R. Muller, and B. E. Wood, "Angular Distribution of Visible and Near IR Radiation Reflected from CO_2 Cryodeposits," *AIAA Journal*, **7**, 2274-2280, 1969.

4. A. S. Toporets, "Specular Reflection from a Rough Surface," *Optics and Spectroscopy,* **16**, 54-58, 1964.
5. T. K. Chinmayanandam, "On the Specular Reflection from Rough Surfaces," *Physical Review,* **13**, 96-101, 1919.
6. K. E. Torrance and E. M. Sparrow, "Off-Specular Peaks in the Directional Distribution of Reflected Thermal Radiation," *Journal of Heat Transfer, Trans. ASME,* Series C, **88**, 223-230, 1966.
7. N. O. Voishvillo, "Reflection of Light by a Rough Glass Surface at Large Angles of Incidence of the Illuminating Beam," *Optics and Spectroscopy,* **22**, 517-520, 1967.
8. R. Graf and J. Paulon, "E'tude Radiocristallographique de la Structure d'un Cryodepot de CO_2." T. P. No. 589, Office National D'Etudes et de Rechereches Aerospatiales, Chatillon, France, 1968.
9. L. G. Dowell and A. P. Rinfret, "Low-Temperature Forms of Ice as Studied by X-Ray Diffraction," *Nature,* **188**, 1144-1148, 1960.

Acknowledgment

The research reported in this paper was sponsored by the Arnold Engineering Development Center, Air Force Systems Command, under Contract No. F40600-72-C-0003 with ARO, Inc.

18

CRYOGENIC VACUUM SEALS AND RF WINDOWS FOR A 94 GHz MASER

L. O. Hoppie, R. E. Hayes, and N. W. Baer

Vespel SP-1 was used in the construction of a 94 GHz traveling wave maser to make reusable tube and flange type cryogenic vacuum seals and as a rf window-vacuum seal combination. This design permitted the maser to be operated successfully with no waveguide going from room temperature to the cryogenic region containing the maser crystal. Thus the large amount of noise usually added by the input waveguide of masers operating at this frequency was eliminated, resulting in a noise temperature of only a few degrees.

18-1. Maser System Design Criterion

The objective of this project was to design and construct a maser, operating in the vicinity of 94 GHz, which would be a prototype of a practical amplifier for use in radio astronomy and/or space communications. The design criterion required simply a stable, high gain, efficient, reliable, and economical maser with low noise figure and high detectivity. It was felt that all these goals could be attained by the successful employment of quasi-optical techniques, a cryogenic detector, and a low-frequency pumping scheme.

The potential advantages of a maser employing quasi-optical techniques and a cryogenic detector become apparent when one considers the noise sources of the two conventional masers shown in Figure 18-1. The amplifier noise temperatures, T_{amp},* of the active maser crystals used in either type of maser have the same

*For definition of noise temperature see for example A. E. Siegman, *Microwave Solid State Masers,* McGraw-Hill, 1964.

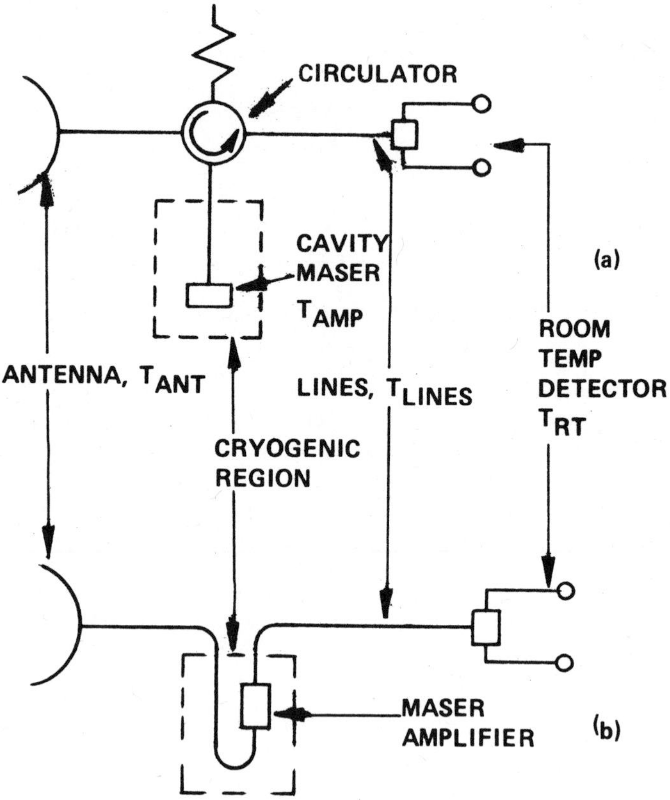

Fig. 18-1. Noise sources of (a) cavity maser and (b) traveling wave maser

ultimate value, namely, the magnitude of the spin temperature. This is in turn close to the lattice temperature. The primary difference in noise temperatures of the two types of masers arises from the microwave structure required to process the signal. Since the masers are assumed to have identical antennas, lines, and room temperature detectors, the noise added to each system by these elements is the same, but the circulator employed with most cavity masers can add as much as a few tens of degrees of additional noise. Since a circulator is not required with a traveling wave maser, this additional noise is eliminated.

The equivalent input noise temperatures of the maser systems shown in Figure 18-1 (a) and (b) are thus given by

$$T_{cavity} = T_{ant.} + T_{circ} + T_{input\ line} + T_{amp} + \frac{1}{G}\left(T_{output\ line} + T_{R.T.\ det}\right) \quad (1)$$

and

$$T_{tr\ wave} = T_{cavity} - T_{circ} \quad (2)$$

Here G is the power gain of the maser and $T_{R.T.\ det}$ is the noise temperature of the room temperature detector. It is necessary to divide the noise temperature of the output line and the detector by the maser gain in order to obtain their contribution to the equivalent input noise temperature of the system.

By contrast the traveling wave maser employing quasi-optical techniques and a cryogenic detector is shown in Figure 18-2. Much of the input line noise is eliminated and the detector noise reduced substantially, and it is possible to reduce the noise temperature of a quasi-optical traveling wave maser to the value

$$T_{quasi\text{-}optical} = T_{ant.} + T_{amp} + \frac{T_{cryo\ det}}{G} \quad (3)$$

Typical values for the above noise temperatures are:

T_{ant} = 7°K

T_{circ} = 20°K

$T_{input\ line}$ = $T_{output\ line}$ = 182°K

Note: This value is based on two feet of waveguide having a loss of 2 db/foot.

T_{amp} = 4.2°K

$T_{R.T.\ det}$ = 1000°K

$T_{cryo\ det}$ = 4.2°K

The results obtained using the above values in equations 1-3 are shown in Figure 18-3. The advantage of the quasi-optical maser is obvious, particularly for higher microwave frequencies where waveguide loss is in fact on the order of two db/foot.

A low-frequency pumping scheme was considered to be important because the current state of the art of millimeter wave

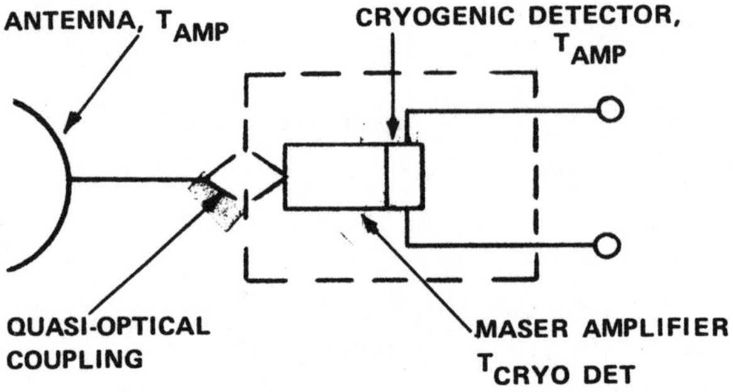

Fig. 18-2. Noise sources of a quasi-optical maser

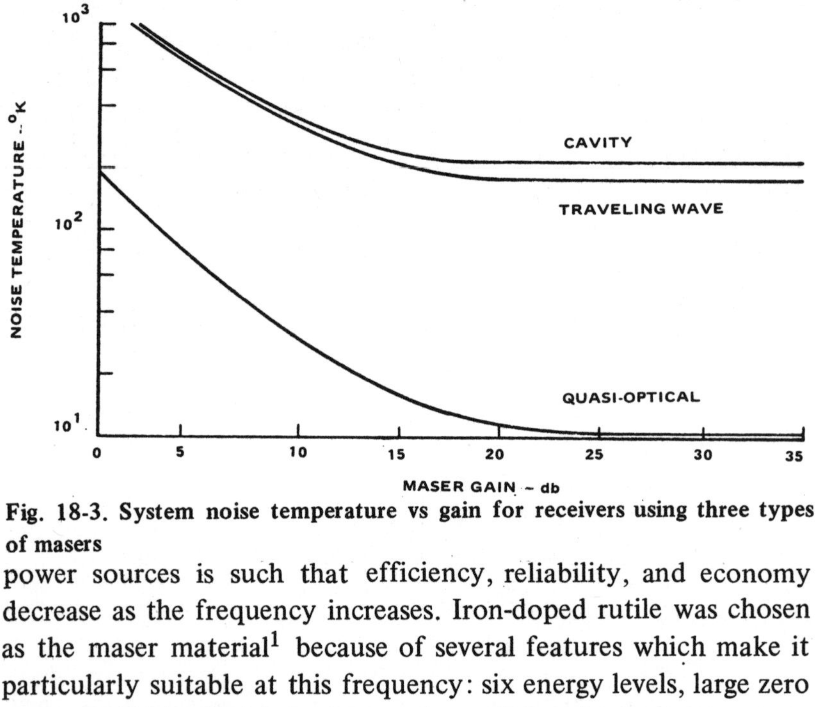

Fig. 18-3. System noise temperature vs gain for receivers using three types of masers

power sources is such that efficiency, reliability, and economy decrease as the frequency increases. Iron-doped rutile was chosen as the maser material[1] because of several features which make it particularly suitable at this frequency: six energy levels, large zero magnetic field splitting, and a very large dielectric constant.

The six energy levels allow much freedom in the choice of a pumping scheme, including low-frequency pumping schemes. The large zero magnetic field splitting maintains the required dc magnetic field at reasonably small values, and the large dielectric constant of rutile makes an external slowing structure unnecessary, since such a high dielectric material serves as its own slowing structure. The temperature dependence of the relative permittivity and the magnetic field dependence of the allowed energy levels (for a particular crystalline orientation) are shown in Figures 18-4 and 18-5, respectively.

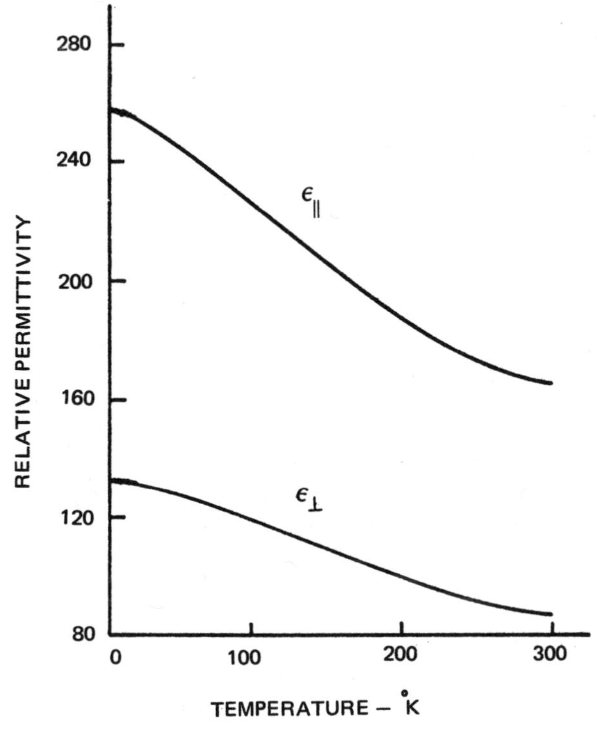

Fig. 18-4. Relative permittivity ot rutile for electric field along c-axis ($\epsilon_{\parallel}$) or perpendicular to c-axis ($\epsilon_{\perp}$)

A simplified energy level diagram of the low-frequency pumping scheme employed in the maser system is shown in Figure 18-6. All six levels are employed in this scheme; for levels (separated by the pump frequency) are used to obtain a low spin density in the $S' = -1/2$ state, and two levels (also separated by the pump frequency) are used to increase the spin density in the $S' = +3/2$ state. Stimulated emission at the signal frequency then occurs between the $S' = +3/2$ and $S' = -1/2$ states.

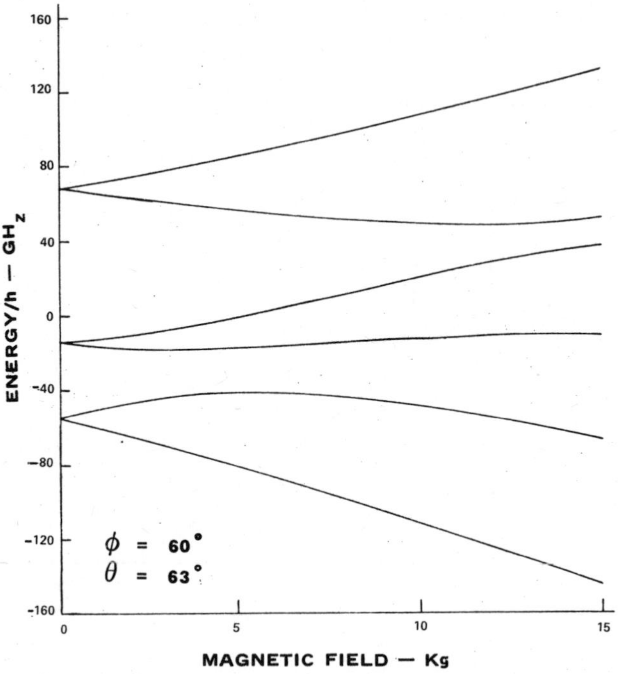

Fig. 18-5. Allowed energy levels in iron-doped rutile

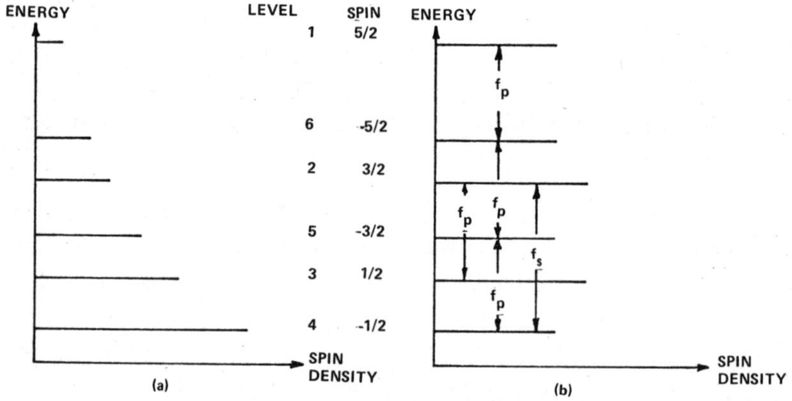

Fig. 18-6. Simplified energy level diagram for case 1—H_o=7.25 Kg, ϕ=34°, f_s=94.6 GHz, f_p=62.3 GHz. (a) no pump signal, (b) with pump signal

18-2. Signal Processing Components

As shown in Figure 18-7, the maser crystal, matching sections, and detector are mounted in a square cross-sectional waveguide structure of overall length 1.125" and 0.0625" on each side. To the bottom of this waveguide is attached a cryogenic horn and lens arrangement. The entire assembly is then attached to gimbals which can be activated from outside the dewar to make small adjustments on the crystal's orientation with respect to the dc magnetic field produced by a superconducting magnet.

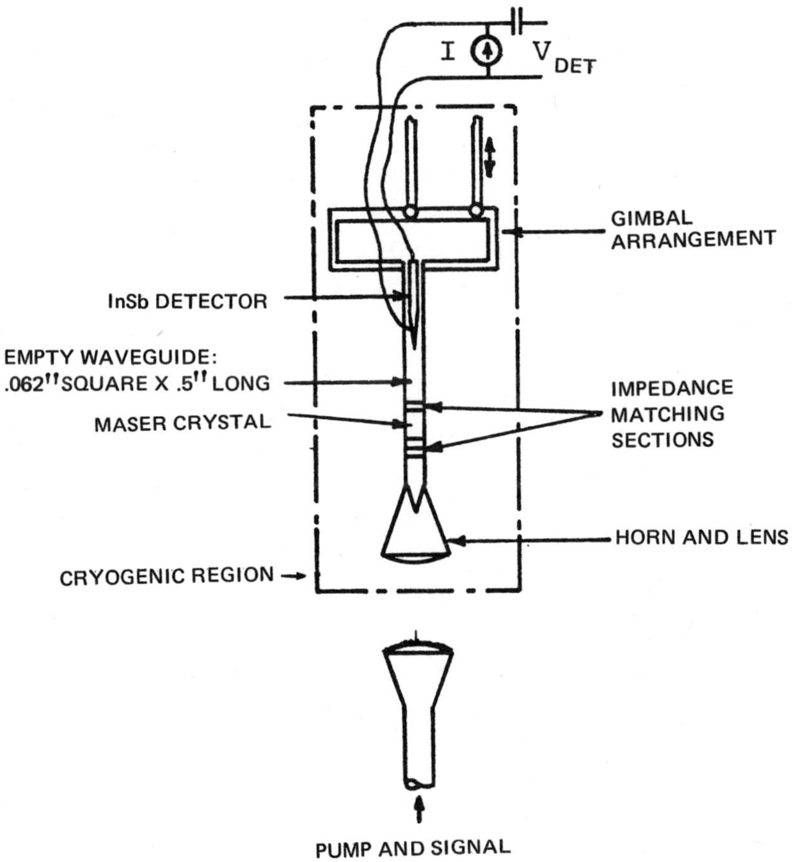

Fig. 18-7. Quasi-optical maser design

The maser crystal and matching sections fill the cross section of the waveguide. Pump power requirements and the amount of pump power available determined the length of the maser crystal. It was cut to have the proper orientation with the gimbals in the undisplaced position. The empty section of waveguide between the crystal and the detector is also 0.0625" on a side and consequently has a cut-off frequency of 94.49 GHz. The length of empty guide (0.5 inches) was chosen so that any pump power leaving the maser is attenuated by approximately 160 db before reaching the detector.

As indicated in Figure 18-7, two matching sections in the form of impedance transformers are employed to match (a) dielectric loaded waveguide to the maser crystal; (b) maser crystal to the empty waveguide. Both matching sections are made from Emerson and Cumings HI-K. This material is an epoxy into which powdered $Ti0_2$ is mixed in varying amounts to give virtually any dielectric constant desired. Dissipation factors for HI-K are quoted by the manufacturers as less than 0.002.

The required relative dielectric constant and the actual values used are listed in Table 18-1.

Table 18-1. Required and Actual Values of ϵ/ϵ_o for Matching Sections

Matching Section	*Required ϵ/ϵ_o*	*Actual ϵ/ϵ_o*
(a) loaded guide-crystal	5.0	5.0
(double quarter-wave)	14.4	14.0
(b) crystal-empty waveguide	7.27	7.0

Two types of indium antimonide detectors were considered for this application: point contact diodes between the substrate and metallic whiskers, and bulk type bolmeters.[2] Although the point contact type was indeed a feasible detection scheme, it was decided to employ the bulk type detector. Much higher responsivities can be realized with this type, and the mechanical difficulties associated with point contact diodes are avoided.

The geometry of the bulk detector which was used is shown in Figure 18-8. By tapering the indium antimonide as shown, it serves as its own dielectric tapered matching section since in operation $\omega\epsilon >> \sigma$. The bulk conductivity of indium antimonide depends upon the power dissipated in the crystal, and envelope detection is accomplished by monitoring the voltage across a crystal biased at constant current.

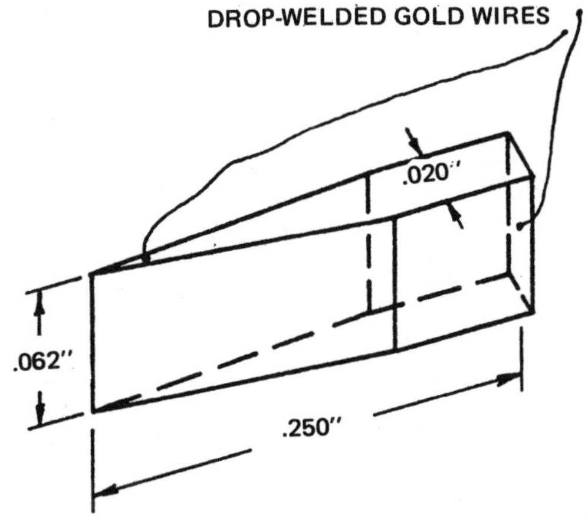

Fig. 18-8. Geometry of bulk InSb detector

18-3. Cryogenic Seals and RF Windows

Figure 18-9 shows the assembled dewar for the maser system. It was decided to design the dewar so that it could be completely disassembled and any section used in some other cryogenic application. This necessitated the incorporation of removable and reusable vacuum seals which would operate at cryogenic temperatures. Three distinctly different seals were developed to accomplish this. All of them were made from Vespel SP-1, a material developed by DuPont for use as a high vacuum bakeable seal.[3] The three seals which were used are discussed separately below.

Nitrogen Fill Tube Seal. The nitrogen chamber can be removed or adjusted through the use of the tubular seals shown in

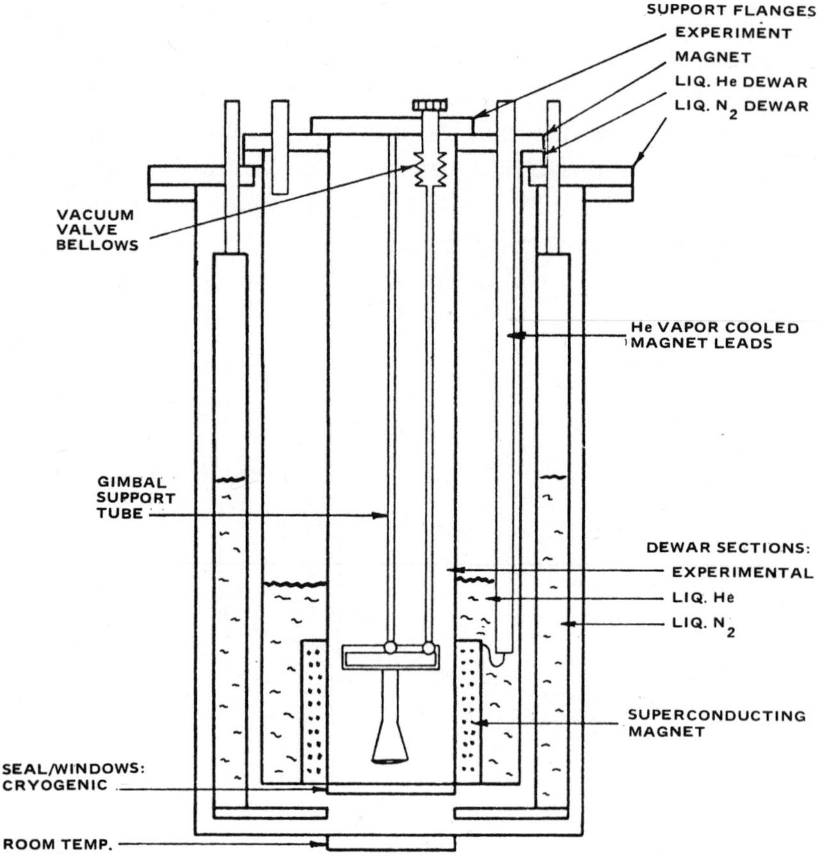

Fig. 18-9. Quasi-optical maser dewar

Figure 18-10. During the filling of the nitrogen chamber, the seal approaches the temperature of liquid nitrogen. The tubular seal and the tube must provide a tight slip-fit during assembly of the seal. By tightening the sealing nut, the Vespel is compressed between the seal retainer and the tube, thus forming the seal and supporting the nitrogen chamber.

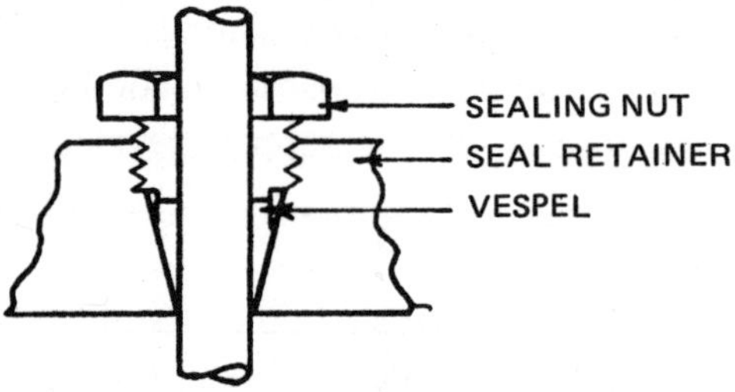

Fig. 18-10. Vespel Tubular Seal

Seals of this type have successfully operated during numerous transfers of liquid nitrogen and have been adjusted on several occasions.

Flange Seal. The seal between the liquid helium and liquid nitrogen dewar flanges (see Figure 18-9) can become quite cold when liquid helium is being transferred. In fact, before a Vespel seal was employed here, conventional "O"-rings failed because of contraction on three occasions during the helium transfer. Since the Vespel flange seal was incorporated, no such vacuum failures occurred.

The flange seal developed is shown in Figure 18-11. The trapazoidal cross section was chosen to minimize the actual sealing area while providing rigidity to the ring itself. The height of the ring should be such that a ten percent deformation per seal surface occurs when the two flanges are bolted together. Thus, in this application, the ring is machined to exceed the groove depth by 20 percent.

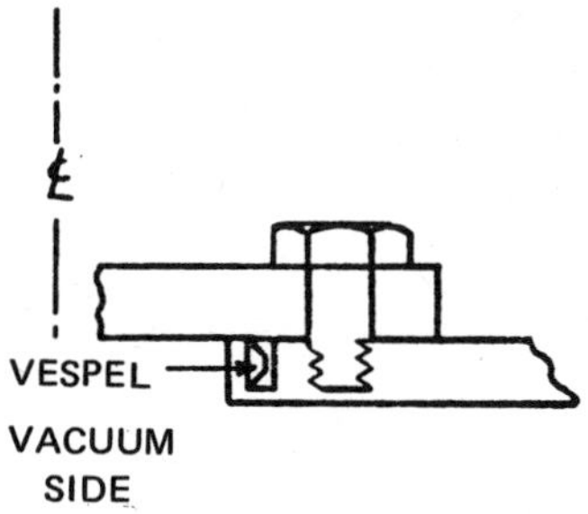

Fig. 18-11. Vespel Flange Seal

If a design should call for a long seal of this type, care must be taken to allow for the thermal contraction of Vespel (approximately 1.0 percent contraction on going from room temperature to liquid helium temperature).

Window Seal. Vespel's dielectric constant of 3.5 and dissipation factor of 0.002 make it a suitable material for rf windows, lenses, etc. In the application shown in Figure 18-12, the Vespel serves as both window and vacuum seal and operates at temperatures near liquid helium.

This seal has been cycled from room temperature to liquid helium temperatures and back on numerous occasions and has shown no signs of failure.

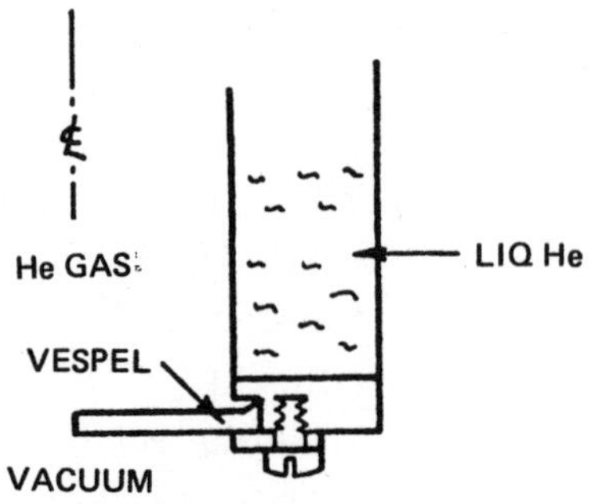

Fig. 18-12. Vespel Seal and rf Window

18-4. Results and Conclusions

The theoretical and experimental results of the maser system are summarized in Table 18-2.

This maser system demonstrated that the quasi-optical approach can indeed be applied to traveling wave maser systems operating in the millimeter wavelength region. Two primary advantages of employing this approach at these frequencies were realized:

1. The input waveguide which is normally required in conventional maser systems was eliminated. As was shown, the advantages of employing conventional maser systems at these frequencies are normally only marginal due to the noise added by the input waveguide. When this noise source is eliminated, receivers with masers become quite attractive.
2. The necessity of constructing a slow wave structure, normally required in conventional traveling wave maser systems, was eliminated. At the millimeter wavelength frequencies, slow wave structures become quite difficult to construct due to the small dimensions and close tolerances required. By propagating the signal through a bulk piece of maser material having a high dielectric constant, the slow wave structure problem is eliminated.

The primary difficulty encountered during this work was development of a reusable cryogenic rf window which would form

Table 18-2. Properties of the 94 GHz Quasi-Optical Traveling-Wave Maser

SIGNAL DATA	Central Frequency	94.587GHz*	
	Electronic gain, max.	6.24 db*	Gain (in db) proportional to crystal length
	Linewidth between half power pts	29.2MHz*	
	Tunability	90MHz*	Low frequency pumping scheme limits tunability
PUMP DATA	Pump frequency	62.3GHz*	
	Required pump power	2.5 mw*	Does not include waveguide or horn losses; required pump power proportional to crystal volume
LOSSES	Double horn & lens	4.5 db*	Can be minimized by proper horn design and/or using antenna
	Windows	<.01 db**	Based on manufacturer's loss tangents and length used
DETECTOR DATA	Responsivity observed, max.	5.5mv/μw**	Measured at a signal power level of 10^{-7} watts. Depends on temperature, power level, and magnetic field.
	Max. theoretical responsivity	6.2 mv/μw**	Assumes signal power level → 0. Based on same temperature and magnetic field as for the max. observed responsivity above.

*measured
**estimated

/more/

Table 18-2. Concluded

NOISE DATA	Amplifier noise temperature	34.0 °K**	
	N.E.P.† of detector	5.22×10^{-13}w**	Assumes noise voltage produced by detector is twice Johnson noise voltage, and uses max. theoretical responsivity.
	N.E.P. of system	1.24×10^{-13}w**	Same as above, and also uses a gain of 6.24 db.

**estimated
†noise equivalent power

a vacuum seal to stainless steel. This problem was solved by using Vespel, an inexpensive, easily machined material which is transparent to rf radiation. The material was machined in such a way that a single piece served as both vacuum seal and rf window.

References

1. M. S. Lin and G. I. Haddad, "Energy Levels and Transition Matrix Elements of Fe^{3+} in $Ti0_2$ (Rutile)," Univ. of Michigan, TR **89**, 1966.
2. F. Arams et al., "Millimeter Mixing and Detection in Bulk InSb," *Proc. of IEEE,* **54**, 612, 1966.
3. Vespel Design Handbook, E. I. duPont de Nemours & Co., Wilmington, Delaware, 1970.

Acknowledgment

This work was sponsored by the National Science Foundation under Grant GK 1723.

Cumulative Subject Index—Volumes 1 through 5

References in the index are in the following form: the first number in parentheses is the volume number; the dash number is the chapter; and the decimal number is the section. The page number is outside the parentheses. Omission of a decimal section reference indicates that most of the paper is on the subject. Page numbers are given for this volume (Volume 5) only.

Cumulative Author Index—Volumes 1 through 5